Food Processing

Food Processing

Advances in Non-Thermal Technologies

Edited by
Kshirod Kumar Dash
Sourav Chakraborty

CRC Press
Taylor & Francis Group
Boca Raton London New York

CRC Press is an imprint of the
Taylor & Francis Group, an **informa** business

First edition published 2021 by
CRC Press
6000 Broken Sound Parkway NW, Suite 300, Boca Raton, FL 33487-2742

and by
CRC Press
2 Park Square, Milton Park, Abingdon, Oxon, OX14 4RN

First edition published by CRC Press 2021

CRC Press is an imprint of Taylor & Francis Group, LLC

ISBN: 978-0-367-75610-9 (hbk)
ISBN: 978-0-367-75615-4 (pbk)
ISBN: 978-1-003-16321-3 (ebk)

Typeset in Times
by SPi Global, India

Contents

Preface vii
List of Figures ix
List of Tables xi
About the Editors xiii
Contributors xv

Chapter 1 High-Pressure-Based Food-Processing Technologies for Food Safety and Quality 1

Musfirah Zulkurnain, Alifdalino Sulaiman, and V.M. Balasubramaniam

Chapter 2 Application of Pulse Electric Fields in Food Processing 27

Sourav Chakraborty, Mausumi Sarma, Swapnil Prashant Gautam, and Purba Prasad Borah

Chapter 3 Recent Advances in Ultrasound Processing of Food 39

Abhinav Tiwari, Animesh Singh Sengar, Anjali H. Kurup, and Ashish Rawson

Chapter 4 Osmotic Dehydration in Food Processing 67

Santanu Malakar, T. Manonmani, Saptashish Deb, and Kshirod Kumar Dash

Chapter 5 Pulsed Light Technology Applied in Food Processing 91

Prasanna Bhalerao, Rishab Dhar, and Snehasis Chakraborty

Chapter 6 Application of Membrane Technology in Food-Processing Industries 123

Ananya Bardhan, Senthilmurugan Subbiah, and Kaustubha Mohanty

Chapter 7 Irradiation Technology for the Food Industry 165

Dipika Trivedi and Anil Kumar Dikshit

Chapter 8 Cryogenic Freezing.. 181

Himani Singh, Murlidhar Meghwal, and Pramod K. Prabhakar

Chapter 9 Nanofiltration: Principles, Process Modeling, and Applications 195

Siddhartha Vatsa, Manibhushan Kumar, Neeraj Ghanghas, Pramod K. Prabhakar, and Murlidhar Meghwal

Chapter 10 Atmospheric Pressure Non-Thermal Plasma in Food Processing 223

Mahreen, Priyanka Prasad, Satyananda Kar, and Jatindra K. Sahu

Index..253

Preface

Today's consumers are highly conscious of the quality and sustainability of the food products they consume. Food science, technology, and engineering have witnessed new advances in thermal and non-thermal food production methods that address the future scope of the field and also its challenges. The adoption of innovative food-processing technologies which both retain the nutrients of food and increase the shelf life of products is likely to further drive the growth of the global market.

Food Processing, which describes newly emerging technology for both academic and industrial research, is divided into two volumes (Advances in Thermal Technologies (Volume I), Advances in Non-Thermal Technologies (Volume II). Volume I covers the major aspects of and emerging developments in thermal operations in food processing, and Volume II covers advances in non-thermal food processing. The principal non-thermal processing technologies, including high-pressure processing, ultrasound, ohmic heating, pulse electric field, pulse light, membrane processing, cryogenic freezing, nanofiltration, and cold plasma processing technology, are covered. Non-thermal operations in food processing represent a viable alternative to thermal operations in order to retain the quality and organoleptic properties of food products. Non-thermal food processing techniques help to retain color, flavor, texture, or nutritional characteristics of the food products with minimal changes to product profile. The development and application of these techniques is having an important and meaningful effect on the food industry, as well as contributing to improved health and well-being for consumers.

The key feature of non-thermal food processing is the development of food products of exceptional quality and safety. The application of various non-thermal processes is set to increase in the coming years, as consumer demand increases for fresh foods and beverages with clean labels and fewer preservatives. By focusing both on basic facts and on recent advances, this book seeks to provide academic and industrial researchers with a wealth of information on newly emerging non-thermal food-processing technologies.

Kshirod Kumar Dash
Sourav Chakraborty

Preface

List of Figures

1.1 High-pressure processing equipment layout 2
1.2 Phase diagram of water and high-pressure freezing pathways 7
1.3 Phase diagram of binary mixture of 30% fully hydrogenated fat and soybean oil 11
2.1 PEF processing unit 29
2.2 Commonly used pulse wave shapes and generic electrical circuits 30
2.3 Configuration of treatment chamber: (a) parallel-plate and (b) coaxial 31
2.4 Breakdown of cell membrane during various stages of electroporation 33
3.1 The acoustic spectrum 40
3.2 Ultrasonic cutting assembly for slicing bread 41
3.3 Formation of pores on the surface of *L. sakei* strain after the application of ultrasound 42
3.4 Combined effects of ultrasound and ozone processing on *E. coli* in water 44
3.5 Action of cavitation bubbles during ultrasound-assisted extraction 46
3.6 Collapse of cavitation bubble and releasing of plant materials in extraction of essential oil from basil 48
3.7 Ultrasound-assisted air-blast freezing system 49
3.8 Cryo-SEM micrographs of potato after (a) immersion freezing and (b) ultrasound-assisted immersion freezing 50
3.9 Changes in the hull during ultrasound-assisted dehulling of pulses 51
3.10 Instability of oil/water interface and the formation of initial droplets 53
3.11 Disruption of large primary droplets in aqueous medium 53
3.12 Double emulsion formed using ultrasonic emulsification 54
3.13 Confocal micrograph of potato strips 57
3.14 SEM images of beef subjected to different ultrasonic treatments 57
4.1 Mass transfer mechanisms during osmotic dehydration 73
4.2 Process flow chart of osmotic dehydration 77
5.1 Batch type of pulsed light processing unit 94
5.2 Continuous type of pulsed light processing unit 95
5.3 Comparison between the peak power of pulsed light and that from continuous UV 98
6.1 Diagrammatic representation of membrane processing for raw milk 131
6.2 Bactocatch process patented by Alfa Laval (Sweden) 133
6.3 Application of membranes in whey processing 135
6.4 Conventional membrane separation process for juice processing 141
7.1 Reactions for water breakdown due to ions and radicals from irradiation 167
7.2 Typical layout of a γ-ray-based irradiation facility 173
7.3 Typical machine for x-ray-based irradiation facility 175

7.4 Typical machine for high-energy electron beam-based irradiation facility 176
7.5 Radura logo for irradiated food 177
8.1 Different phases of food products during freezing 186
8.2 Spiral freezer 188
9.1 Nanofiltration modules: a) plate and frame, b) tubular, c) spiral wound, and d) hollow fiber 199
9.2 Tubular membrane module 200
9.3 Plate-and-frame module for membrane separation 200
9.4 Spiral-wound module 201
9.5 Capillary hollow-fiber module 201
9.6 Two main types of filtration a) dead-end; b) cross-flow 202
9.7 Permeability: Volume flux as a function of the operating pressure 203
9.8 Relationship for obtaining ω and σ for a nanofiltration membrane 203
9.9 Hypothetical NF membrane with dissociated carboxylic groups 206
9.10 A schematic illustration of ions inside a negatively charged membrane and in the bulk 207
9.11 A schematic drawing of the electrochemical double layer at a membrane solution interface 207
9.12 Mass transport in NF systems 208
10.1 Different methods of plasma generation 225
10.2 Classification of atmospheric pressure plasma 225
10.3 Breakdown potential of some gases as a function of potential difference for $d = 10$ cm 227
10.4 Electron and gas temperature as a function of pressure 228
10.5 Schematic of corona discharge 228
10.6 General dielectric barrier discharge (DBD) configurations 229
10.7 Plasma jet configurations: (a) linear field; (b) cross-field 230
10.8 Formation of malondialdehyde (MDA) 232

List of Tables

2.1 Effect of Pulse Electric Field (PEF) on Inactivation of Microorganisms and Enzymes 34
3.1 Recent Studies on Ultrasonic Inactivation of Microbes and Enzymes 45
3.2 Comparison of Different Novel Extraction Methods 47
3.3 Different Ultrasound-Assisted Drying Methods 52
3.4 Effective Ultrasonic Nano-Emulsions Utilized in Food Systems 55
4.1 Different Model Equations Developed During Osmotic Dehydration of Food Products 75
5.1 Microbial Inactivation in Liquid Foods after Pulsed Light Processing 103
5.2 Literature Covering the Effects on Liquid Foods of Processing Technologies in Combination with Pulsed Light 110
6.1 Merits and Demerits of Different Membrane Modules 125
6.2 Classification of Membrane Material Based on Technical Development 128
6.3 Comparison of Polymer and Ceramic Membranes 129
6.4 Applications of Ceramic Membrane in Food Processing 130
6.5 Membrane Processes and Their Application in Food Industries 131
6.6 Composition of Cow's Milk 132
6.7 Membrane Application in Dairy Processing 132
6.8 Separation Range of Pressure-Driven Membrane Process for Fruit Juice 138
6.9 Composition, Nutritive and Medicinal Properties of Wine 143
6.10 Applications of Integrated Membrane Processes in Food Processing 153
7.1 Classification Based on Intensity of Radiation 170
7.2 Classification Based on Different Forms of Radiation Applied in Food 171
7.3 Characteristics of Radiation Sources 171
7.4 Additional Product Costs Due to Irradiation Treatment 178
9.1 Differences between Tight and Loose Nanofiltration Membranes 197
9.2 Morphology Characterization Methods for NF Membranes 205
9.3 Application of Nanofiltration in Food and Allied Industries 214
10.1 Temperature Difference of Atmospheric Pressure Thermal and Non-Thermal Plasmas 226
10.2 A and B Values for the Determination of V_b 227
10.3 Required Electrical Field for Gas Breakdown at Atmospheric Pressure 227
10.4 Electron and Neutral Gas Temperatures for Different Plasma Discharges 230
10.5 Plasma Parameters for Different Plasma Sources 231
10.6 Impact of Atmospheric Pressure Non-Thermal Cold Plasma on Microbial Inactivation, Shelf-Life Extension, and Quality Enhancement of Food Products 238
10.7 Research Findings Evaluating the Effect of Cold Plasma on Food Packaging Material 244

About the Editors

Kshirod Kumar Dash, PhD, is an Associate Professor and Head of the Department of Food Processing Technology, Ghani Khan Choudhury Institute of Engineering and Technology, Malda, West Bengal, India. His previous post was Assistant Professor in the Department of Food Engineering and Technology, Tezpur University, Tezpur, Assam. He obtained his BTech in Agricultural Engineering from the College of Agricultural Engineering and Technology, OUAT, Bhubaneswar, and his MTech in Dairy and Food Engineering, and PhD in Food Process Engineering from IIT Kharagpur, West Bengal, India. His post-doctoral research was carried out at Ohio State University, USA. He has published several research papers in peer-reviewed journals and participated in various national and international conferences. He has been involved in research projects funded by various funding agencies such as DST, MOFPI, and MSME. He is paper coordinator and content writer for Food Technology in the development of the e-course program of ePGPathshala-Inflibnet. Dr Dash has received numerous awards and fellowships in recognition of his teaching and research achievements. His principal current research interests are drying technology, extraction, food process modeling, and heat and mass transfer in food processing.

Sourav Chakraborty, PhD, is currently Assistant Professor in the Department of Food Engineering and Technology, Tezpur University, Tezpur, Assam. He was previously a faculty member in the Department of Agricultural Engineering, Assam University, Silchar. He obtained his BTech in Agricultural Engineering from the North Eastern Regional Institute of Science and Technology (NERIST), and his MTech and PhD degrees in Food Engineering and Technology from Tezpur University. He has expertise in the food process engineering and food process modeling fields. Dr Chakraborty has published various research papers in national and international peer-reviewed journals. He has received honors at numerous international conferences for his groundbreaking research.

Contributors

V.M. Balasubramaniam
The Ohio State University, Department of Food Science and Technology, Department of Food Ag Bio Engineering, Food Safety Engineering Laboratory, Center for Clean Process Technology, Columbus, Ohio, USA

Ananya Bardhan
Department of Chemical Engineering, Indian Institute of Technology, Guwahati, Assam, India

Prasanna Bhalerao
Department of Food Engineering and Technology, Institute of Chemical Technology, Matunga, Mumbai, India

Purba Prasad Borah
Department of Food Engineering and Technology, Tezpur University, Tezpur, Assam, India

Snehasis Chakraborty
Department of Food Engineering and Technology, Institute of Chemical Technology, Matunga, Mumbai, India

Sourav Chakraborty
Department of Food Engineering and Technology, Tezpur University, Tezpur, Assam, India

Kshirod Kumar Dash
Department of Food Processing Technology, Ghani Khan Choudhury Institute of Engineering and Technology (GKCIET), Malda, West Bengal, India

Saptashish Deb
Sant Longowal Institute of Engineering & Technology, Longowal, Sangrur, India

Rishab Dhar
Department of Food Engineering and Technology, Institute of Chemical Technology, Matunga, Mumbai, India

Anil Kumar Dikshit
Environmental Science and Engineering (ESE) Department, Indian Institute of Technology Bombay, Maharashtra, India

Swapnil Prashant Gautam
Department of Food Engineering and Technology, Tezpur University, Tezpur, Assam, India

Neeraj Ghanghas
Department of Food Science and Technology, National Institute of Food Technology Entrepreneurship and Management, Kundli, Sonipat, Haryana, India

Satyananda Kar
Centre for Energy Studies, Indian Institute of Technology Delhi, New Delhi, India

Mahreen
Centre for Energy Studies, Indian Institute of Technology Delhi, New Delhi, India

Manibhushan Kumar
Department of Food Science and Technology, National Institute of Food Technology Entrepreneurship and Management, Kundli, Sonipat, Haryana, India

Anjali H. Kurup
Department of Food Safety and Quality Testing, Indian Institute of Food Processing Technology, Thanjavur, India
Department of Agricultural and Environmental Sciences, Tennessee State University, Nashville, USA

Santanu Malakar
National Institute of Food Technology Entrepreneurship and
Management, Sonipat, Haryana, India

T. Manonmani
Department of Food Engineering, National Institute of Food Technology Entrepreneurship and Management, Sonepat, Haryana,India

Murlidhar Meghwal
Department of Food Science and Technology, National Institute of Food Technology Entrepreneurship and Management, Kundli, Sonipat, Haryana, India

Kaustubha Mohanty
Department of Chemical Engineering, Indian Institute of Technology, Guwahati, Assam, India

Pramod K. Prabhakar
Department of Food Science and Technology, National Institute of Food Technology Entrepreneurship and Management, Kundli, Sonipat, Haryana, India

Priyanka Prasad
Centre for Rural Development and Technology, Indian Institute of Technology Delhi, New Delhi, India

Ashish Rawson
Department of Food Safety and Quality Testing, Indian Institute of Food Processing Technology, Thanjavur, India

Jatindra K. Sahu
Centre for Rural Development and Technology, Indian Institute of Technology Delhi, New Delhi, India

Mausumi Sarma
Department of Food Engineering and Technology, Tezpur University, Tezpur, Assam, India

Animesh Singh Sengar
Department of Food Safety and Quality Testing, Indian Institute of Food Processing Technology, Thanjavur, India

Himani Singh
Department of Food Science and Technology, National Institute of Food Technology Entrepreneurship and Management, Kundli, Sonipat, Haryana, India

Senthilmurugan Subbiah
Department of Chemical Engineering, Indian Institute of Technology, Guwahati, Assam, India

Alifdalino Sulaiman
Universiti Putra Malaysia, Department of Process and Food Engineering, Faculty of Engineering, Selangor, Malaysia

Abhinav Tiwari
Department of Primary Processing, Storage and Handling, Indian Institute of Food Processing Technology, Thanjavur, India
Department of Biosystems Engineering, Price Faculty of Engineering, University of Manitoba, Winnipeg, Canada

Dipika Trivedi
Centre for Technology Alternatives in Rural Areas (CTARA), Indian Institute of Technology Bombay, Maharashtra, India

Siddhartha Vatsa
Department of Agricultural and Environmental Sciences, National Institute of Food Technology Entrepreneurship and Management, Kundli, Haryana, India

Musfirah Zulkurnain
UniversitiSains Malaysia, Food Technology Division, School of Industrial Technology, Penang, Malaysia

1 High-Pressure-Based Food-Processing Technologies for Food Safety and Quality

Musfirah Zulkurnain
Universiti Sains Malaysia, Penang, Malaysia
Alifdalino Sulaiman
Universiti Putra Malaysia, Serdang, Selangor, Malaysia
V.M. Balasubramaniam
Center for Clean Process Technology, The Ohio State University, Columbus, Ohio, USA

CONTENTS

1.1 Introduction 2
1.2 Technology Principles 3
  1.2.1 Isotactic Principle 3
  1.2.2 Le Chatelier–Braun Principle 3
  1.2.3 Transition State Theory 4
  1.2.4 Microscopic Arrangement/Ordering 4
1.3 Application of High-Pressure-Based Technologies for Food Preservation, Safety, and Quality 5
  1.3.1 High-Pressure Pasteurization 5
  1.3.2 Pressure-Assisted Thermal Processing 5
  1.3.3 High-Pressure Freezing and Thawing 6
  1.3.4 High-Pressure Crystallization of Lipids 10
  1.3.5 High-Pressure Extraction 12
  1.3.6 Role of High Pressure in Mass Transfer and Infusion 14
1.4 Conclusions 15
Acknowledgment 16
Nomenclature 16
References 16

1.1 INTRODUCTION

High-pressure processing (HPP) has a long history. It is more than 100 years since Bert Hite first reported the potential application of high pressure as an innovative food preservation technology for milk and other foods (Hite 1899). Since 1987, high-pressure pasteurization has been a commercially feasible process. Companies such as JBT Avure Technologies (USA), Hiperbaric (Spain), Multivac (German), Bao Tou KeFA (China), and Kobe Steel (Japan) manufacture industrial-scale high-pressure equipment with batch sizes ranging from 35 to 525 liters. Laboratory-scale or pilot-scale equipment is also available from 1 mL to 5-liter capacity.

Major components of a high-pressure system include a pressure chamber (or pressure vessel) fitted with top and bottom closures, pressure intensifier, low-pressure pump, decompression valve, pressure medium tank, and heating/cooling jacket to heat or cool down the vessel (Ting 2010) (see Figure 1.1). Although current research seeks to develop continuous high-pressure processing methods for liquid foods, no commercial applications are yet available (Martinez-Monteagudo et al. 2017).

Improved research understanding of the effects of high-pressure on food components and their various physical, chemical, or biological changes has created various potential food-manufacturing applications. High-pressure treatments with or without heat that can inactivate a variety of pathogenic and spoilage vegetative bacteria, yeasts, molds, viruses, and spores, are able to produce microbiologically safe, high-quality foods. In addition, the phase change phenomenon of water and lipids under high pressure opened up a new direction of research seeking to improve the quality of frozen and lipid-based foods. High-pressure processing has also demonstrated its potential to modify bioaccessibility and extraction of bioactive components in food

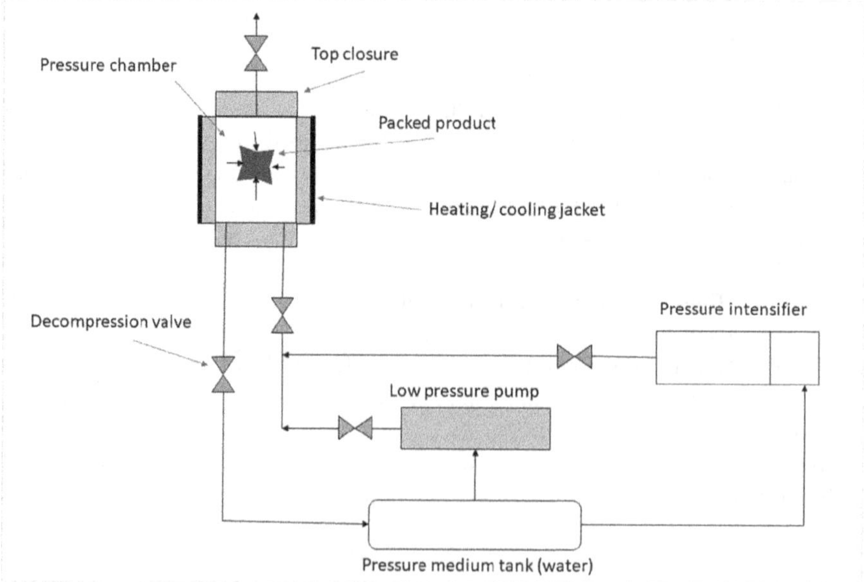

FIGURE 1.1 High-pressure processing equipment layout

matrices. Infusion of food ingredients to improve nutritional value, and the development of "clean-label" products are recent examples of the potential industrial applications of high-pressure processing technology.

1.2 TECHNOLOGY PRINCIPLES

1.2.1 Isostatic Principle

The isostatic principle demonstrates that a force transmitted to the surface of a fluid is equally distributed across the surface with which the fluid is in contact. Hydrostatic describes the forces or pressure equilibrium of a fluid system. During high-pressure food processing, the food product is pre-packaged in flexible containers and suspended in pressuring fluid (typically water). This cause the pressure to be instantaneously and homogeneously distributed throughout the sample volume regardless of its geometry and size. A pressure treatment of 300–600 MPa at ambient temperatures can transiently reduce the volume of water by 10–17% (Farkas and Hoover 2000). Products with a large number of air pockets or delicate structures, such as mushrooms, leafy greens, and marshmallows, where applying pressure could deform its structure and shape as the air escapes from the voids while the food is pressurized, may not be good candidate products for pressure treatment.

1.2.2 Le Chatelier–Braun Principle

The Le Chatelier–Braun principle states that if a process in equilibrium is disturbed, the process responds in such a way as to reduce the disturbance. HPP causes a decrease in the volume of the system it acts upon and avoids an increase in volume. Hence, any phenomenon (chemical reaction, molecular configuration, phase change) that leads to a decrease in volume during processing is favored, based on this principle.

For example, for a simple chemical reaction, the kinetics of conversion of a reaction of A and B to C with intermediate $AB^{\neq}$ can be stated as follows:

$$A + B \leftrightarrow AB^{\neq} \leftrightarrow C \tag{1.1}$$

For this reaction, the pressure (p), temperature (T), system's free enthalpy (ΔG), thermal energy (ΔE), volume (ΔV), and entropy (ΔS) can be linked by the following equation:

$$\Delta G = \Delta E + p\Delta V - T\Delta S \tag{1.2}$$

Under isothermal condition, the kinetics of this reaction can be explained by:

$$\Delta V^{\neq} = V^{\neq} - V_A = \left(\frac{\delta \Delta G^{\neq}}{\delta p}\right)_T = -RT\left(\frac{\delta \ln k}{\delta p}\right)_T \tag{1.3}$$

where k is the reaction rate constant and R is the universal gas constant ($R = 8.314$ J/mol·K). $\Delta G^{\neq}$ and $\Delta V^{\neq}$ relate the changes in free activation enthalpy and activation volume, respectively. $V^{\neq}$ represents the volume of the activated system, while V_A represents the volume before activation (Pfister 2001; Martinez-Monteagudo et al. 2014). A positive ΔV implies a shift toward reactants at higher pressures. Depending on the mechanism, some reactions may be increased or reduced by pressure. Positive ΔV favors bond cleavage and negative ΔV favors bond formation.

The reaction kinetics for isothermal conditions as a function of pressure are governed by Van't Hoff-like equations (Martinez-Monteagudo et al. 2014):

$$\left(\frac{\delta \ln K}{\delta P}\right)_T = -\frac{\Delta V^o}{RT} \tag{1.4}$$

$$\left(\frac{\delta \ln k}{\delta P}\right)_T = -\frac{\Delta V^{\neq}}{RT} \tag{1.5}$$

where K and k are equilibrium reaction constant and the reaction rate constant, respectively. V^o is the reaction volume. For a given biochemical or chemical reaction, the effect of pressure favors negative reaction volume and the reaction pathway that has negative activation volume (Martinez-Monteagudo et al. 2014).

1.2.3 Transition State Theory

Reaction rates of elementary chemical reactions are described by transition state theory, which is related to the Le Chatelier–Braun principle. This theory states that, for a pressure-assisted process, the reaction rates depend on the intermediate complex of the reaction. If the activation volume of the intermediate complex changes from the reactants' components, the reaction rate may be faster or slower.

1.2.4 Microscopic Arrangement/Ordering

At a given temperature, the degree of microscopic ordering of molecules increases with pressure. This principle shows that pressure limits rotational, vibrational, and translational movement of molecules, and thus causes molecules to be better arranged. This principle explains the antagonist effect of pressure and temperature on molecular structure and chemical reactions (Balny and Masson 1993). Pressure will inevitably alter the volume of a food product as it is applied. On the molecular level, pressure alters the distance between molecules, and thus affects the interaction between molecules. A short distance between molecules is also the strongest; covalent bonding (0.2 nm) is the most stable under pressure as this type of bonding is not altered much by high pressure. Hence, small molecules like most vitamins, antioxidants, and salts are not affected by pressure lower than 1,500 MPa (Mozhaev et al. 1994). Unlike covalent bonds, hydrogen bonding, electrostatic, and hydrophobic interactions are very distance dependent with large working distances of >2 nm (Martinez-Monteagudo and Saldana 2014).

1.3 APPLICATION OF HIGH-PRESSURE-BASED TECHNOLOGIES FOR FOOD PRESERVATION, SAFETY, AND QUALITY

Currently, HPP is used in production of products such as fruit juice, guacamole, jellies, dips, salsas, meat and poultry, seafood, and ready-to-eat meals. These products are all commercially available, mainly in Australia, Canada, China, India, Japan, Europe, and the United States (Nguyen and Balasubramaniam 2011; Evelyn and Silva 2019; Houška and Pravda 2017). High-pressure pasteurization is performed in the industry at a pressure of 400–600 MPa, at chilled or ambient temperatures; high-pressure sterilization (referred to in the literature as pressure-assisted thermal processing) carried out at >600 MPa pressure with a temperature 90–120°C is more common for resistant food pathogens and spoilage spores.

1.3.1 High-Pressure Pasteurization

Pressure at near ambient temperature or low/chilled temperatures around 400–600 MPa has been shown to inactivate some food vegetative pathogens and spoilage, producing excellent safe, fresh-like, higher-quality food products with an extensive shelf life (Farkas and Hoover 2000; Evelyn and Silva 2019; Lee and Oey 2017; Sánchez-Moreno and De Ancos 2017). However, processing under these conditions has limited effectiveness against spoilage enzymes and spores (Evelyn and Silva 2019; Sulaiman et al. 2015; Hendrickx et al. 1998). HPP causes damage to the microbial cell membrane, affecting permeability and ion exchange. The inactivation of microorganisms is dependent on many factors, such as the type of microorganism, food composition, pH, and water activity. Gram-negative bacteria can generally be more easily inactivated by pressure treatment than gram-positive bacteria. High pressure denatures proteins through conformational structure changes which affect the quaternary, tertiary, and secondary structures of enzymes. This structure change occurs because HPP only affects non-covalent bonds (e.g., hydrogen bonds in enzymes) but does not break covalent bonds (e.g., peptide bonds in enzymes) (Hendrickx et al. 1998; Mozhaev et al. 1996). Enzymes may be activated or inactivated depending on the magnitude of the pressure during processing. The origin of the enzymes (food types, fruit/vegetable cultivar), pressure, temperature, and time of treatment affect the outcome of HPP. However, a higher level of pressure, temperature, and time is expected to give higher inactivation or lower residual activity of the enzyme after processing (Silva and Sulaiman 2019). High-pressure pasteurization treatment is not sufficient to inactivate bacterial spores. Hence, food products are stored and distributed under refrigeration. Since spores are not inactivated by high-pressure pasteurization, any deviation from refrigerated storage temperature (i.e., temperature abuse during handling, distribution, and storage) is a food safety concern.

1.3.2 Pressure-Assisted Thermal Processing

Pressure-assisted thermal processing (PATP) is a technique that employs 600 MPa at a temperature between 90 and 120°C for commercial sterilization of pre-heated,

packaged low-acid foods. PATP has been shown to inactivate harmful bacterial spores (Matser et al. 2004; Sizer et al. 2002; Ahn et al. 2007; Daryaei and Balasubramaniam 2013). Some of the resistant spoilage spore formers of interest for PATP are *Clostridium botulinum, Clostridium sporogenes, C. tyrobutylicum, Thermoanaerobacterium thermosaccharolyticum, Bacillus amyloliquefaciens*, and *B. sphaericus*. Two of the strains that have shown resistance toward PATP are *T. thermosaccharolyticum* and *B. amyloliquefaciens* (Ahn et al. 2007; de Lamo-Castellvi et al. 2010). Ahn et al. (2007) showed that the *B. amyloliquefaciens* species in deionized water can be inactivated up to 7- to 8-log in only 1 min when processed at 700 MPa at 121°C.

It is known that the target pathogen spore former in low-acid food is *Clostridium botulinum* (Carlin et al. 2000; Silva and Evelyn 2018). This pathogen produces a potent and fatal neurotoxin (Brown 2000). According to the US Food and Drug Administration, *C. botulinum* types A, B, E, and F have been associated with key human foodborne botulism cases (FDA 2001). *C. sporogenes* is used as surrogate for *C. botulinum* as this species has high heat resistance and thus provides a good estimation of reduction under specific treatments. PATP at 700–900 MPa and temperatures from 80 to 100°C at less than 5 mins have been shown to achieve sterilization standards in ground beef (Zhu et al. 2008). As PATP processing relies on high pressure and temperature for a short time, this technique preserves overall food quality compared to retorted products (Matser et al. 2004; Juliano et al. 2006; Nguyen and Balasubramaniam 2011).

It is worth noting the importance of monitoring and controlling temperature distribution during product pre-heating as well during PATP treatment. The thermal distribution depends on the food properties and packaging (e.g., compressibility, heat of compression, thermal conductivity, specific heat, density) as well as the pressure-transmitting fluid used during processing (Park et al. 2016). Often mathematical modeling is employed to evaluate the extent of thermal non-uniformity within a given processed volume (Grauwet et al. 2016).

PATP is not currently a commercial process even though the US Food and Drug Administration has issued a "letter of no objection" in response to two applications to sterilize low-acid foods (Stewart et al. 2016). However, PATP has shown great potential as a good preservation technique for a variety of low-acid foods, ready-to-eat meals, soups, and sauces.

1.3.3 High-Pressure Freezing and Thawing

Bridgman (1912) first reported the phase transition and volume change of pure water under pressure. The application of high pressure in freezing foods has drawn attention due to the formation of smaller ice crystals that increase quality preservation in frozen foods. As shown in Figure 1.2, the phase diagram of pure water describes the existence of different types of ice crystal structures at different combinations of pressures and temperatures. It is well recognized that ice formed at atmospheric pressure (Ice I) has a density lower than water and its volume increases during freezing. Ice I is hexagonal in structure and contains much empty space that causes mechanical damage and is responsible for considerable textural, color, and

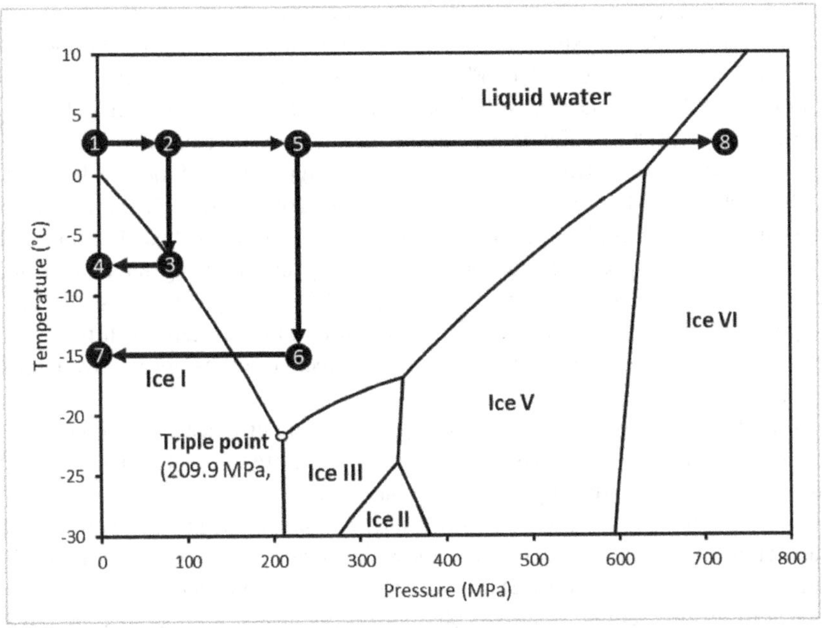

FIGURE 1.2 Phase diagram of water and high-pressure freezing pathways: pressure-assisted freezing (1-2-3-4), pressure-shift freezing (1-2-5-6-7), and pressure-induced freezing above 0 °C (1-2-5-8)

organoleptic alterations after thawing. The shape and size of ice crystals, governed by the kinetics of freezing, is recognized to contribute to structural damage of frozen foods (Duckworth 1975).

The Clausius–Clapeyron equation describes the phase transition lines in the phase diagram as follows:

$$\frac{dT_m}{dP} = \frac{\Delta V.T_m}{\Delta H} \tag{1.6}$$

where T_m is melting temperature and P is pressure. The enthalpy change, ΔH, is always negative irrespective of the modification of the solid state. At low pressure below 210 MPa, the slope of the liquid–Ice I equilibrium line is negative, thus increase in pressure is accompanied by a positive volume change, ΔV. As pressure opposes the reactions accompanied with volume increment, pressurization within this region results in melting of Ice I. At higher pressure levels, the liquid–ice slopes become positive, resulting in denser ice phases. The Ice II, III, V, and VI exist at pressures of 213, 209.9, 350.1, and 632.4 MPa at much higher densities of 1.17, 1.14, 1.23, and 1.31 g/cm^3 due to the bending ability of hydrogen bonds (Fletcher 1970). These different ice polymorphs can be exploited to reduce structural impact on frozen foods.

The formation of ice crystals during freezing happens in the two consecutive steps of nucleation and crystal growth. Nucleation is governed by the degree of supercooling when product temperature falls below the liquid–solid equilibrium point in the

phase diagram and is strongly affected by pressure. The temperature range for initiating nucleation under pressure can be determined from the phase diagram (Otero and Sanz 2003; Schlüter et al. 2004). Phase diagrams of different food models such as sodium chloride and sucrose solutions (Guignon et al. 2005), and gelatin gels (Chevalier et al. 2000a; Guignon et al. 2008) have been established in order to exploit high-pressure freezing (Knorr et al. 1998). Phase diagrams of foods such as potato (Schlüter et al. 2004), tomato paste (Denys et al. 2000), broccoli (Guignon et al. 2008), tuna and surimi (Takai et al. 1991) have also been established. The fundamental and technical aspects of high-pressure freezing and thawing have been documented by the SAFE ICE project (Urrutia et al. 2007). High-pressure vessels must be equipped with thermostatic circuits or immersed in a thermostatic bath for pressurization under sub-zero conditions. However, the technology is not yet available for commercial use.

Three methods of high-pressure freezing have been introduced, based on different freezing pathways: pressure-assisted freezing (PAF), pressure-shift freezing (PSF), and pressure-induced freezing (PIF) (Urrutia Benet et al. 2004). For PAF (pathway 1-2-3-4 in Figure 1.2), the phase transition happens under constant pressure when the temperature of the food product falls below the corresponding freezing point. PSF (pathway 1-2-5-6-7 in Figure 1.2) involves pressurization and cooling under pressure in the non-freezing region, where the phase transition happens upon pressure release. For PIF (pathway 1-2-5-8 in Figure 1.2), phase transition happens at high pressure above 632.4 MPa and continues at constant pressure. Although theoretically possible, PIF is not practical as the frozen products should be maintained and thawed under pressure to achieve the benefits of high-density ice. In addition to improving ice crystal characteristics, high-pressure freezing has the potential to alter the freezing rate.

Compared to atmospheric freezing, PAF shortens phase transition time: the liquid–solid equilibrium is reached quickly as the degree of supercooling increases with increase in pressure levels (Schlüter et al. 2004). Crystallization under constant pressure is identical to atmospheric crystallization, where the freezing rate depends on the sample size due to the existence of a temperature gradient. Crystal structures of Ice I formed under constant pressure were described as needle-shaped, radially oriented and observing size increment from the surface to the center of the frozen products (Lévy et al. 1999). Formation of ice crystals of smaller volumes (Ice III, Ice IV, Ice V) at high pressure levels requires a higher degree of supercooling to initiate nucleation (Fuchigami and Teramoto 1997; Molina-García et al. 2004; Schlüter et al. 2004). For example, Ice III needs supercooling of at least 15°C (Luscher et al. 2005). However, these ices revert back to Ice I on the release of pressure, thus the benefit of texture improvement was not significant compared to atmospheric crystallization (Fuchigami and Teramoto 1997). To avoid Ice I, thawing of the product must be conducted under the same pressure, leaving no trace of damage on muscle fibers (Molina-García et al. 2004).

PSF is the most effective and most rapid of the high-pressure freezing methods (Fernández et al. 2006a). By exploiting the positive volume change of Ice I, the ice nucleation during pressure release is quasi-instantaneous throughout the product, independent of its volume, according to the isostatic principle, with crystal growth completed at atmospheric pressure (Luscher et al. 2005). PSF produced homogeneous

nucleation with regular ice crystals distributed throughout the product, which reduced drip losses after thawing compared to PAF and conventional freezing (Alizadeh et al. 2007a; Chevalier et al. 2000b). The degree of supercooling during rapid depressurization increases with increase in pressure and decrease in temperature (Otero et al. 1998). The amount and size of the ice crystal increase with an increase in the freezing rate (Fernández et al. 2006a; Zhu et al. 2005). It is suggested that the optimum onset point for PSF with maximum degree of supercooling is the triple point in the phase diagram of water (Schlüter et al. 2004). PSF preserves textural integrity of plant tissues (Luscher et al. 2005; Urrutia-Benet et al. 2007; Fuchigami et al. 2006; Fernández et al. 2006b; Otero et al. 1998; Otero et al. 2000) and reduces drip loss in animal tissues after thawing, compared to atmospheric freezing (Hong and Choi 2016; Choi et al. 2016; Hansen et al. 2003; Chevalier et al. 2000b; Chevalier et al. 2000c; Alizadeh et al. 2007a, 2007b; Zhu et al. 2003; Tironi et al. 2007). However, pressure above 200 MPa increased toughness and drip loss in meat and sea foods (Fernandez-Martin et al. 2000; Zhu et al. 2005; Chevalier et al. 2000b; Chevalier et al. 2000c; Alizadeh et al. 2007a), and texture deformation of cheddar and mozzarella cheeses (Johnston 2000) due to protein denaturation.

Moreover, high-pressure freezing is also reported to have increased microbial inactivation in both PAF (Hayakawa et al. 1998; Park et al. 2008) and PSF treatments (Ballestra et al. 2010; Choi et al. 2008; Park et al. 2008; Picart et al. 2004, 2005; Préstamo et al. 2007) compared to atmospheric freezing. PSF has been shown to inactivate polyphenoloxidase and peroxidase enzymes at high pressure levels although it is not able to inactivate the enzymes completely (Préstamo et al. 2004; Préstamo et al. 2005; Urrutia-Benet et al. 2007).

On the other hand, high-pressure thawing takes advantage of the negative slope of the liquid–Ice I of the phase diagram to transform Ice I into water at a lower temperature. The reduction in the melting point of ice with an increase in pressure up to 210 MPa increases the temperature difference between the product and surrounding water, significantly increasing the driving force of melting. This process increases the thawing rate to one-third of conventional thawing, which allows better retention of food quality (Alizadeh et al. 2007b; Chevalier et al. 1999; Zhu et al. 2004). There are two different methods of high-pressure thawing: pressure-assisted thawing (PAT) and pressure-induced thawing (PIT). In PAT, pressurization up to 100 MPa is followed by temperature increments under constant pressure (pathway 4-3-2-1 in Figure 1.2). Under PIT, pressurization above 200 MPa initiates melting in food samples and the melting process reaches completion during pressure holding (pathway 7-6-5-1 in Figure 1.2). High-pressure thawing significantly reduces thawing and cooking drips in muscle tissues although protein denaturation was evident at high pressure levels above 200 MPa. The combination of high pressure and low temperature also resulted in microbial inactivation as studied in frozen strawberries by Eshtiaghi and Knorr (1996) and selected sea foods by Le Bail et al. (2002). Depending on the food matrix and process parameters applied, high-pressure thawing benefits vegetable tissues more than muscle tissues. Alizadeh et al. (2007b) reported that the impact of high-pressure freezing on the quality of muscle tissues was greater than high-pressure thawing, suggesting that their interactions are crucial to retaining the quality attributes of muscle tissues.

1.3.4 High-Pressure Crystallization of Lipids

Lipids are the most pressure-sensitive biological components. Hydrophobic interactions governing lipid assembly have low activation volume, which can be easily overcome by pressure (Rivalain et al. 2010). Relative to water, lipids have higher compressibility and heat of compression values due to their larger free volume (Min et al. 2010). Increase in pressure enhances the formation of van der Waals forces due to the volume reduction reaction that accelerates the solidification of lipids. The effect of pressure on lipid food is utilized to improve product quality, and to save time and energy (Nosho et al. 2002; Sevdin et al. 2018; Yasuda and Mochizuki 1992; Zulkurnain et al. 2017).

Lipids, either in the form of emulsions or in bulk, play an important role in providing texture and organoleptic properties of foods, for example the creaminess, mouth feel, plasticity, uniform appearance, glossiness, and brittleness of chocolate. In many edible fat applications, the fat crystal's morphologies, polymorphism, distribution, number, and size determine the suitability of the fat for different purposes (Marangoni et al. 2012). Similar to ice crystal formation, lipid crystallization is governed by consecutive stages of nucleation and crystal growth, with supersaturation required in the formation of the new solid phase from the lipid melt (Himawan et al. 2006). Unlike ice, some varieties of triacylglycerols have a broad melting profile (e.g., milk fat in the range of −8 to 50°C) that requires tempering, a repeated process of heating and cooling to stabilize the crystal structures and prevent product defects such as sandiness, blooming, cracking and dull appearances from post-crystallization. Moreover, the fat phase in foods is frequently found in a dispersed form in emulsions such as ice cream, whipped toppings, and mayonnaise. The fat crystallization in emulsified systems is different from bulk systems that usually require tedious tempering processes and a high degree of supercooling to be transformed into a crystalline state (Coupland 2002).

Understanding the crystallization dynamics of lipids under pressure, one of the key constituents in food, is essential to efficient running of these processes (Zulkurnain et al. 2016b). Figure 1.3 is a typical phase diagram of a lipid. According to the Clausius–Clapeyron relation (Eq. 1.6), dT_m/dP denotes the equilibrium of the solid–liquid boundary of the phase transition diagram. For lipids, the phase boundary coefficient, dT_m/dP, is always reported as positive, showing a non-linear relation (Ferstl et al. 2011; Hiramatsu et al. 1989; Zulkurnain et al. 2019). This is because the volume (ΔV) and enthalpy (ΔH) changes that correspond to lipid melting are always positive. Thus, pressure treatments accelerate phase transition of lipids by increasing their melting temperatures and degree of supercooling as the temperature gradient (ΔT) between melting temperature, T_m, and temperature of lipid medium under pressure (T_c) is extended. The increase in driving force for lipid crystallization with increase in pressure is observed in the formation of a homogeneous distribution of small crystals compared to atmospheric crystallization (Zulkurnain et al. 2017).

Knowledge of the thermodynamics and kinetics of lipid crystallization under pressure will enable us to obtain the desired crystal structures with different functional applications in food (Zulkurnain et al. 2016b). During high-pressure treatment, the volume reduction reaction establishes supersaturation conditions that act as the

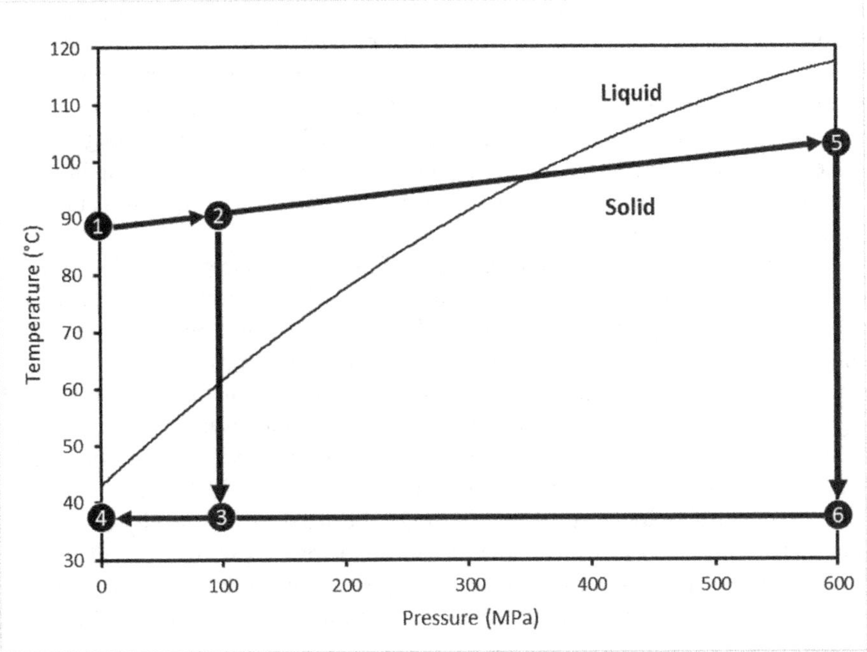

FIGURE 1.3 Phase diagram of binary mixture of 30% fully hydrogenated fat and soybean oil. Arrows show pathways of lipid crystallization under pressure during isobaric cooling at 100 MPa (1-2-3-4), and during adiabatic compression at 600 MPa (1-5-6-4)

driving force for nucleation (Greiner et al. 2012). It has been suggested that the microscopic reordering of lipid molecules facilitates phase transition under pressure (Adrjanowicz et al. 2014; Moritoki et al. 1997). Pressure treatment reduces the system's free energy, as the entropy of the system decreases, resulting in increase in density and interactions between molecules (Heremans 1982; Siegoczynski et al. 1989).

Different pathways taken during pressurization may result in the formation of different crystal structures (Zulkurnain et al. 2017, 2019). Figure 1.3 summarizes crystallization pathways of lipid mixtures during high-pressure treatment resulting in the onset of phase transition during isobaric cooling (see pathway 1-2-3-4 in Figure 1.3) at 100 MPa and adiabatic compression (see pathway 1-5-6-4 in Figure 1.3). Isobaric cooling of lipid is observed in gradual crystal growth from heterogeneous nucleation resulting in broad crystal size distribution similar to atmospheric crystallization (Kościesza et al. 2010; Ferstl et al. 2011; Zulkurnain et al. 2017). However, the microstructure of the fat crystals was reported to be smaller in size than in atmospheric crystallization, suggesting that volume reduction reaction with an increase in pressure levels limits crystal growth and increases the rate of crystallization (Zulkurnain et al. 2017).

On the other hand, the onset of crystallization induced during adiabatic compression was instantaneous, regardless of pressure levels, resulting in a modified crystal

structure with smaller particle size distribution compared to isobaric cooling (Zulkurnain et al. 2019). Complete removal of crystal memory prior to pressurization is essential and is achieved by keeping the initial temperature higher than the melting temperature of the lipid mixture. The seed crystals that form at atmospheric conditions prior to pressurization may have the properties of heterogeneously large crystal make-up and affect the formation of lipid crystals under pressure, resulting in a final crystal structure with large crystal size distribution similar to the crystals that form at atmospheric conditions (Zulkurnain et al. 2017).

High-pressure treatments favor formation of the most stable polymorphic form of fat crystal (Sevdin et al. 2018; Zulkurnain et al. 2017; Zulkurnain et al. 2019). Fat crystals formed under different conditions show differences in their crystal properties with amplification from nanostructure to micro- and macrostructural level, but resemblance in their polymorphic properties (Zulkurnain et al. 2019). Formation of high numbers of small-sized crystals at the microstructural level amplifies the functional properties at macrostructural levels. Alipid gel formed of a high density of small crystals created under high pressure showed strong mechanical properties, albeit having a 10% lower solid fraction compared with gels formed by crystallization at atmospheric pressure. The solid mass fraction of a binary fat mixture affected crystallization behaviors during adiabatic compression more than the low solid fraction, which could be due to forced convection during compression (Tefelski et al. 2011).

High-pressure treatment has potential as a processing aid for optimizing functional properties of lipid foods. Among successful applications were chocolate processing at 50–200 MPa (Yasuda and Mochizuki 1992), improving consistency of plastic fats at 10–150 MPa (Nosho et al. 2002), ice cream up to 500 MPa (Huppertz et al. 2011) and, recently, food emulsions such as solid lipid nanoparticles (SLNs) (Sevdin et al. 2018). These high-pressure treatments effectively induce, accelerate, and control crystallization of emulsion systems (Buchheim et al. 1996; Buchheim and El-Nour 1992; Dumay et al. 1996; Huppertz et al. 2011), which can be useful for development of triglyceride nanodispersions (Blümer and Mäder 2005; Sevdin et al. 2018). High-pressure treatment enhances homogeneity and decreases crystal-size distribution, delivering excellent mechanical properties to the final lipid products (Zulkurnain et al. 2016a). High-pressure processing accelerates lipid crystallization, saves time and energy, and improves the product quality of lipid foods.

1.3.5 High-Pressure Extraction

High-pressure treatment is gaining attention as a green technology for the extraction of bioactive compounds from plant materials. Among these are bioactive compounds of varying polarities, such as flavonoids, polyphenols, ginsenosides, anthocyanins, lycopene, caffeine, salidroside, corilagin, and momordicosides (Jun 2013). The extraction of specific bioactive compounds requires the plant cell walls to be broken to release the intracellular compounds and purify the bioactive compounds (Scepankova et al. 2018). In conventional methods, heat and physical treatments like ultrasonication, microwaves, pulsed electric field, and ohmic heating that are used to disrupt the tissue structure can degrade thermo-sensitive bioactive compounds.

High-pressure treatments increase membrane permeability of plant cells due to the large pressure gap between the inner and outer cell membrane, and disrupt the integrity of the bilayer membrane structure (Knorr 1992). A low pressure level of 100 MPa was enough to rupture the cell walls of onions (Butz et al. 1994). High-pressure treatments also disrupt non-covalent interactions within macromolecules, such as cell structure, proteins, starch, enzymes, and cell membranes which assist deformation of different plant compartments (Prasad et al. 2009). For example, high pressure applied to oil-bearing materials prior to aqueous enzymatic extraction decreased the emulsifying capacity of the proteins, increasing recovery of free oil (Mat Yusoff et al. 2017). The denaturation of the protein's structures also lowered protein solubility, making peptide bonds more susceptible to hydrolysis by proteolytic enzymes in pressure-treated oil-bearing materials.

Moreover, the disruption of the plant's structural integrity facilitates solvent penetration into the cell to dissolve the bioactive compounds (Yan and Xi 2017; Toepfl et al. 2006). Pressure increases the mass transfer rate and solubility of bioactive components (Richard 1992). Consequently, the rates of extraction and availability of bioactive molecules are enhanced by high-pressure treatments, especially from matrices which are difficult to release. It is known that pressure has the ability to reduce intracellular pH (Samaranayake and Sastry 2013). This contributes to the extraction of a more stable form of anthocyanins at low pH (below 4) in acylated form (Corrales et al. 2008). High-pressure treatments also decrease the dielectric constant of water and solvents that assist the release of phenolic compounds and most stable anthocyanins in acylated form, which are less polar (Corrales et al. 2008).

Besides extraction efficiency for maximum yield, the extraction process also aims to enhance the bioactivity of the bioactive components for highest functional properties. The key advantage of high-pressure treatment lies in the preservation of low molecular weight compounds, such as vitamins, antioxidants, and aroma compounds, which are stabilized by the covalent bonds due to their low compressibility. The extraction of phenolic compounds from various plant materials using HPP has been shown to have great advantages, with higher antioxidant activities and enhanced bioactivity due to the non-barosensitivity of their structures (Briones-Labarca et al. 2013; Corrales et al. 2008; Oey et al. 2008; Patras et al. 2009; Prasad et al. 2010; Shinwari and Rao 2018; Tokusoglu et al. 2010; Xi and Shouqin 2007).

In contrast to thermal extraction, which disrupts covalent bonds and consequently the functionality of the bioactive compounds, high pressure affects only non-covalent interactions like hydrogen, electrostatic, van der Waals, and hydrophobic bonds (Balasubramaniam et al. 2015). For example, bioactive compounds may lose their provitamin A activity after extraction of carotenoids, due to isomerization of the carotenoids to *cis*-isomers with heat treatment (Knockaert et al. 2011; Schieber and Carle 2005). High-pressure extraction increased the bioaccessibility of carotenoids without influencing the isomerization of carotenoids because pressure does not affect covalent bonds (Knockaert et al. 2013; Oey et al. 2008). Low temperature has also proven to enhance extract quality because of improved diffusivity and lower viscosity of solvent under pressure (Prasad et al. 2010; Knockaert et al. 2011). The extraction of lycopene and carotenoids from tomato juices increased up to 62% and 56%, respectively, following pressurization at low initial temperature of 4°C (Hsu 2008).

The bioactivity of extracts obtained from high-pressure treatments of saffron at lower temperature exhibited 28% higher efficiency in inhibiting cancer cell growth compared to conventional extraction methods (Shinwari and Rao 2018).

1.3.6 Role of High Pressure in Mass Transfer and Infusion

High-pressure treatment has been applied to accelerate the infusion of small molecule solutes and bioactive compounds into food matrices. Traditionally, infusion of solute into food matrices occurs through osmotic dehydration, which is a slow mass transfer process. Rastogi and Niranjan (1998) were the first to demonstrate that high-pressure pretreatment significantly increased osmotic dehydration of pineapple slices. The efficiency of the mass transfer from HPP was 3–5 times higher than with alternative methods like vacuum impregnation, pulsed electric field, and ultrasound pretreatments (Karwe et al. 2016). Sopanangkul et al. (2002) reported that the infusion of sucrose in potato cylinders increased eight-fold when treated with high pressure. High-pressure-assisted dehydration is gaining attention due to the significant reduction in time and energy that it offers, while improving product quality. Moreover, osmotic dehydration under high-pressure conditions also reduces the risk of microbial spoilage and increases shelf life of products.

As discussed earlier, high-pressure treatments increase cell permeability due to the damaging of the cell wall structures. Pressure-driven transport is another mechanism that accelerates the uptake of solute under high pressure, as suggested by Mahadevan et al. (2015). An increase in pressure increased the amount of solute infused (Mahadevan et al. 2015; Gosavi et al. 2019). Further, Vatankhah and Ramaswamy (2017) confirmed the need for a porous structure for an efficient liquid in-flow under pressure. They showed that blanching reduced solute uptake in apple cubes under pressure as the modeled liquid in-flow was limited by the collapsed pores. Kamat et al. (2018) also reported that acidified vegetable pickles treated under pressure at 65°C retained better quality attributes without blanching. Moreover, the infusion of viscous fluid into porous biomaterial has been shown to be effective under pressure (Vatankhah and Ramaswamy 2019).

HPP facilitates counter-current mass transport of water or solutes into porous biomaterials from a surrounding hypertonic solution such as syrup or brine (Dash et al. 2019; Rastogi and Niranjan 1998). A higher rate of dehydration has been reported in high-pressure-treated plant materials such as kiwifruit (Dalla Rosa et al. 1997), pineapple (Rastogi and Niranjan 1998), potato (Sopanangkul et al. 2002), strawberries (Nuñez-Mancilla et al. 2013; Taiwo et al. 2003; Nuñez-Mancilla et al. 2011), banana (Verma et al. 2014), and ginger (Dash et al. 2019) compared to treatment by conventional osmotic dehydration. Increase in pressure levels increased water loss and solid gain in the plant materials, ultimately reaching total permeability (Rastogi et al. 2000). In muscle tissue, high-pressure treatments accelerate diffusion of salt into the meat where the cell membrane acts as a semi-permeable film. This effect is reported in turkey breast (Villacíset al. 2008) and jumbo squid (Lemus-Mondaca et al. 2018).

Increase in pressure reduces the temperature dependence of the osmotic dehydration process. Low-temperature osmotic dehydration under high pressure retained higher nutrients and color (Nuñez-Mancilla et al. 2013) and prevented a browning

reaction (Verma et al. 2014). High pressure modifies cell integrity and increases permeability for mass transfer, resulting in higher rate of water loss compared to the rate of solid gain in ginger slices observed at all pressure levels (Dash et al. 2019). For solid enrichment, a solute of smaller molecular weight such as glucose is an effective osmotic agent under high pressure compared to fructose and sucrose, which are the most effective osmotic agents for water loss (Dash et al. 2019).

Likewise, high-pressure treatment also facilitates grain hydration. Muthukumarappan and Gunasekaran (1992) showed that high pressure increased water infusion into corn grains, which shortened soaking time. Ahromrit et al. (2006) studied the application of high pressure up to 600 MPa combined with elevated temperatures up to 70°C on hydration of glutinous rice. The overall water uptake of the glutinous rice increased with increase in pressure and temperature up to 300 MPa and 60°C, above which the effective diffusion coefficient decreased due to starch gelatinization. High-pressure infusion of polished rice with natural thiamine caused some gelatinization with increase in temperature and pressure, which provided advantages over parboiling, such as shortened processing time, increased palatability, and maintenance of the grain shape (Balakrishna and Farid 2020).

Moreover, HPP is shown to reduce thermal softening of plant tissues by activation of pectin methyl esterase enzyme (PME) at high pressure levels (Fraeye et al. 2010; Terefe et al. 2014). The controlled activation of PME has been exploited for tissue firming by increasing pectin crosslinking in the presence of divalent cations such as calcium (Duvetter et al. 2005). Calcium infusion during HPP further improved the textural damage of plant tissues from pressure-assisted thermal processing (Rastogi et al. 2008, 2010; Techakanon and Barrett 2017; Yu et al. 2018). A decrease in the degree of methylation (DM) requires calcium to form calcium bridges into an egg-box structure that reduces their susceptibility to thermal depolymerization by β-elimination reaction (Krall and Mcfeeters 1998; Sila et al. 2004). As well as texture improvement, Gosavi et al. (2019) suggested high-pressure infusion as a method of food fortification. Significantly large amounts of calcium were successfully infused into baby carrots under pressure at 14% higher than the reference daily intake (RDI). This infusion was achieved by five 15-minutecy cles at moderate pressure (350 MPa). No alteration in the hardness of the texture of the high-pressure-infused baby carrots was observed with the increase in pressure cycling. The extent of PME and calcium infusion has been shown to be influenced by cell permeability and the extent of volume reduction of the cells (Gosavi et al. 2019).

Dourado et al. (2020) introduced the application of high pressure for diffusion of asparaginase into raw potato sticks, which successfully reduced acrylamide formation in fried potato up to 47%. The high-pressure impregnation of asparaginase enzymes into the potato sticks also resulted in leaching of soluble solid and significant moisture loss from the fried potato. Asparaginase treatment or high-pressure treatment alone were not significant in reducing the acrylamide levels.

1.4 CONCLUSIONS

Like thermal processing, high pressure can be used as a tool in various food-processing applications, including ensuring food safety, blanching to control enzymes, extraction,

infusion, and crystallization. This enables food processors to develop various foods with minimal degradation of their nutritional, textural, and sensory qualities.

Pressure accelerates various physical, chemical, and biological reactions associated with volume reduction reaction. By understanding the underlying mechanisms and kinetics of the biochemical (i.e., bond formation, bond cleavage, enzymatic reactions) and physical reactions (i.e., phase transition, fluid in-flow, mass transfer) under pressure, this technique can be used to achieve the desired quality improvement of foods. Pressure, temperature, and pressure holding time are key process parameters that can be manipulated to offer various solutions to satisfy consumer demand for "clean-label" food products.

ACKNOWLEDGMENT

Author V.M. Balasubramaniam (VMB) is Professor of Food Engineering and Director of The Ohio State University Food Safety Engineering Laboratory, Center for Clean Food Process Technology Development, Columbus, Ohio. VMB acknowledges financial support from USDA NIFA, USDA HATCH program, OARDC, and participating food processors and equipment providers. Author Musfirah Zulkurnain acknowledges financial support from the Department of Higher Education, Malaysia and University Sains Malaysia. Author Alifdalino Sulaiman acknowledges financial support from Universiti Putra Malaysia (Geran Putra: GP-IPM and GP-IPS). References to commercial products or trade names are made with the understanding that no endorsement or discrimination by contributing universities is implied.

NOMENCLATURE

E	thermal energy (J-kg^{-1})
G	free enthalpy (J-kg^{-1})
$G^{\neq}$	free activation enthalpy (J-kg^{-1})
H	enthalpy (J-kg^{-1})
K	equilibrium reaction constant
k	reaction rate constant
p	pressure (MPa)
R	gas constant (J/mol·K)
S	entropy (J-K^{-1})
T	temperature (°C)
T_m	melting temperature (°C)
V	volume (m^3)
$V^{\neq}$	activated volume (m^3)
V_A	volume before activation (m^3)
V^o	reaction volume (m^3)

REFERENCES

Adrjanowicz, K., A. Grzybowski, K. Grzybowska, J. Pionteck, and M. Paluch. 2014. Effect of high pressure on crystallization kinetics of van der Waals liquid: An experimental and theoretical study. *Crystal Growth & Design*, 14: 2097–2104.

Ahn, J., V.M. Balasubramaniam, and A.E. Yousef. 2007. Inactivation kinetics of selected aerobic and anaerobic bacterial spores by pressure-assisted thermal processing. *International Journal of Microbiology*, 113: 321–329.

Ahromrit, A., D.A. Ledward, and K. Niranjan. 2006. High pressure induced water uptake characteristics of Thai glutinous rice. *Journal of Food Engineering*, 72(3): 225–233.

Alizadeh, E., N. Chapleau, M. De Lamballerie, and A. Le Bail. 2007a. Effect of different freezing processes on the microstructure of Atlantic salmon (Salmo salar) fillets. *Innovative Food Science & Emerging Technologies*, 8(4): 493–499.

Alizadeh, E., N. Chapleau, M. De Lamballerie, and A. Le Bail. 2007b. Effects of freezing and thawing processes on the quality of Atlantic salmon (Salmo salar) fillets. *Journal of Food Science*, 72(5): E279–E284.

Balakrishna, A.K. and M. Farid. 2020. Enrichment of rice with natural thiamine using high-pressure processing (HPP). *Journal of Food Engineering*, 283: 110040.

Balasubramaniam, V.M., D. Farkas, and E.J. Turek. 2008. Preserving foods through high-pressure processing. *Food Technology*, 62(11): 32–38.

Balasubramaniam, V.M., S.I. Martínez-Monteagudo, and R. Gupta. 2015. Principles and application of high-pressure-based technologies in the food industry. *Annual Review of Food Science & Technology*, 6: 435–462.

Ballestra, P., C. Verret, C. Cruz, A. Largeteau, G. Demazeau, and A. El Moueffak. 2010. High pressure inactivation of Pseudomonas in black truffle – Comparison with *Pseudomonas fluorescens* in tryptone soya broth. *High Pressure Research*, 30(1): 104–107.

Balny, C. and P. Masson. 1993. Effects of high pressure on proteins. *Food Reviews International*, 9(4): 611–628.

Blümer, C. and K. Mäder. 2005. Isostatic ultra-high-pressure effects on supercooled melts in colloidal triglyceride dispersions. *Pharmaceutical Research*, 22: 1708–1715.

Bridgman, P.W. 1912. Water in the liquid and five solid forms under pressure. *Proceedings of the American Academy of Arts and Sciences*, 47: 411–558.

Briones-Labarca, V., C. Giovagnoli-Vicuña, P. Figueroa-Alvarez, I. Quispe-Fuentes, and M. Pérez-Won. 2013. Extraction of β-carotene, vitamin C and antioxidant compounds from *Physalis peruviana* (Cape Gooseberry) assisted by high hydrostatic pressure. *Food & Nutrition Sciences*, 4: 109–118.

Brown, K.L. 2000. Control of bacterial spores. *British Medical Bulletin*, 56(1): 158–171.

Buchheim, W. and A.A. El-Nour. 1992. Induction of milkfat crystallization in the emulsified state by high hydrostatic pressure. *Lipid/Fett*, 94: 369–373.

Buchheim, W., M. Schütt, and E. Frede. 1996. High pressure effects on emulsified fats. In: R. Hayashi and C. Balny (eds), *High Pressure Bioscience and Biotechnology*. Amsterdam: Elsevier, pp. 331–333.

Butz, P., W.D. Koller, B. Tauscher, and S. Wolf. 1994. Ultra-high pressure processing of onions: Chemical and sensory changes. *LWT - Food Science & Technology*, 27: 463–467.

Carlin, F., H. Girardin, M.W. Peck, S.C. Stringer, G.C. Barker, A. Martínez, A. Fernandez, P. Fernandez, W.M. Waites, S. Movahedi, F. van Leusden, M. Nauta, R. Moezelaar, M.D. Torre, and S. Litman. 2000. Research on factors allowing a risk assessment of spore-forming pathogenic bacteria in cooked chilled foods containing vegetables: A FAIR collaborative project. *International Journal of Food Microbiology*, 60(2–3): 117–135.

Cheftel, J.C. 1995. High-pressure, microbial inactivation and food preservation. *Food Science & Technology International*, 1(2–3): 7590.

Chevalier, D., A. Le Bail, J.M. Chourot, and P. Chantreau. 1999. High pressure thawing of fish (whiting): Influence of the process parameters on drip losses. *LWT - Food Science and Technology*, 32: 25–31.

Chevalier, D., A. Le Bail, and M. Ghoul. 2000a. Freezing and ice crystals formed in a cylindrical food model: Part II. Comparison between freezing at atmospheric pressure and pressure-shift freezing. *Journal of Food Engineering*, 46(4): 287–293.

Chevalier, D., M. Sentissi, M. Havet, and A. Le Bail. 2000c. Comparison of air-blast and pressure shift freezing on Norway lobster quality. *Journal of Food Science*, 65(2): 329–333.

Chevalier, D., A. Sequeira-Munoz, A. Le Bail, B.K. Simpson, and M. Ghoul. 2000b. Effect of freezing conditions and storage on ice crystal and drip volume in turbot (Scophthalmus maximus): Evaluation of pressure shift freezing vs. air-blast freezing. *Innovative Food Science & Emerging Technologies*, 1(3): 193–201.

Choi, M.J., S.G. Min, and G.P. Hong. 2008. Effect of high pressure shift freezing process on microbial inactivation in dairy model food system. *International Journal of Food Engineering*, 4(5): 1556–3758.

Choi, M.J., S.G. Min, and G.P. Hong. 2016. Effects of pressure-shift freezing conditions on the quality characteristics and histological changes of pork. *LWT - Food Science & Technology*, 67: 194–199.

Corrales, M., S. Toepfl, P. Butz, D. Knorr, and B. Tauscher. 2008. Extraction of anthocyanins from grape by-products assisted by ultrasonics, high hydrostatic pressure or pulsed electric fields: A comparison. *Innovative Food Science & Emerging Technologies*, 9: 85–91.

Coupland, J.N. 2002. Crystallization in emulsions. *Current Opinion in Colloid & Interface Science*, 7: 445–450.

Dalla Rosa, M., F. Bressa, D. Mastrocola, and G. Carpi. 1997. Evaluation of mass transfer kinetics in kiwifruit slices during osmotic dehydration under high pressure treatments. In: E. Sfakiotakis and J. Porlingis (eds), *Proceedings of the Third International Symposium on Kiwifruit*, vol. 444. Thessaloniki, Greece: International Society for Horticultural Science (ISHS), pp. 655–662.

Daryaei, H. and V.M. Balasubramaniam. 2013. Kinetics of *Bacillus coagulans* spore inactivation in tomato juice by combined pressure-heat treatment. *Food Control*, 30(1): 168–175.

Dash, K.K., V.M. Balasubramaniam, and S. Kamat. 2019. High pressure assisted osmotic dehydrated ginger slices. *Journal of Food Engineering*, 247: 19–29.

de Lamo-Castellvi, S., W. Ratphitagsanti, V.M. Balasubramaniam, and A.E. Yousef. 2010. Inactivation of *Bacillusamyloliquefaciens*spores by a combination of sucrose laurate and pressure-assisted thermal processing. *Journal of Food Protection*, 73(11): 2043–2052.

Delgado, A., L. Kulisiewicz, C. Rauh, and R. Benning. 2010. Basic aspects of phase changes under high pressure. *Annals of the New York Academy of Sciences*, 1189: 16–23.

Denys, S., A.M. Van Loey, and M.E. Hendrickx. 2000. A modeling approach for evaluating process uniformity during batch high hydrostatic pressure processing: Combination of a numerical heat transfer model and enzyme inactivation kinetics. *Innovative Food Science & Emerging Technologies*, 1(1): 5–19.

Dourado, C., C.A. Pinto, S.C. Cunha, S. Casal, and J.A. Saraiva. 2020. A novel strategy of acrylamide mitigation in fried potatoes using asparaginase and high pressure technology. *Innovative Food Science and Emerging Technologies*, 60: 102310.

Duckworth, R.B. 1975. *Water Relations of Foods*. Food Science and Technology Monographs. Academic Press, New York.

Dumay, E., C. Lambert, S. Funtenberger, and J.C. Cheftel. 1996. Effects of high pressure on the physicochemical characteristic of dairy creams and model oil/water emulsions. *LWT - Food Science & Technology*, 29: 606–625.

Duvetter, T., I. Fraeye, T. Van Hoang, S. Van Buggenhout, I. Verlent, C. Smout, A.M. Van Loey, and M.E. Hendrickx. 2005. Effect of pectin methylesterase infusion methods and processing techniques on strawberry firmness. *Journal of Food Science*, 70(6): S383–S388.

Eshtiaghi, M. and D. Knorr. 1996. High hydrostatic pressure thawing for the processing of fruit preparations from frozen strawberries. *Food Technology*, 10(2): 143–148.

Evelyn and F.V.M. Silva. 2019. Heat assisted HPP for the inactivation of bacteria, moulds and yeasts spores in foods: Log reductions and mathematical models. *Trends in Food Science & Technology*, 88(March): 143–156.

Farkas, D.F. and D.G. Hoover. 2000. High pressure processing. *Journal of Food Science*, 65: 47–64.

FDA (U.S. Food and Drug Administration). 2001. *BAM: Clostridium botulinum*. https://www.fda.gov/Food/FoodScienceResearch/LaboratoryMethods/ucm070879.htm.

Fernández, P.P., L. Otero, B. Guignon, and P.D. Sanz. 2006a. High-pressure shift freezing versus high-pressure assisted freezing: Effects on the microstructure of a food model. *Food Hydrocolloids*, 20(4): 510–522.

Fernández, P.P., G. Préstamo, L. Otero, and P. Sanz. 2006b. Assessment of cell damage in high-pressure-shift frozen broccoli: Comparison with market samples. *European Food Research & Technology*, 224(1): 101–107.

Fernandez-Martin, F., L. Otero, M.T. Solas, and P.D. Sanz. 2000. Protein denaturation and structural damage during high-pressure-shift freezing of porcine and bovine muscle. *Journal of Food Science*, 65: 1002–1008.

Ferstl, P., C. Eder, W. Ruß, and A. Wierschem. 2011. Pressure induced crystallization of triacylglycerides. *High Pressure Research*, 31: 339–349.

Ferstl, P., S. Gillig, C. Kaufmann, C. Dürr, C. Eder, A. Wierschem, and W. Russ. 2010. Pressure-induced phase transitions in triacylglycerides. *Annals of the New York Academy of Sciences*, 1189: 62–67.

Fletcher, N.H. 1970. *The Chemical Physics of Ice*. Cambridge: Cambridge University Press.

Fraeye, I., G. Knockaert, S. Van Buggenhout, T. Duvetter, M. Hendrickx, and A. Van Loey. 2010. Enzyme infusion prior to thermal/high pressure processing of strawberries: Mechanistic insight into firmness evolution. *Innovative Food Science & Emerging Technologies*, 11(1): 23–31.

Fuchigami, M., N. Kato, and A.I. Teramoto. 2006. High-pressure-freezing effects on textural quality of carrots. *Journal of Food Science*, 62(4): 804–808.

Fuchigami, M. and A. Teramoto. 1997. Structural and textural changes in kinu-tofu due to high-pressure-freezing. *Journal of Food Science*, 62(4): 828–832 and 837.

Galazka, V.B. and D.A. Ledward. 1995. Developments in high pressure food processing. In: A. Turner (ed.), *Food Technology International Europe*. London: Sterling Publications International, pp. 123–125.

George, J.M., T.S. Selvan, and N.K. Rastogi. 2016. High-pressure-assisted infusion of bioactive compounds in apple slices. *Innovative Food Science & Emerging Technologies*, 33: 100–107.

Gosavi, N.S., D. Salvi, and M.V. Karwe. 2019. High pressure-assisted infusion of calcium into baby carrots, part I: Influence of process variables on calcium infusion and hardness of the baby carrots. *Food & Bioprocess Technology*, 12(2): 255–266.

Grauwet, T., I. Van der Plancken, L. Vervoort, M. Hendrickx, and A. Van Loey. 2016. High-pressure processing uniformity. In: V.M. Balasubramaniam, G.V. Barbosa-Cánovas, and H. Lelieveld (eds), *High Pressure Processing of Food*. New York: Springer, pp. 253–268.

Greiner, M., A.M. Reilly, and H. Briesen. 2012. Temperature- and pressure-dependent densities, self-diffusion coefficients, and phase behavior of monoacid saturated triacylglycerides: Toward molecular-level insights into processing. *Journal of Agricultural & Food Chemistry*, 60: 5243–5249.

Guignon, B., L. Otero, A. Molina-García, and P. Sanz. 2005. Liquid water-ice I phase diagrams under high pressure: Sodium chloride and sucrose models for food systems. *Biotechnology Progress*, 21(2): 439–445.

Guignon, B., J. Torrecilla, L. Otero, A. Ramos, A. Molina-García, and P. Sanz. 2008. The initial freezing temperature of foods at high pressure. *Critical Reviews in Food Science & Nutrition*, 48(4): 328–340.

Hansen, E., R.A. Trinderup, M. Hviid, M. Darré, and L.H. Skibsted. 2003. Thaw drip loss and protein characterization of drip from air-frozen, cryogen-frozen, and pressure-shift-frozen pork longissimus dorsi in relation to ice crystal size. *European Food Research & Technology*, 218(1): 2–6.

Hayakawa, K., Y. Ueno, S. Kawamura, T. Kato, and R. Hayashi. 1998. Microorganism inactivation using high-pressure generation in sealed vessels under sub-zero temperature. *Applied Microbiology & Biotechnology*, 50(4): 415–418.

Heinz, V., and R. Buckow. 2010. Food preservation by high pressure. *Journal of Consumer Protection & Food Safety*, 5: 73–81.

Hendrickx, M., L. Ludikhuyze, I. Van den Broeck, and C. Weemaes. 1998. Effects of high pressure on enzymes related to food quality. *Trends in Food Science & Technology*, 9(5): 197–203.

Heremans, K. 1982. High pressure effects on proteins and other biomolecules. *Annual Review of Biophysics & Bioengineering*, 11: 1–12.

Himawan, C., V.M. Starov, and A.G.F. Stapley. 2006. Thermodynamic and kinetic aspects of fat crystallization. *Advances in Colloid & Interface Science*, 122: 3–33.

Hiramatsu, N. and T. Inoue, M. Suzuki, and K. Sato. 1989. Pressure study on thermal transitions of oleic acid polymorphs by high-pressure differential thermal analysis. *Chemistry & Physics of Lipids*, 51: 47–53.

Hite, B. 1899. The effect of pressure in the preservation of milk. *West Virginia Agricultural Experimental Station Bulletin*, 58: 15–35.

Hong, G.-P. and M.-J. Choi. 2016. Comparison of the quality characteristics of abalone processed by high-pressure sub-zero temperature and pressure-shift freezing. *Innovative Food Science & Emerging Technologies*, 33: 19–25.

Houška, M. and P. Pravda. 2017. Examples of commercial fruit and vegetable juices and smoothies cold pasteurized by high pressure. In: M. Houška and F.V.M. Silva (eds), *High Pressure Processing of Fruit and Vegetable Products*. Boca Raton, FL: CRC Press, pp. 147–154.

Hsu, K. 2008. Evaluation of processing qualities of tomato juice induced by thermal and pressure processing. *LWT - Food Science & Technology*, 41: 450–459.

Huppertz, T., M.A. Smiddy, H.D. Goff, and A.L. Kelly. 2011. Effects of high pressure treatment of mix on ice cream manufacture. *International Dairy Journal*, 21: 718–726.

Johnston, D.E. 2000. The effects of freezing at high pressure on the rheology of cheddar and mozzarella cheeses. *Milchwissenschaft*, 55(10): 559–562.

Juliano, P., M. Toldrág, T. Koutchma, V.M. Balasubramaniam, S. Clark, J.W. Mathews, C.P. Dunne, G. Sadlerand, and G.V. Barbosa-Cánovas. 2006. Texture and water retention improvement in high-pressure thermally treated scrambled egg patties. *Journal of Food Science*, 71(2): 52–61.

Jun, X. 2013. High-pressure processing as emergent technology for the extraction of bioactive ingredients from plant materials. *Critical Reviews in Food Science & Nutrition*, 53: 837–852.

Kamat, S.S., K.K. Dash, and V.M. Balasubramaniam. 2018. Quality changes in combined pressure-thermal treated acidified vegetables during extended ambient temperature storage. *Innovative Food Science & Emerging Technologies*, 49: 146–157.

Karwe, M.V., D. Salvi, and N.S. Gosavi. 2016. High pressure–assisted infusion in foods. *Reference Module in Food Science*, 1–6.

Knockaert, G., A. De Roeack, L. Lemmens, S. Van Buggenhout, M. Hendrickx, and A. Van Loey. 2011. Effect of thermal and high pressure processes on structural and health-related properties of carrots (*Daucus carota*). *Food Chemistry*, 125: 903–912.

Knockaert, G., S.K. Pulissery, L. Lemmens, S. Van Buggenhout, M. Hendrickx, and A. Van Loey. 2013. Isomerisation of carrot β-carotene in presence of oil during thermal and combined thermal/high pressure processing. *Food Chemistry*, 138(2–3): 1515–1520.

Knorr, D. 1992. High pressure effects on plant derived foods. In: C. Balny, R. Hayashi, K. Herman, and P. Masson (eds), *High Pressure and Biotechnology*, vol. 24. Montrouge, France: John Libbey Euro-text, pp. 211–217.

Knorr, D., O. Schlüter, and V. Heinz. 1998. Impact of high hydrostatic pressure on phase transition of foods. *Food Technology*, 52(9): 42–45.

Kościesza, R., L. Kulisiewicz, and A. Delgado. 2010. Observations of a high-pressure phase creation in oleic acid. *High Pressure Research*, 30: 118–123.

Krall, S.M. and R.F. Mcfeeters. 1998. Pectin hydrolysis: Effect of temperature, degree of methylation, p H, and calcium on hydrolysis rates. *Journal of Agricultural & Food Chemistry*, 46(4): 1311–1315.

Le Bail, A., L. Boillereaux, A. Davenel, M. Hayert, T. Lucas, and J.Y. Monteau. 2003. Phase transition in foods: Effect of pressure and methods to assess or control phase transition. *Innovative Food Science & Emerging Technologies*, 4: 15–24.

Le Bail, A., D. Mussa, J. Rouillé, H.S. Ramaswamy, N. Chapleau, M. Anton, M. Hayert, L. Boillereaux, and D. Chevalier. 2002. High pressure thawing. Application to selected sea-foods. *Progress in Biotechnology*, 19(C): 563–570.

Lee, P.Y. and I. Oey. 2017. Sensory properties of high-pressure treated fruit and vegetable juices. In: M. Houška and F.V.M. Silva (eds), *High Pressure Processing of Fruit and Vegetable Products*. Boca Raton, FL: CRC Press, pp. 121–134.

Lemus-Mondaca, R., C. Zambra, F. Marín, M. Pérez-Won, and G. Tabilo-Munizaga. 2018. Mass transfer kinetic and quality changes during high-pressure impregnation (HPI) of jumbo squid (Dosidicus gigas) slices. *Food & Bioprocess Technology*, 11(8): 1516–1526.

Lévy, J., E. Dumay, E. Kolodziejczyk, and J.C. Cheftel. 1999. Freezing kinetics of a model oil-in-water emulsion under high pressure or by pressure release. Impact on ice crystals and oil droplets. *LWT - Food Science and Technology*, 32(7): 396–405.

Luscher, C., O. Schluter, and D. Knorr. 2005. High pressure low temperature processing of foods: Impact on cell membranes, texture, color and visual appearance of potato tissue. *Innovative Food Science & Emerging Technologies*, 6(1): 59–71.

Mahadevan, S., N. Nitin, D. Salvi, and M.V. Karwe. 2015. High-pressure enhanced infusion: Influence of process parameters. *Journal of Food Process Engineering*, 38: 601–612.

Marangoni, A.G., N. Acevedo, F. Maleky, E. Co, F. Peyronel, G. Mazzanti, B. Quinn, and D. Pink. 2012. Structure and functionality of edible fats. *Soft Matter*, 8: 1275–1300.

Martinez-Monteagudo, S.I., M.G. Gänzle, and M.D.A. Saldaña. 2014. High-pressure and temperature effects on the inactivation of *Bacillus amyloliquefaciens*, alkaline phosphatase and storage stability of conjugated linoleic acid in milk. *Innovative Food Science & Emerging Technologies*, 26: 59–66.

Martinez-Monteagudo, S.I., S. Kamat, N. Patel, G. Konuklar, N. Rangavajla, and V.M. Balasubramaniam. 2017. Improvements in emulsion stability of dairy beverages treated by high pressure homogenization: A pilot-scale feasibility study. *Journal of Food Engineering*, 193: 42–52.

Martinez-Monteagudo, S.I. and M.D.A. Saldana. 2014. Chemical reactions in food systems at high hydrostatic pressure. *Food Engineering Reviews*, 6(4): 105–127.

Mat Yusoff, M., M.H. Gordon, O. Ezeh, and K. Niranjan. 2017. High pressure pre-treatment of Moringa oleifera seed kernels prior to aqueous enzymatic oil extraction. *Innovative Food Science & Emerging Technologies*, 39: 129–136.

Matser, A.M., B. Krebbers, R.W. van den Berg, and P.V. Bartels. 2004. Advantages of high pressure sterilisation on quality of food products. *Trends in Food Science & Technology*, 15(2): 79–85.

McInerney, J.K., C.A. Seccafien, C.M. Stewart, and A.R. Bird. 2007. Effects of high pressure processing on antioxidant activity, and total carotenoid content and availability, in vegetables. *Innovative Food Science & Emerging Technologies*, 8: 543–548.

Min, S., S.K. Sastry, and V.M. Balasubramaniam. 2010. Compressibility and density of select liquid and solid foods under pressures up to 700 MPa. *Journal of Food Engineering*, 96: 568–574.

Molina-García, A.D., L. Otero, M.N. Martino, N.E. Zaritzky, J. Arabas, J. Szczepek, and P.D. Sanz. 2004. Ice VI freezing of meat: Supercooling and ultrastructural studies. *Meat Science*, 66(3): 709–718.

Moritoki, M., N. Nishiguchi, and S. Nishida. 1997. Features of the high-pressure crystallization process in industrial use. In: G. Botsaris and K. Toyokura (eds), *Separation and Purification by Crystallization*. Washington, DC: ACS Publications, pp. 136–149.

Mozhaev, V.V., K. Heremans, J. Frank, P. Masson, and C. Balny. 1994. Exploiting the effects of high hydrostatic pressure in biotechnological applications. *Trends in Biotechnology*, 12(12): 493–501.

Mozhaev, V.V. and R. Lange, E.V. Kudryashova, and C. Balny. 1996. Application of high hydrostatic pressure for increasing activity and stability of enzymes. *Biotechnology & Bioengineering*, 52(2): 320–331.

Muthukumarappan, K. and S. Gunasekaran. 1992. Above-atmospheric hydration of corn. *Transaction of the ASAE*, 35(6): 1885–1889.

Nguyen, L.T. and V.M. Balasubramaniam. 2011. Fundamentals of food processing using high pressure. In: H.Q. Zhang, G.V. Barbosa-Cánovas, V.M. Balasubramaniam, C.P. Dunne, D.F. Farkas, and J.T.C. Yuan (eds), *Nonthermal Processing Technologies for Food*. Blackwell Publishing Ltd, pp. 3–19.

Nosho, Y., S. Hachimoto, M. Kato, and K. Suzuki. 2002. A novel pressure technology for the production of margarine. In: R. Winter (ed.), *Advances in High Pressure Bioscience and Biotechnology II*. Berlin: Springer Press, pp. 447–451.

Nuñez-Mancilla, Y., M. Pérez-Won, E. Uribe, A. Vega-Gálvez, V. Arias, G. Tabilo-Munizaga, V. Briones-Labarca, R. Lemus-Mondaca, and K. Di Scala. 2011. Modelling mass transfer during osmotic dehydration of strawberries under high hydrostatic pressure conditions. *Innovative Food Science & Emerging Technologies*, 12: 338–343.

Nuñez-Mancilla, Y., M. Pérez-Won, E. Uribe, A. Vega-Gálvez, and K. Di Scala. 2013. Osmotic dehydration under high hydrostatic pressure: Effects on antioxidant activity, total phenolics compounds, vitamin C and colour of strawberry (Fragaria vesca). *LWT - Food Science & Technology*, 52: 151–156.

Oey, I., I. Van der Plancken, A. Van Loey, and M. Hendrickx. 2008. Does high pressure processing influence nutritional aspects of plant based food systems? *Trends in Food Science & Technology*, 19: 300–308.

Oh, J.-H. and B.G. Swanson. 2006. Polymorphic transitions of cocoa butter affected by high hydrostatic pressure and sucrose polyesters. *Journal of the American Oil Chemists' Society*, 83: 1007–1014.

Otero, L., M. Martino, N. Zaritzky, M. Solas, and P.D. Sanz. 2000. Preservation of microstructure in peach and mango during high-pressure-shift freezing. *Journal of Food Science*, 65(3): 466–470.

Otero, L. and P.D. Sanz. 2003. High pressure-assisted and high pressure-induced thawing: Two different processes. *Food Engineering & Physical Properties*, 68: 2523–2528.

Otero, L., M.T. Solas, P.D. Sanz, C. de Elvira, and J.A. Carrasco. 1998. Contrasting effects of high-pressure-assisted freezing and conventional air-freezing on eggplant tissue microstructure. *Zeitschrift für Lebensmitteluntersuchung und-Forschung A*, 206(5): 338–342.

Park, S.H., G.P. Hong, S.G. Min, and M.J. Choi. 2008. Combined high pressure and subzero temperature phase transition on the inactivation of *Escherichia coli* ATCC 10536. *International Journal of Food Engineering*, 4(4): 1556–3758.

Park, S.H., L.T. Nguyen, S. Min, V.M. Balasubramaniam, and S.K. Sastry. 2016. *In situ* thermal, volumetric and electrical properties of food matrices under elevated pressure and the techniques employed to measure them. In: V.M. Balasubramaniam, G.V. Barbosa-Cánovas, and H. Lelieveld (eds), *High Pressure Processing of Food*. New York: Springer, pp. 97–121.

Patras, A., N.P. Brunton, S. Da Pieve, and F. Butler. 2009. Impact of high pressure processing on total antioxidant activity, phenolic, ascorbic acid, anthocyanin content and colour of strawberry and blackberry purées. *Innovative Food Science & Emerging Technologies*, 10(3): 308–313.

Pfister, M.K.H. 2001. Influence of high pressure treatment on chemical alterations in foods: A literature review. *Bg VV-Hefte*, Germany.

Picart, L., E. Dumay, J.-P. Guiraud, and J.C. Cheftel. 2004. Microbial inactivation by pressure-shift freezing: Effects on smoked salmon mince inoculated with *Pseudomonas fluorescens, Micrococcus luteus* and *Listeria innocua. LWT - Food Science & Technology*, 37(2): 227–238.

Picart, L., E. Dumay, J.-P. Guiraud, and J.C. Cheftel. 2005. Combined high pressure sub-zero temperature processing of smoked salmon mince: Phase transition phenomena and inactivation of *Listeria innocua. Journal of Food Engineering*, 68(1): 43–56.

Prasad, K.N., B. Yang, J. Shi, C. Yu, M. Zhao, S. Xue, and Y. Jiang. 2010. Enhanced antioxidant and antityrosinase activities of longan fruit pericarp by ultra-high-pressure-assisted extraction. *Journal of Pharmaceutical & Biomedical Analysis*, 51(2): 471–477.

Prasad, K.N., E. Yang, C. Yi, M. Zhao, and Y. Jiang. 2009. Effects of high pressure extraction on the extraction yield, total phenolic content and antioxidant activity of longan fruit pericarp. *Innovative Food Science & Emerging Technologies*, 10: 155–159.

Préstamo, G., L. Palomares, and P.D. Sanz. 2004. Broccoli (Brasica oleracea) treated under pressure-shift freezing process. *European Food Research & Technology*, 219: 598–604.

Préstamo, G., L. Palomares, and P.D. Sanz. 2005. Frozen foods treated by pressure shift freezing: Proteins and enzymes. *Journal of Food Science*, 70(1): S22–S27.

Préstamo, G., A. Pedrazuela, B. Guignon, and P.D. Sanz. 2007. Synergy between high-pressure, temperature and ascorbic acid on the inactivation of Bacillus cereus. *European Food Research & Technology*, 225(5–6): 693–698.

Rastogi, N.K., A. Angersbach, and D. Knorr. 2000. Synergistic effect of high hydrostatic pressure pretreatment and osmotic stress on mass transfer during osmotic dehydration. *Journal of Food Engineering*, 45: 25–31.

Rastogi, N.K., L.T. Nguyen, and V.M. Balasubramaniam. 2008. Effect of pretreatments on carrot texture after thermal and pressure-assisted thermal processing. *Journal of Food Engineering*, 88(4): 541–547.

Rastogi, N.K., L.T. Nguyen, B. Jiang, and V.M. Balasubramaniam. 2010. Improvement in texture of pressure-assisted thermally processed carrots by combined pretreatment using response surface methodology. *Food & Bioprocess Technology*, 3: 762–771.

Rastogi, N.K. and K. Niranjan. 1998. Enhanced mass transfer during osmotic dehydration of high pressure treated pineapple. *Journal of Food Science*, 63(3): 508–511.

Richard, J.S. 1992. *High Pressure Phase Behaviour of Multicomponent Fluid Mixtures*. Amsterdam: Elsevier.

Rivalain, N., J. Roquain, and G. Demazeau. 2010. Development of high hydrostatic pressure in biosciences: Pressure effect on biological structures and potential applications in biotechnologies. *Biotechnology Advances*, 28(6): 659–672.

Roßbach, A., L.A. Bahr, S. Gäbel, A.S. Braeuer, and A. Wierschem. 2019. Growth rate of pressure-induced triolein crystals. *Journal of the American Oil Chemists' Society*, 96: 25–33.

Samaranayake, C.P. and S.K. Sastry. 2013. In-situ p H measurement of selected liquid foods under high pressure. *Innovative Food Science & Emerging Technologies*, 17: 22–26.

Sánchez-Moreno, C. and B. De Ancos. 2017. High-pressure processing effect on nutrients and their stability. In: M. Houška and F.V.M. Silva (eds), *High Pressure Processing of Fruit and Vegetable Products*. Boca Raton, FL: CRC Press, pp. 85–104.

Sarker, M.R., S. Akhtar, J.A. Torres, and D. Paredes-Sabja. 2015. High hydrostatic pressure-induced inactivation of bacterial spores. *Critical Reviews in Microbiology*, 41(1): 18–26.

Scepankova, H., M. Martins, L. Estevinho, I. Delgadillo, and J.A. Saraiva. 2018. Enhancement of bioactivity of natural extracts by non-thermal high hydrostatic pressure extraction. *Plant Foods for Human Nutrition*, 73: 253–267.

Schieber, A. and R. Carle. 2005. Occurrence of carotenoid cis-isomers in food: Technological, analytical, and nutritional implications. *Trends in Food Science & Technology*, 16: 416–422.

Schlüter, O., G. Urrutia Benet, V. Heinz, and D. Knorr. 2004. Metastable states of water and ice during pressure-supported freezing of potato tissue. *Biotechnology Progress*, 20(3): 799–810.

Sevdin, S., B. Ozel, U. Yucel, M.H. Oztop, and H. Alpas. 2018. High hydrostatic pressure induced changes on palm stearin emulsions. *Journal of Food Engineering*, 229: 65–71.

Shinwari, K.J. and P.S. Rao. 2018. Thermal-assisted high hydrostatic pressure extraction of nutraceuticals from saffron (Crocus sativus): Process optimization and cytotoxicity evaluation against cancer cells. *Innovative Food Science & Emerging Technologies*, 48(7): 296–303.

Siegoczynski, R.M. and J. Jedrzejewski, R. Wisniewski. 1989. Long time relaxation effect of liquid castor oil under high pressure conditions. *High Pressure Research*, 1: 225–233.

Sila, D.N., C. Smout, T.S. Vu, and M.E. Hendrickx. 2004. Effects of high-pressure pretreatment and calcium soaking on the texture degradation kinetics of carrots during thermal processing. *Journal of Food Science*, 69: E205–E211.

Silva, F.V.M. and Evelyn. 2018. High pressure processing effect on microorganisms in fruit and vegetable products. In: M. Houškaand and F.V.M. Silva (eds), *High Pressure Processing of Fruit and Vegetable Products*. Boca Raton, FL: CRC Press, pp. 343–378.

Silva, F.V.M. and A. Sulaiman. 2019. Polyphenoloxidase in fruit and vegetables: Inactivation by thermal and non-thermal processes. In: L. Melton, F. Shahidiand, and P. Varelis (eds), *Encyclopedia of Food Chemistry*, vol. 2, Amsterdam: Elsevier, pp. 287–301.

Sizer, C., V.M. Balasubramaniam, and E. Ting. 2002. Validating high-pressure processes for low-acid foods. *Food Technology*, 56(2): 36–42.

Sopanangkul, A., D.A. Ledward, and K. Niranjan. 2002. Mass transfer during sucrose infusion into potatoes under high pressure. *Journal of Food Science*, 67: 2217–2220.

Stewart, C.M., C.P. Dunne, and L. Keener. 2016. Pressure-assisted thermal sterilization validation. In: V.M. Balasubramaniam, G.V. Barbosa-Cánovas, and H. Lelieveld (eds), *High Pressure Processing of Food*. New York: Springer, pp. 687–716.

Sulaiman, A., Soo, M.J., Yoon, M.M.L., Farid, M., and Silva, F.V.M. 2015. Modeling the polyphenoloxidase inactivation kinetics in pear, apple and strawberry purees after high pressure processing. *Journal of Food Engineering*, 147: 89–94.

Taiwo, K.A., M.N. Eshtiaghi, B.I.O. Ade-Omowaye, and D. Knorr. 2003. Osmotic dehydration of strawberry halves: Influence of osmotic agents and pretreatment methods on mass transfer and product characteristics. *International Journal of Food Science & Technology*, 38: 693–707.

Takai, R., T.T. Kozhima, and T. Suzuki. 1991. Low temperature thawing by using high-pressure, 1951–1955. In: *Proceeding of the XVII International Congress of Refrigeration*. Montreal: Elsevier.

Techakanon, C. and D.M. Barrett. 2017. The effect of calcium chloride and calcium lactate pretreatment concentration on peach cell integrity after high-pressure processing. *International Journal of Food Science & Technology*, 52: 635–643.

Tefelski, D.B., L. Kulisiewicz, A. Wierschem, A. Delgado, A.J. Rostocki, and R.M. Siegoczyński. 2011. The particle image velocimetry method in the study of the dynamics of phase transitions induced by high pressures in triolein and oleic acid. *High Pressure Research*, 31(1): 178–185.

Terefe, N.S., R. Buckow, and C. Versteeg. 2014. Quality-related enzymes in fruit and vegetable products: Effects of novel food processing technologies, Part 1 – High-pressure processing. *Critical Reviews in Food Science & Nutrition*, 54(1): 24–63.

Tironi, V., A. Le Bail, and M. de Lamballerie. 2007. Effects of pressure-shift freezing and pressure-assisted thawing on sea bass (*Dicentrarchus labrax*) quality. *Journal of Food Science*, 72(7): C381–C387.

Toepfl, S., A. Mathys, V. Heinz, and D. Knorr. 2006. Review: Potential of high hydrostatic pressure and pulsed electric fields for energy efficient and environmentally friendly food processing. *Food Reviews International*, 22(4): 405–423.

Tokusoglu, O., H. Alpas, and F. Bozoglu. 2010. High hydrostatic pressure effects on mold flora, citrinin mycotoxin, hydroxytyrosol, oleuropein phenolics and antioxidant activity of black table olives. *Innovative Food Science & Emerging Technologies*, 11: 250–258.

Urrutia Benet, G., O. Schlüter, and D. Knorr. 2004. High pressure-low temperature processing. Suggested definitions and terminology. *Innovative Food Science & Emerging Technologies*, 5: 413–427.

Urrutia, G., J. Arabas, K. Autio, S. Brul, M. Hendrickx, A. Kakolewski, D. Knorr, A. Le Bail, M. Lille, A.D. Molina-García, A. Ousegui, P.D. Sanz, T. Shen Van, and S. Buggenhout. 2007. SAFE ICE: Low-temperature pressure processing of foods – Safety and quality aspects, process parameters and consumer acceptance. *Journal of Food Engineering*, 83: 293–315.

Urrutia-Benet, G., T. Balogh, J. Schneider, and D. Knorr. 2007. Metastable phases during high-pressure low-temperature processing of potatoes and their impact on quality-related parameters. *Journal of Food Engineering*, 78(2): 375–389.

Vatankhah, H. and H.S. Ramaswamy. 2017. Dynamics of fluid migration into porous solid matrix during high pressure treatment. *Food & Bioproducts Processing*, 103: 122–130.

Vatankhah, H. and H.S. Ramaswamy. 2019. High pressure impregnation (HPI) of apple cubes: Effect of pressure variables and carrier medium. *Food Research International*, 116: 320–328.

Verma, D., N. Kaushik, and P.S. Rao. 2014. Application of high hydrostatic pressure as a pretreatment for osmotic dehydration of banana slices (Musa cavendishii) finish-dried by dehumidified air drying. *Food Bioprocess Technology*, 7(5): 1281–1297.

Villacís, M.F., N.K. Rastogi, and V.M. Balasubramaniam. 2008. Effect of high pressure on moisture and Na Cl diffusion into turkey breast. *LWT - Food Science & Technology*, 41(5): 836–844.

Wilson, D.R., L. Dabrowski, S. Stringer, R. Moezelaar, and T.F. Brocklehurst. 2008. High pressure in combination with elevated temperature as a method for the sterilisation of food. *Trends in Food Science & Technology*, 19(6): 289–299.

Xi, J. and Z. Shouqin, 2007. Antioxidant activity of ethanolic extracts of propolis by high hydrostatic pressure extraction. *International Journal of Food Science & Technology*, 42(11): 1350–1356.

Yan, L. and J. Xi. 2017. Micro-mechanism analysis of ultrahigh pressure extraction from green tea leaves by numerical simulation. *Separation & Purification Technology*, 180: 51–57.

Yasuda, A. and K. Mochizuki. 1992. The behavior of triglycerides under high pressure: The high pressure can stably crystallize cocoa butter in chocolate. *High Pressure Biotechnology*, 224: 255–259.

Yu, Y., X. Jiang, H.S. Ramaswamy, S. Zhu, and H. Li. 2018. High pressure processing treatment of fresh-cut carrots: Effect of presoaking in calcium salts on qualityparameters. *Journal of Food Quality*, Article ID 7863670. doi:10.1155/2018/7863670

Zhao, Y., R.A. Flores, and D.G. Olson. 1998. High hydrostatic pressure effects on rapid thawing of frozen beef. *Journal of Food Science*, 63(2): 272–275.

Zhu, S., A. Le Bail, and H.S. Ramaswamy. 2003. Ice crystal formation in pressure shift freezing of Atlantic salmon (*salmo salar*) as compared to classical freezing methods. *Journal of Food Processing Preservation*, 27: 427–444.

Zhu, S., F. Naim, M. Marcotte, H. Ramaswamy, and Y. Shao. 2008. High-pressure destruction kinetics of *Clostridiumsporogenes*spores in ground beef at elevated temperatures. *International Journal of Food Microbiology*, 126(1–2): 86–92.

Zhu, S., H.S. Ramaswamy, and A. Le Bail. 2005. High-pressure calorimetric evaluation of ice crystal ratio formed by rapid depressurization during pressure-shift freezing of water and pork muscle. *Food Research International*, 38(2): 193–201.

Zhu, S., H.S. Ramaswamy, and B.K. Simpson. 2004. Effect of high-pressure versus conventional thawing on color, drip loss and texture of Atlantic salmon frozen by different methods. *LWT - Food Science & Technology*, 37(3): 291–299.

Zulkurnain, M., V.M. Balasubramaniam, and F. Maleky. 2017. Thermal effects on lipids crystallization kinetics under high pressure. *Crystal Growth & Design*, 17: 4835–4843.

Zulkurnain, M., V.M. Balasubramaniam, and F. Maleky. 2019. Effects of lipid solid mass fraction and non-lipid solids on crystallization behaviors of model fats under high pressure. *Molecules*, 24(15): 2853.

Zulkurnain, M., F. Maleky, and V.M. Balasubramaniam. 2016a. High pressure crystallization of binary fat blend: A feasibility study. *Innovative Food Science & Emerging Technologies*, 38: 302–311.

Zulkurnain, M., F. Maleky, and V.M. Balasubramaniam. 2016b. High pressure processing effects on lipids thermophysical properties and crystallization kinetics. *Food Engineering Reviews*, 8: 393–413.

2 Application of Pulse Electric Fields in Food Processing

Sourav Chakraborty, Mausumi Sarma, Swapnil Prashant Gautam, and Purba Prasad Borah
Tezpur Univesity, Tezpur, Assam, India

CONTENTS

2.1 Introduction 27
2.2 Principles of Pulse Electric Field Processing 28
2.3 PEF System Components 29
2.3.1 Power Supply 29
2.3.2 High-Power Capacitor 29
2.3.3 Switches 30
2.3.4 High-Voltage Pulse Generator 30
2.3.5 Treatment Chamber 31
2.4 Factors Affecting the Outcomes of Pulse Electric Field Treatment 31
2.4.1 Technological Factors 32
2.4.2 Biological Factors 32
2.4.3 Media Factors 32
2.5 Modeling of the Inactivation Rate 33
2.6 Application of Pulse Electric Field in Food Processing 33
2.6.1 Inactivation of Microorganisms 33
2.6.2 Processing of Milk 34
2.6.3 Processing of Eggs 35
2.6.4 Processing of Juice and Soaps 35
2.7 Conclusion 35
References 36

2.1 INTRODUCTION

Consumer demand for fresh and natural food products has reached a high level and is continuing to grow, making preservation of natural quality during food processing a key concern. The thermal processes that are mainly used in food processing to kill contaminating microorganisms have the drawback of degrading the taste, color, flavor, and nutritional quality of the finished product. To overcome these demerits, cold pasteurization methods have been developed in which microorganisms are killed

without the application of heat, while natural quality attributes are retained. Pulse electric field (PEF) is one such method used for pasteurization and possibly sterilization, in combination with other processing hurdles such as microfiltration with large pore ≥1.2 μm, ultraviolet based irradiation, high-intensity light pulses, etc. (Jeyamkondan et al., 1998).

PEF processing is a novel non-thermal preservation method for producing food with excellent sensory and nutritional quality, and shelf life. It is suitable for preserving liquid and semi-liquid foods, removing microorganisms and producing functional constituents, and is also used for cell hybridization and electro fusion in the areas of genetic engineering and biotechnology. PEF technology is regarded as superior to traditional heat treatment because it causes fewer detrimental changes to the sensory and physical properties of food (Quass, 1997; Qin et al., 1995). It is suggested for the pasteurization of foods such as juices, milk, yogurt, soups, and liquid eggs (Vega-Mercado et al., 1997; Bendicho et al., 2003; Puértolas et al., 2013). In high-intensity pulsed electric field (HIPEF) processing, food is placed between two electrodes and pulses of high voltage (typically 20–80 kV/cm) are applied to it. PEF treatment is performed for a very short time (less than 1 s) at ambient, sub-ambient, or slightly above ambient temperature, minimizing energy loss and undesirable changes in sensory properties.

2.2 PRINCIPLES OF PULSE ELECTRIC FIELD PROCESSING

The basic principle of PEF is application of short pulses of high electric fields with an intensity of 10–80 kV/cm for a very short duration. Pulse electric currents are passed through food products placed between a set of electrodes. PEF processing time is calculated by multiplying the number of pulses by the effective pulse duration. The electric field generated by the application of high voltage causes microbial inactivation of food products. The electric field can be applied in various forms: exponentially decaying, square wave, bipolar, or oscillatory pulses (Jeyamkondan et al., 1998).

Microorganisms are protected from the surrounding environment by cell membranes acting as semi-permeable barriers that help control the metabolic activities of the cell. With the application of high-voltage PEF, the cell membrane is disrupted and microorganisms are inactivated by its electromechanical instability. Microbial cells that lose the ability to grow and divide in a nutrient medium are regarded as deactivated.

PEF processing uses a combination of high-voltage power source, an energy storage capacitor bank, a charging current limiting resistor, a switch to discharge energy from the capacitor across the food, and a treatment chamber. High voltage electrical pulses are received by the electrodes and passed to the food product placed between them. The food product experiences force per unit charge which causes irreversible dielectric cell membrane breakdown in microorganisms, and interaction with the charged molecules of food (Fernandez-Díaz et al., 2000; Zimmermann, 1986). An oscilloscope is used to observe pulse waveform. The high-voltage DC generator converts voltage from a utility line (110 V) to high-voltage AC followed by rectification to high-voltage DC. The capacitor stores energy from the power source and discharges it by means of an electrical switch through the treatment chamber to generate an

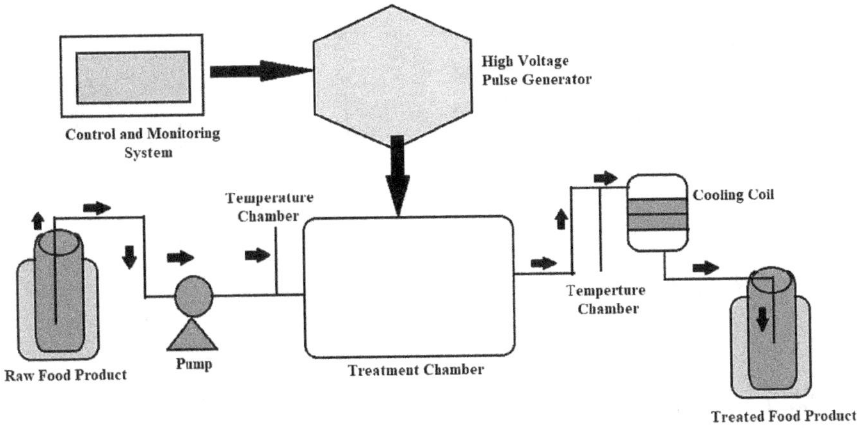

FIGURE 2.1 PEF processing unit.
Source: Mohamed and Eissa (2012)

electric field in the food material. The bank of capacitors is charged from the DC power source obtained from amplified and rectified regular AC. In continuous systems a pump conveys the food through the treatment chamber. A chamber cooling system may also be in use to control food temperature during treatment. High-voltage and high-current probes measure the voltage and current delivered to the chamber (Ho et al., 1995; Barbosa-Cánovas et al., 1999; Floury et al., 2006; Amiali et al., 2006). Figure 2.1 illustrates the PEF processing unit (Ortega-Rivas et al., 1998).

2.3 PEF SYSTEM COMPONENTS

The PEF processing system consists of power supply, capacitor bank, switch, high-voltage repetitive pulser, treatment chamber, cooling system, voltage- and current-measuring devices, control unit, and data acquisition system.

2.3.1 Power Supply

A high-voltage pulse generator, using either an ordinary source of direct current or a capacitor, supplies high-voltage pulses of the required intensity, shape, and duration to the system. The capacitor is generally charged with high-frequency AC inputs that provide a command charge with higher repetitive rates than the DC power supply (Zhang et al., 1996). Figure 2.2 shows commonly used pulse wave shapes and the generic electrical circuits for a PEF system.

2.3.2 High-Power Capacitor

Storage capacitors and on-off switches are the main components of the high-power source. Due to their relatively high ohmic power consumption compared with capacitors, inductors are less commonly used. The energy stored in the capacitors is used

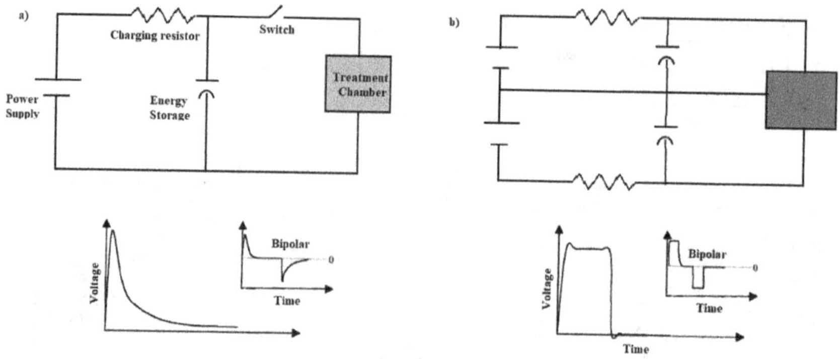

FIGURE 2.2 Commonly used pulse wave shapes and generic electrical circuits: (a) Mono-polar exponential decaying circuits and possible waveform; (b) Mono-polar square circuit and possible waveform.
Source: Mohamed and Eissa (2012)

to generate electric or magnetic fields. Electric fields accelerate charged particles, leading to thermal, chemical, mechanical, electromagnetic wave, or breakdown effects. Electromagnetic fields transfer energy in the form of electromagnetic waves. Typical examples are X-ray, microwaves, and laser beam generation (Weise, 2001).

2.3.3 Switches

The discharge switch is important for efficient control of the PEF system. High-power switching systems are used to connect elements between storage device and load. Inpulse-forming elements, the properties of the switches control rise time, shape, and amplitude of the generator output pulse. Closing switches are used for generators with a capacitive storage device, while opening switches are used for inductive storage devices (Bluhm, 2006). Solid-state semiconductor switches are seen as the future of high-power switching (Bartos, 2000). Two main groups of switches are currently available: ON switches and ON/OFF switches. ON switches fully discharge the capacitor and turn off after the completion of discharge, and can handle high voltages at lower cost than ON/OFF switches. However, they have a shorter life and low repetition rates. Types of ON switch include ignitron, gas spark gap, trigatron, and thyratron. ON/OFF switches provide control over the pulse generation process with partial or complete discharge of capacitors: examples include the gate turn-off (GTO) thyristor, insulated gate bipolar transistor (IGBT), and symmetrical gate commutated thyristor (SGCT).

2.3.4 High-Voltage Pulse Generator

The high-voltage pulse generator provides electrical pulses of the desired voltage, shape, and duration with the help of a pulse-forming network (PFN). PFN is defined as an electrical circuit that combines a number of components: one or more DC power supplies, a charging resistor, a capacitor bank formed by two or more

units connected in parallel, one or more switches, and pulse-shaping inductors and resistors. Charging of the capacitor bank to the desired voltage is done by the DC power supply. With the help of this device, the AC power from the utility line (50–60 Hz) is converted into high-voltage AC power and then rectified to high-voltage DC power (Zhang et al., 1995).

2.3.5 Treatment Chamber

The treatment chamber is one of the most important and complicated components in the PEF processing system. This chamber helps to keep the treated product inside during pulsing. Uniformity of the PEF process mainly depends on the design of the treatment chamber. Applied electric fields exceeding the field strength of the food products induce spark breakdown. Treatment chambers are mainly grouped together in batch or continuous systems. Batch systems are generally used to handle static volumes of solid or semi-solid foods.

Treatment chambers are of two types: parallel-plate and coaxial (Figure 2.3). Parallel-plate chambers are used for batch processing, whereas coaxial chambers are used for continuous processing, in which a higher inactivation rate is observed (Qin et al., 1998).

2.4 FACTORS AFFECTING THE OUTCOMES OF PULSE ELECTRIC FIELD TREATMENT

Where PEF technology is used for pasteurization, its efficacy against pathogenic and spoilage food-borne microorganisms must be predicted. Knowledge of the critical factors affecting microbial inactivation, and an understanding of PEF inactivation kinetics and mechanisms, are therefore required. The lethality factors responsible for controlling the efficiency of PEF technology can be classified as technological,

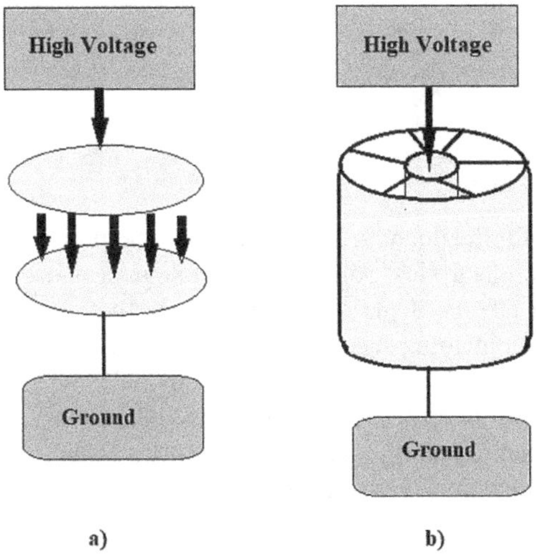

FIGURE 2.3 Configuration of treatment chamber: (a) parallel-plate and (b) coaxial.
Source: Mohamed and Eissa (2012)

biological, and media factors. Type of equipment, processing parameters, target microorganism, and medium conditions are also relevant factors.

2.4.1 Technological Factors

The microbial inactivation rate during PEF processing is affected by various technological factors, including field strength, treatment time, treatment temperature, pulse shape, type of microorganism, growth stage of microorganism, and characteristics of the treatment substrate.

Electric field strength (EFS) is an important parameter for PEF processing. Uniform distribution of EFS is required for the efficient inactivation of microorganisms in food. No microbial inactivation is observed at electric field intensity of 4–8 kV/cm; intensity in the range of 12–45 kV/cm is generally required for inactivation of microorganisms (Peleg, 1995). Pulses are the main factors that distinguish PEF treatment from other microbial inactivation technologies. Generally square or exponential-shape pulses are used for inactivation of microorganisms. Treatment time is defined as the effective time required to subject the microorganism to the applied field strength. The number and width of pulses affect treatment time. The lethality of the treatment is ultimately controlled by the EFS, and the number and width of pulses. Treatment time is calculated by multiplying the pulse number by pulse duration (Sale and Hamilton, 1967; Jayaram et al., 1992; Jayaram et al., 1993; Barsotti and Cheftel, 1999; Wouters et al., 2001).

2.4.2 Biological Factors

Among biological factors, individual characteristics of target microorganisms and their physiological and growth states affect the PEF treatment process. Various intrinsic parameters of the microorganism such as size, shape, species, or growth state determine its susceptibility to PEF inactivation. In PEF treatment, greater resistance by gram-positive than gram-negative bacteria is observed. Higher sensitivity is observed for yeast than for bacteria (Sale and Hamilton, 1967; Hulsheger et al., 1983; Zhang et al., 1994).

2.4.3 Media Factors

The efficiency of PEF treatment is also affected by the properties of the food being treated between the electrodes. Although technological and biological factors are very important, the effects of pH, temperature, resistivity, and composition of the enzyme- or protein-containing medium or food system are also important for controlling the performance of PEF treatment (Barsotti and Cheftel, 1999).

The effectiveness of microbial inactivation, as well as recovery of injured microbial cells and their subsequent growth during PEF treatment, is strongly influenced by the physical and chemical characteristics of food products, which include conductivity, resistivity, dielectric properties, ionic strength, pH, and composition (Wouters et al., 2001; Ho et al., 1995; Grahl and Märkl, 1996; Martín et al., 1997). When foods are subjected to a high-intensity pulse electric field, temperature is also important (Jayaram et al., 1993). The lethality of PEF treatment generally increases with the

processing temperature. A cooling device should be attached to the PEF treatment unit to maintain the proper temperature without degradation of nutritional, sensory, and functional properties (Wouters et al., 2001).

2.5 MODELING OF THE INACTIVATION RATE

Inactivation of *Escherichia coli* in a static treatment chamber based on an exponential pulse generator was studied by Hulsheger and Niemann (1980). With the application of 10 pulses of 30-ms pulse width and 20-kV/cm electric-field strength, 3-log reduction of *E. coli* was observed. A linear relationship between the inactivation (log reduction) of *E. coli* and the EFS was obtained. Treatment of food material with electric fields below a threshold value of 3 kV/cm did not show a lethal effect on *E. coli*. A mathematical model for the survival rate as a function of electric-field strength and treatment time was developed by Hulsheger et al. (1981):

$$s = \left(\frac{t}{t_c}\right)^{\left[-(E-E_c)/k\right]} \tag{2.1}$$

Here, s = survival ratio (ratio of number of microorganisms present in the food after treatment and initial number of microorganisms present before the treatment), t = treatment time, t_c = critical treatment time, which is a threshold value above which inactivation occurs (ms), E = electric field strength (kV/cm), E_c = critical electric field strength and k = specific constant for a microorganism.

2.6 APPLICATION OF PULSE ELECTRIC FIELD IN FOOD PROCESSING

2.6.1 Inactivation of Microorganisms

PEF treatment has a lethal effect on various vegetative bacteria, mold, and yeast. The cell membranes of vegetative microorganisms are broken by expanding existing pores (electroporation) or by creating new ones with the application of high-voltage pulses. Reversibility or irreversibility of pore formation depends on various factors such as field intensity, pulse duration, and number of pulses (Figure 2.4). Due to this, small molecules can easily penetrate through the membrane of the PEF-treated cell,

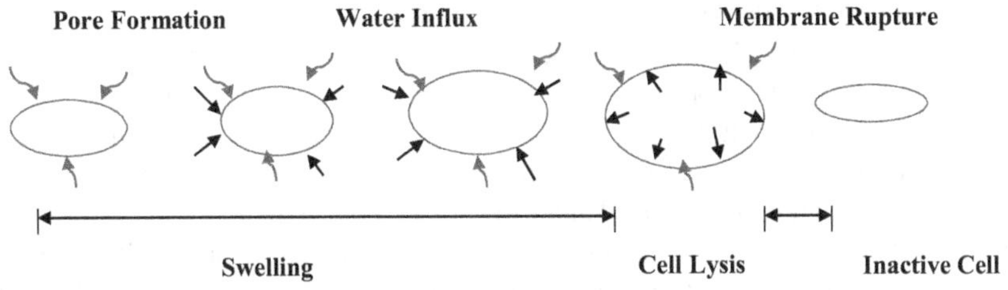

FIGURE 2.4 Breakdown of cell membrane during various stages of electroporation.
Source: Kumar et al. (2016)

TABLE 2.1
Effect of Pulse Electric Field (Pef) on Inactivation of Microorganisms and Enzymes

Food Product	Reduction in Log Cycle	Target Microorganism	Type of Treatment Vessel	Process Condition
Raw skimmed milk	2.6	*Listeria innocua*	Coaxial	Temperature: 15–28 °C Voltage: 50 kV/cm Time: 2 μs
Pasteurized whole milk	3.0–4.0	*Listeria monocytogenes*	Cofield flow	Temperature: 10–50 °C Voltage: 30 kV/cm Time: 1.5 μs
Yoghurt	2	*Lactobacillus brevis*	Parallel plates	Temperature: 50 °C Voltage: 1.8 kV/μm
Pea soup	5.3	*B. subtilis* spores	Coaxial	Temperature: <5.5 °C Voltage: 3.3 V/μm Time: 2 μs Frequency: 4.3 Hz Total pulses: 30

Source: Kumar et al. (2016)

resulting in swelling and eventual rupture of the membrane. Table 2.1 illustrates the effect of the PEF system on inactivation of microorganisms and enzymes.

2.6.2 Processing of Milk

PEF processing improves shelf life and overall quality of milk and milk products. Fernandez-Molina et al. (2005) reported shelf life of 2 weeks for PEF-processed skimmed milk treated at 40 kV/cm with 30 pulses (exponential decaying pulses) for a treatment time of 2 μs. Increased shelf life of 22 days was observed for skimmed milk treated at 80°C for 6 s followed by PEF treatment at 30 kV/cm. During the PEF treatment of skimmed milk, the processing temperature did not exceed 28°C.

Shelf life of homogenized milk can also be enhanced by the application of PEF treatment. Dunn and Pearlman (1987) studied shelf life of homogenized milk (inoculated with *Salmonella* Dublin) treated at 36.7 kV/cm with 40 pulses over 25 min, and reported no growth of *Salmonella* Dublin after storage at 7–9°C for 8 days. Bacterial population for naturally occurring milk increases to 10^7 cfu/mL, as compared to 4×10^2 cfu/mL for PEF-treated milk. Less flavor degradation and no chemical or physical changes in milk quality attributes for cheese making was reported in further studies by Dunn (1996). For milk subjected to 2 steps of 7 pulses and 1 step of 6 pulses with an electric field of 40 kV/cm, shelf life of 2 weeks at refrigerated temperature can be observed. No significant changes in physical, chemical, and sensory attributes were observed between heat-pasteurized and PEF-treated milk (Qin et al., 1995). Calderon-Miranda (1998) reported inactivation of *Listeria innocua* suspended in skimmed milk and its subsequent sensitization to nisin. After PEF treatments at 30, 40, or 50 kV/cm, the microbial population of *L. innocua* was reduced by 2.5-log cycles. The lethality effect of PEF was reported as a function of field intensity and treatment time.

2.6.3 Processing of Eggs

Processing of eggs with the application of PEF treatment helps to improve shelf life with minimal changes to quality attributes. Dunn and Pearlman (1987) treated egg in a static parallel electrode treatment chamber with a 2-cm gap using 25 exponentially decaying pulses with peak voltages of around 36 kV. With the addition of potassium sorbate and citric acid as preservatives, tests were conducted on liquid eggs and heat-pasteurized liquid egg products. Extension in shelf life between the regular heat-pasteurized sample and the PEF-treated one was observed at both low (4°C) and high (10°C) refrigeration temperatures.

Further, PEF treatment studies conducted by Qin et al. (1995) and Martín-Belloso et al. (1997) showed a reduction in viscosity along with an increase in color of liquid eggs compared to freshly treated ones. No significant changes in sensory properties were observed between PEF-treated and freshly treated egg products. Greater strength was observed in sponge cake prepared with PEF-treated eggs than in a cake made with normally treated ones. The difference in strength was attributed to lower expansion after baking with PEF-treated eggs.

2.6.4 Processing of Juice and Soaps

Vega-Mercado et al. (1996) reported extended shelf life of 4 weeks at refrigeration temperature by treating pea soup with two steps of 16 pulses at 35 kV/cm. No significant changes in the physical and chemical properties or sensory attributes of the pea soup directly after PEF processing or during the four weeks of storage at refrigeration temperatures were observed.

Simpson et al. (1995) reported an increase in the shelf life of apple juice from 21 to 28 days with PEF treatment. The apple juice was treated with PEF at 50 kV/cm for 10 pulses with a pulse width of 2 µs. No changes in ascorbic acid, sugars, and sensory properties were observed in the PEF-treated juice samples.

PEF can also be effectively applied for the extraction of juice from fruits. Grimi (2009) reported extraction of juice from Chardonnay white grape using PEF with two pressure conditions. Juice yield with PEF treatment was increased by 67–75% compared to the control sample without any treatment. Jarupan Kuldiloke et al. (2008) applied PEF treatment to the extraction of juice from sugar cane. They reported higher juice yield from a PEF-treated sample than from heat-treated and untreated sugar cane. Lower energy consumption for the disintegration of sugar cane was observed for the PEF-treated samples. Jemai and Vorobiev (2006) applied PEF to cold juice extraction from sugar beet cossettes. They reported higher purity of the juice in case of PEF treatment based extraction process.

2.7 CONCLUSION

Pulse electric field (PEF) is one of the most promising non-thermal processing methods for the preservation of food products. PEF technology involves the application of very short pulses (micro to milliseconds) to liquid or semi-solid food placed between two electrodes at electric field intensities ranging from 10 to 80 kV/cm.

For the application of high-intensity pulsed electric fields, short time pulses are generated between two parallel-plate electrodes enclosing a dielectric material. The PEF processing system consists of a high-power pulse generator, treatment cell, and voltage- and current-measuring devices. During PEF treatment, parallel-plate chambers are used for batch processing, whereas coaxial chambers are used for continuous processing. In the case of coaxial chamber-based continuous processing systems, a higher inactivation rate is observed.

PEF processing inactivates microorganisms present in food products with minimum changes to their physical, sensory, and functional properties. Exposure of food products to high electric field pulses develops pores in the cell membrane of the microorganism either by enlarging the existing pores or creating new ones. Hence, it can be concluded that PEF processing is a pioneering approach to the preservation of food material.

REFERENCES

Amiali, M. and M.O. Ngadi, J.P. Smith and V.G. Raghavan. 2006. Inactivation of Escherichia coli O157: H7 and Salmonella enteritidis in liquid egg white using pulsed electric field. *J. Food Sci.* 71(3): 88–94.

Barbosa-Cánovas, G.V. and M.M. Góngora-Nieto, U.R. Pothakamury and B.G. Swanson. 1999. Fundamentals of high-intensity pulsed electric fields (PEF). In G.V. Barbosa-Cánovas et al. (eds), *Preservation of Foods with Pulsed Electric Fields*. San Diego: Academic Press, pp. 1–19, 76–107, 108–155.

Barsotti, L. and J.C. Cheftel. 1999. Food processing by pulsed electric fields. II. Biological aspects. *Food Rev. Inter.* 15(2): 181–213.

Bartos, F. 2000. SPS/IPC/Drives' 99: More coverage online at www.controleng.com. *Control Engineering*, 47(3), 32–33.

Bendicho, S., G.V. Barbosa-Cánovas and O. Martín. 2003. Reduction of protease activity in simulated milk ultrafiltrate by continuous flow high intensity pulsed electric field treatments. *J. Food Sci.* 68(3): 952–957.

Bluhm, H. 2006. *Switches. Pulsed Power Systems: Principles and Applications*, Springer, 83–133.

Dunn, J. 1996. Pulsed light and pulsed electric field for foods and eggs. *Poul. Sci.* 75(9): 1133–1136.

Dunn, J.E. and J.S. Pearlman. 1987. U.S. Patent No. 4,695,472. Washington, DC: U.S. Patent and Trademark Office.

Fernandez-Díaz, M.D., L. Barsotti, E. Dumay and J.C. Cheftel. 2000. Effects of pulsed electric fields on ovalbumin solutions and dialyzed egg white. *J. Agri. Food Chem.* 48(6): 2332–2339.

Fernandez-Molina, J.J., G.V. Barbosa-Canovas, and B.G. Swanson. 2005. Skim milk processing by combining pulsed electric fields and thermal treatments. *J. Food Process. Preserv.*, 29(5–6): 291–306.

Floury, J., N. Grosset, N. Leconte, M. Pasco, M.N. Madec and R. Jeantet. 2006. Continuous raw skim milk processing by pulsed electric field at non-lethal temperature: Effect on microbial inactivation and functional properties. *Le Lai.* 86(1): 43–57.

Grahl, T. and H. Märkl. 1996. Killing of microorganisms by pulsed electric fields. *App. Micro. Biotech.* 45(1–2): 148–157.

Grimi, N., N.I. Lebovka, E. Vorobiev and J. Vaxelaire. 2009. Effect of a pulsed electric field treatment on expression behavior and juice quality of chardonnay grape. *Food Biophy.* 4(3): 191–198.

Ho, S.Y., G.S. Mittal, J.D. Cross and M.W. Griffiths. 1995. Inactivation of Pseudomonas fluorescens by high voltage electric pulses. *J. Food Sci.* 60(6): 1337–1340.
Hulsheger, H. and E.G. Niemann. 1980. Lethal effects of high-voltage pulses on E. coli K12. *Rad. Env. Biophy.* 18(4): 281–288.
Hulsheger, H., J. Potel and E.G. Niemann. 1981. Killing of bacteria with electric pulses of high field strength. *Rad. Env. Biophy.* 20(1): 53–65.
Hulsheger, H., J. Potel and E.G. Niemann. 1983. Electric field effects on bacteria and yeast cells. *Rad. Env. Biophy.* 22(2): 149–162.
Jayaram, S., G.S.P. Castle and A. Margaritis. 1992. Kinetics of sterilization of Lactobacillus brevis cells by the application of high voltage pulses. *Biotech. Bioeng.* 40(11): 1412–1420.
Jayaram, S., G.S.P. Castle and A. Margaritis. 1993. The effects of high field DC pulse and liquid medium conductivity on survivability of Lactobacillus brevis. *App. Micro. Biotech.* 40(1): 117–122.
Jemai, A.B. and E. Vorobiev. 2006. Pulsed electric field assisted pressing of sugar beet slices: Towards a novel process of cold juice extraction. *Biosys. Eng.* 93(1): 57–68.
Jeyamkondan, S., D.S. Jayas and R.A. Holley. 1998, September. *Pasteurization of foods by pulsed electric fields at high voltages*. In *North Central ASAE Meeting*, Brookings, South Dakota, USA (pp. 24–26).
Kuldiloke, J., M.N. Eshtiaghi, C. Neatpisarnvanit and T. Uan-On. 2008. Application of high electronic field pulses for sugar cane processing. *Cur. App. Sci. Tech.* 8(2): 75–83.
Kumar, S., N. Agarwal and P.K. Raghav. 2016. Pulsed electric field processing of foods – A review. *Inter. J. Eng. Res. Mod. Edu.* 1(1): 111–1118.
Martín, O., B.L. Qin, F.J. Chang, G.V. Barbosa-Cánovas and B.G. Swanson. 1997. Inactivation of Escherichia coli in skim milk by high intensity pulsed electric fields. *J. Food Proc. Eng.* 20(4): 317–336.
Martín-Belloso, O., H. Vega-Mercado, B.L. Qin, F.J. Chang, G.V. Barbosa-Cánovas and B.C. Swanson. 1997. Inactivation of Escherichia coli suspended in liquid egg using pulsed electric fields. *J. Food Process. Preserv.*, 21(3): 193–208.
Miranda, M.L.C. 1998. *Inactivation of Listeria innocua by pulsed electric fields and nisin.* Doctoral dissertation, Washington State University.
Mohamed, M.E. and A.H.A. Eissa. 2012. Pulsed electric fields for food processing technology. *Struc. Func. Food Eng.* 11: 275–306.
Peleg, M. 1995. A model of microbial survival after exposure to pulsed electric fields. *J. Sci. Food Agri.* 67(1): 93–99.
Puértolas, E., O. Cregenzán, E. Luengo, I. Álvarez and J. Raso. 2013. Pulsed-electric-field-assisted extraction of anthocyanins from purple-fleshed potato. *Food Chemistry*, 136(3–4): 1330–1336.
Qin, B.L. and G.V. Barbosa-Cánovas, B.G. Swanson, P.D. Pedrow and R.G. Olsen. 1998. Inactivating microorganisms using a pulsed electric field continuous treatment system. *IEEE Trans. Indus. App.* 34(1): 43–50.
Qin, B.L., F.J. Chang, G.V. Barbosa-Cánovas and B.G. Swanson. 1995. Nonthermal inactivation of Saccharomyces cerevisiae in apple juice using pulsed electric fields. *LWT - Food Sci. Tech.* 28(6); 564–568.
Quass, D.W. 1997. *Pulsed electric field processing in the food industry. A status report onpulsed electric field.* Palo Alto, CA: Electric Power Research Institute. CR-109742. pp. 23–35.
Sale, A.J.H. and W.A. Hamilton. 1967. Effects of high electric fields on microorganisms: I. Killing of bacteria and yeasts. *Biochimica et Biophysica Acta (BBA)-General Subjects*, 148(3): 781–788.
Simpson, M.V., G.V. Barbosa-Cánovas and B.G. Swanson. 1995. The combined inhibitory effect of lysozyme and high voltage pulsed electric fields on the growth of Bacillus subtilis spores. In *IFT Annual Meeting: Book of Abstracts* (Vol. 267).

Vega-Mercado, H., O. Martín-Belloso, F.J. Chang, G.V. Barbosa-Cánovas and B.G. Swanson. 1996. Inactivation of Escherichia coli and Bacillus subtilis suspended in pea soup using pulsed electric fields. *J. Food Proc. Preser.* 20(6): 501–510.

Vega-Mercado, H., O. Martín-Belloso, B.L. Qin, F.J. Chang, M.M. Góngora-Nieto, G.V. Barbosa-Cánovas and B.G. Swanson. 1997. Non-thermal food preservation: Pulsed electric fields. *Tre. Food Sci. Tech.* 8(5): 151–157.

Weise, T.H. 2001. *Overview on pulsed power applications*. In *Proceedings of the International Conference on Pulsed Power Applications*, March 27–29, 2001, Gelsenkirchen, Germany.

Wouters, P.C., I. Alvarez and J. Raso. 2001. Critical factors determining inactivation kinetics by pulsed electric field food processing. *Tre. Food Sci. Tech.* 12(3–4): 112–121.

Zhang, Q., G.V. Barbosa-Cánovas and B.G. Swanson. 1995. Engineering aspects of pulsed electric field pasteurization. *J. Food Eng.* 25(2): 261–281.

Zhang, Q., F.J. Chang, G.V. Barbosa-Cánovas and B.G. Swanson. 1994. Inactivation of microorganisms in a semisolid model food using high voltage pulsed electric fields. *LWT - Food Sci. Tech.* 27(6): 538–543.

Zhang, Q., B.L. Qin, G.V. Barbosa-Cánovas, B.G. Swanson and P.D. Pedrow. 1996. U.S. Patent No. 5,549,041. Washington, DC: U.S. Patent and Trademark Office.

Zimmermann, U. 1986. Electrical breakdown, electropermeabilization and electrofusion. In: *Reviews of Physiology, Biochemistry and Pharmacology*, vol. 105. Berlin, Heidelberg: Springer, pp. 175–256.

3 Recent Advances in Ultrasound Processing of Food

Abhinav Tiwari
Indian Institute of Food Processing Technology, Thanjavur, Tamil Nadu, India
University of Manitoba, Winnipeg, MB, Canada

Animesh Singh Sengar
Indian Institute of Food Processing Technology, Thanjavur, Tamil Nadu, India

Anjali H. Kurup
Indian Institute of Food Processing Technology, Thanjavur, Tamil Nadu, India
Tennessee State University, Nashville, TN, USA

Ashish Rawson
Indian Institute of Food Processing Technology, Thanjavur, Tamil Nadu, India

CONTENTS

3.1 Introduction 40
3.2 Ultrasonic Cutting 41
3.3 Microbial and Enzyme Inactivation 41
3.4 Extraction 44
3.5 Freezing 48
3.6 Dehulling 50
3.7 Drying 51
3.8 Ultrasonic or Sono-Emulsification/Homogenization 53
3.9 Structural Modification 56
3.10 Tempering 58
3.11 Treatment of Food Industry Waste Water 59
3.12 Conclusion 59
References 59

3.1 INTRODUCTION

The term "acoustic" relates to the science of elastic waves. There are a wide range of applications from communication to engineering, from life and earth sciences to arts and many more. Applications use different parts of the acoustic spectrum to suit their requirements (Figure 3.1). Ultrasound is applied in food processing for extraction, modification, microbial inactivation, shelf-life extension, quality improvement, and quality evaluation.

Ultrasonication can be used in operations such as cutting, microbial and enzyme inactivation, extraction, freezing, dehulling, drying, emulsification/homogenization, structure modification, tempering, and treatment of food industry waste water.

Ultrasound can be defined as sound waves having frequency higher than the human hearing range. It was first generated in the year 1881 by means of the piezoelectric effect. An alternating voltage was applied to the specific plane of sodium potassium tartrate tetrahydrate crystals (Wang et al. 2018). Ultrasound can also be produced by mechanical vibration using magneto-strictive methods, whistles, etc. (Mackersie et al. 2005). These mechanical vibrations can be amplified to power a sonotrode which converts mechanical vibrations to ultrasound. The sonotrode applies ultrasound directly to the product to be treated.

Ultrasonic waves create cavitation bubbles which on collapse create high temperatures and pressures. High acoustic pressure also creates microjets/microstreaming and causes shear stress which leads to mechanical disturbance in the food structure and cleavage of chemical bonds. Detailed reviews have been published by O'Donnell et al. (2010), Chemat and Khan (2011), and Chemat et al. (2017b) related to enzyme inactivation, extraction and preservation using ultrasound in various ways. This chapter aims to provide insight into recent advances in ultrasonic food-processing techniques.

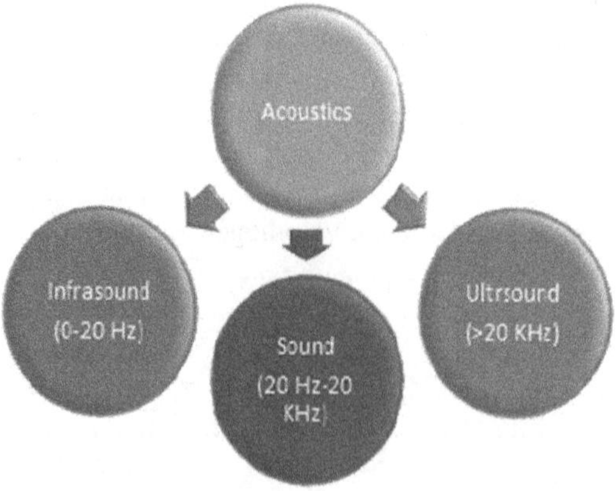

FIGURE 3.1 The acoustic spectrum

3.2 ULTRASONIC CUTTING

In cutting various foods ultrasound can supplement cutting blade movement to achieve better quality than that achieved by conventional methods. The cutting tool motion is actually chopping rather than cutting. A chain of elements are synchronized to achieve a perfect cut in the separation zone, depending on the mount angle of the cutting tool (Figure 3.2). Such arrangements give less crumbling, squeezing, debris and smearing. The aim is to obtain perfect thin slices, better than those produced by conventional cutters in line cutting of bread, dough or cakes (Liu et al. 2015).

Ultrasonic cutting improves hygiene by reducing adherence of food components to the blades. This reduces the chance of microbial growth during long runs. Improved cutting precision and motion reproducibility reduces losses and cracking, thus enhancing standardization of the dimensions and weight of every part being cut (Chemat and Khan, 2011).

3.3 MICROBIAL AND ENZYME INACTIVATION

Controlling microbial growth is key in extending food shelf stability and the extent to which it can be consumed before deterioration. Thermal inactivation, in the form of pasteurization and sterilization, has hitherto been used to control microbes and enzymes causing deterioration during food storage. However, heat treatment may cause deterioration of biomolecules and essential nutrients. Ultrasound processing can be seen as an alternative to heat treatment. It can help retain food quality whilst limiting the growth of harmful microorganisms (Janghu et al., 2017).

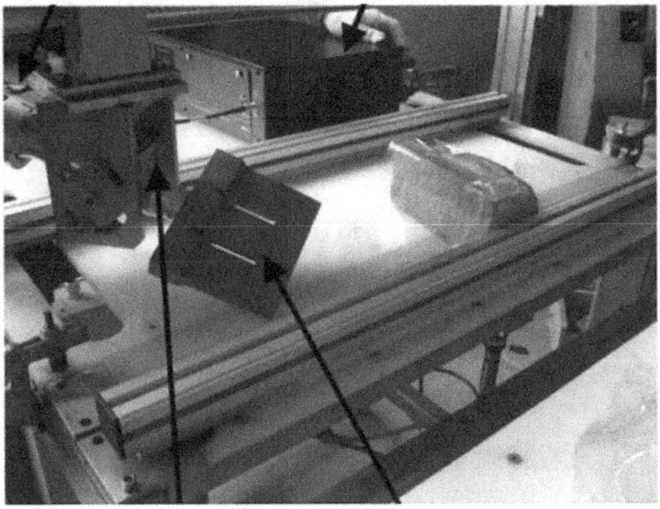

FIGURE 3.2 Ultrasonic cutting assembly for slicing bread.
Source: Adapted from Liu et al. (2015)

The principal mechanism by which ultrasound inactivates microbes and enzymes is based on the physical/mechanical forces associated with cavitation. A cycle of bubble formation, growth, and collapse creates localized hotspots which in turn produce high temperatures. The surrounding cells absorb heat and are inactivated. Along with the mechanical damage, the creation of free radical and electronically excited species may cause oxidative damage which can, to some extent, kill microbes. The earliest literature report of microbial inactivation using ultrasound came from Princeton University, USA, in 1929. Sonication at 375 kHz was used for the reduction of *Bacillus fischeri*. Later, higher-power ultrasound was applied to treat milk and fruit juices until the late 1970s. It provided extended shelf life and retained quality. It was then explained that ultrasound causes cell wall thinning, which ultimately compromised the function of the cytoplasmic membrane. But the versatility of microorganisms meant that ultrasound alone failed on several occasions to restrict microbial growth. It was therefore tested together with heat and pressure to enhance the lethality of the overall process. Ultrasound treatment at moderately elevated temperatures (50–70°C), known as thermosonication, and pressures (200–700 kPa), known as manosonication, caused apparent increases incavitation phenomena and the energetics of bubble collapse.

Extensive research has concluded that ultrasonic inactivation basically follows first-order kinetics but deviations can be observed (Rodríguez-Calleja et al. 2006; Arroyo et al. 2011a, 2011b, 2012a, 2012b). Microscopy observation of treated samples confirmed that inactivation was due to structural damage of the microbes (Figure 3.3). Variation in ultrasound intensity directly influenced the degree of structure disruption. *E. coli* and *S. cerevisiae* have been the most widely studied

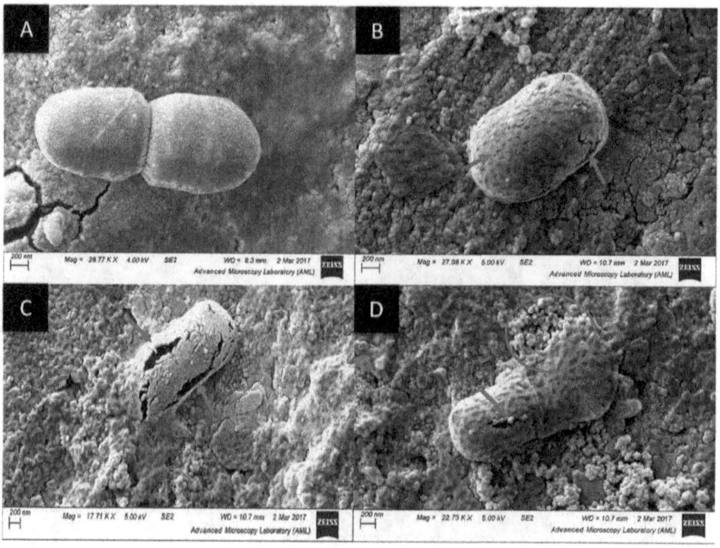

FIGURE 3.3 Formation of pores on the surface of *L. sakei* strain after the application of ultrasound.
Source: Adapted from Ojha et al. (2018)

microorganisms for the application of ultrasound. Flow cytometry and electron microscopy (SEM and TEM) later revealed DNA damage and damage to the internal structure (Bermúdez-Aguirre et al. 2011; Moody et al. 2014; Birmpa et al. 2013; Marchesini et al. 2015). The inactivation rate during ultrasonic treatment can be expressed as the D-value as a function of treatment temperature and time. Ultrasonic treatment of tomato juice and milk, as reported by Adekunte et al. (2010) and Juliano et al. (2014), revealed shoulders in the survival curves for the inactivation of *Pichia fermentans* and *C. sakazakii*. This may be explained as an effect of shock waves causing the disintegration of cell aggregates.

Guimarães et al. (2018) used high-intensity ultrasound at powers ranging from 200 to 600 W to inactivate aerobic mesophilic heterophobic bacteria, yeast, and mold in a prebiotic whey beverage. It was observed that increased ultrasonic power substantially affected the microbial inactivation in short times. It also influenced the zeta potential, thus maintaining the stability of the beverage. Similar observations were reported by Ferrario et al. (2015) who treated apple juice using ultrasound and pulsed light. A 3-log reduction in the population of *Alicyclobacillus* spores and *Saccharomyces cerevisiae* was observed. Combination of the treatment with pulsed light (PL) enhanced inactivation of the spore and the yeast to 6.4 and 5.8 log reduction, respectively. The combined treatment gave the juice a lifetime under ambient conditions of 15 days following treatment. Another recent study by Michelino et al. (2018) describes microbial inactivation in coriander, a major spice, using a combined treatment of ultrasound with supercritical carbon dioxide. Inactivation was reportedly achieved during the CO_2 pressurization phase of the treatment. Mesophilic bacteria were reportedly reduced by up to 4 log cycles and mesophilic spores were reduced by 2 log cycles. Apart from microbial inactivation, this process also led to a considerable loss in the water content of the coriander which enhanced dehydration, making it shelf stable.

Inactivation of enzymes using ultrasound has previously been reviewed by O'Donnell et al. (2010), Islam et al. (2014), Delgado-Povedano and De Castro (2015), and Chemat et al. (2017a). The mechanisms involved are quite complex and need to be monitored precisely. They include removal of lipids from lipoproteins, fragmentation of polypeptides, and even oxidation of aromatic residues. The effects of ultrasound on the enzymes widely used in the food industries are clear, particularly enzymes which cause deterioration in plant-derived foodstuffs, juices, and liquid milk. Most studies suggest that exposure to ultrasound is insufficiently potent for the inactivation of enzymes when used alone, but when combined with heat and pressure, it can emerge as a standalone treatment (Figure 3.4).

Treating milk for the inactivation of alkaline phosphatase (AP) is the prime focus of milk processors. Cameron et al. (2009) described the use of a 20 kHz ultrasound system at a power of 750 W for the inactivation of AP at low temperature (24–26°C). This was the first report of successful inactivation using ultrasound, but scaling up the treatment for mainstream milk processing remains an issue.

Lipase on the other hand is secreted by bacteria even in refrigerated milk. Lipase catalyzes the hydrolysis of milk triglycerides. A study by Vercet et al. (1997) is the main literature describing the use of manothermosonication for the inactivation of lipase. D-values were calculated for experiments conducted at 450 kPa for 1.27 min.

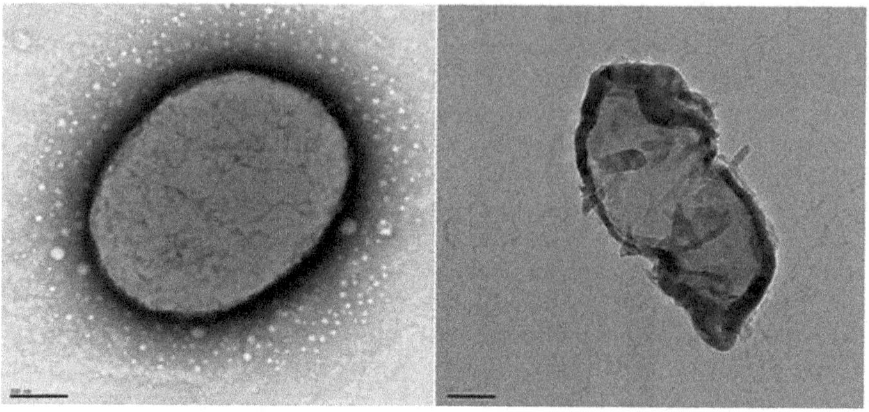

FIGURE 3.4 Combined effects of ultrasound and ozone processing on ***E. coli*** in water.
Source: Adapted from Al-Hashimi et al. (2015)

Vercet et al. (2002) studied the effect of manothermosonication at pressures ranging from 350 to 500 kPa. No added effects were observed when the pressure was increased to around 600 kPa. The D-values reduced by 20% when the temperature was almost 140°C.

The effect of thermosonication on pectin methylesterase (PME) was reported by Aadil et al. (2015) and Saeeduddin et al. (2015) in grape and pear juice, respectively. The inactivation of PME is important to maintain cloud stability in the juices during storage. The treatment retained ascorbic acid and other important phenolic compounds with no change in the pH, acidity and soluble solid content.

The most widely found enzymes in food systems are the peroxidases that cause oxidation of a large group of natural compounds. They are difficult to inactivate due to their thermal stability, and degrade the flavor and color of the foodstuff. Ultrasound treatment is reported to bring about conformational change in the structure of peroxidases and in their prosthetic groups. Ultrasonically induced cell damage and extraction might increase the activity of the enzyme (Terefe et al. 2016). Similar implications were observed in the case of polyphenol oxidase (PPO) by Saeeduddin et al. (2015) for a treatment at 95°C for 2 min. More recently, Huang et al. (2015) and Silva et al. (2015) reported the inactivation of PPO in Satsuma, mandarin, and apple juice, respectively, using thermosonication. Some recent studies of the ultrasonic inactivation of microbes and enzymes are given in Table 3.1.

3.4 EXTRACTION

Ultrasound has proved to be an innovative and energy-efficient method for the extraction of essential bioactive compounds. The shortcomings of existing extraction processes are longer extraction time, higher energy input (about 70%), lower extraction efficiency, increased use of chemicals, and their adverse effect on human health. These drawbacks limit the use of conventional and chemical methods of extraction.

TABLE 3.1
Recent Studies on Ultrasonic Inactivation of Microbes and Enzymes

Microbe/ Enzyme	Medium	Treatment	Specification	Reduction (log)	References
Bacillus subtilis	Black pepper; tapioca starch	Airborne acoustic ultrasound	170 W for 0–120 min	2	Charoux et al. (2019)
Geobacillus spp. A. flavithermus	Water	Ultrasonication combined with H_2O_2 and NaOH	20 kHz, from 0.5 to 20 min	7	Palanisamy et al. (2019)
Staphylococcus spp. Streptococci; Lactobacilli	Sheep's milk	Sonication	20 kHz, 130 W for 4–8 min	4	Balthazar et al. (2019)
Alicyclobacillus acidoterrestris	Apple juice	Thermosonication	35 kHz, 120–480 W	4.8–5.5	Tremarin et al. (2019)
Bacillus subtilis	Chinese bayberry juice	Sonication with mild heat	20 kHz, 900 W at 63 °C	5	Li et al. (2019)
E. coli Staphylococcus aureus	Orange juice	Sono-photodynamic inactivation	20 kHz at 35 °C with blue light emission	4.2; 2.35	Bhavya and Hebbar (2019)
PPO PME	Peach juice	Ultrasound and high pressure homogenization	–	50%	Yildiz (2019)
Peroxidase	Golden berry puree	Ultrasonication and thermal pasteurization	20 kHz, 1000 W	90%	Etzbach et al. (2019)
Invertase	Sucrose	Ultrasonication	20 kHz, 22 W/L	Enhanced activity by 27%	de Souza Soares et al. (2019)

Ultrasonication produces cavitation bubbles, which on rupture generate microjets and shear stress on the cell cuticles. These disrupt cell wall structure and enhance the mass transfer rate of target materials into the extraction solvent. Cavitation collapse near the cell wall causes cell disruption (Figure 3.5), thus improving solvent penetration into the cell structure and increasing the mass transfer rate (Wu et al. 2016). Shock waves created during bubble explosion produce shear forces between the cells which can increase swelling and hydration. Structural modification of the plant tissue ensues and larger pores in the cell wall are formed. These effects are the main causes of the higher diffusion rate of intracellular materials.

Most of the studies suggest that probe-type ultrasound system sare more effective as the horn comes into direct contact with the foodstuff. Bath-type systems are less effective as there is a transmitting medium between the ultrasound generator and the foodstuff. This reduces the energy available for extraction and diminishes the extraction yield (Table 3.2).

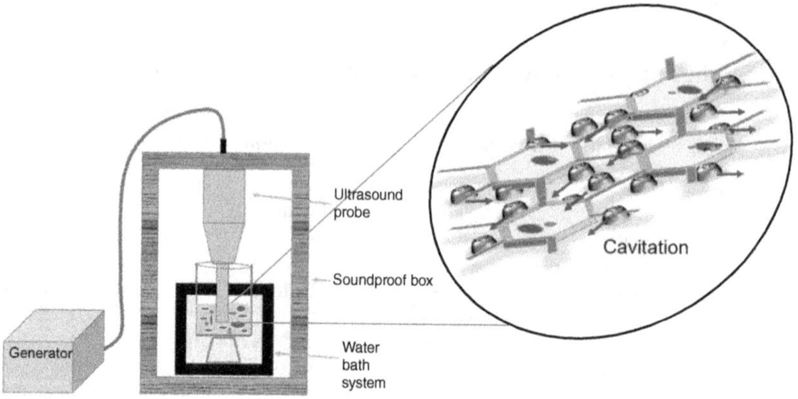

FIGURE 3.5 Action of cavitation bubbles during ultrasound-assisted extraction

Ultrasound-assisted extraction of pectin from food industry processing waste is of increasing interest to reduce the carbon footprint (Sengar et al. 2020). Pomegranate processing produces large amounts of by-products, mainly peel (78%) and seeds (22%) (Moorthy et al. 2015). Extraction of pectin from pomegranate peels used a 20 kHz ultrasound device with an immersed probe (2 cm diameter, flat tip) and maximum power of 130 W. The process variable sex plored were solid–liquid ratio, pH, temperature, and extraction time. An outstanding correlation between the predicted and experimental values was obtained. The author sconcluded that a solid–liquid ratio of 1:18 g/mL with a processing time of 29 min at 62°C gave the highest extraction yield.

Krishnan et al. (2015) have studied the feasibility of ultrasound-assisted extraction of essential oil from rice bran. The maximum oil yield of 10.8% was obtained by extraction at 93% power for 29 min. Higher extraction efficiency was obtained by ultrasound-assisted extraction with ethanol as solvent, when compared with conventional extraction using ethanol and hexane as solvent. The authors reported that ultrasound-treated samples had higher concentrations of unsaturated fatty acids, compared to those present in conventionally extracted oil. The unsaturated fatty acid percentage was higher for ethanol-extracted oil than for oil obtained using hexane as solvent. The results clearly show the superior extraction efficiency of ultrasound-assisted extraction and reportedly it has the potential to replace the conventional solvent extraction method.

Figure 3.6 illustrates the ultrasound-assisted extraction of essential oils from basil leaves (Naik et al., 2021). Collapse of ultrasound cavitation bubbles, followed by the creation of microstreaming, leads to extraction of the target material from the plant cell structure. Other novel extraction methods for essential compounds include microwave-assisted extraction, accelerated solvent extraction and supercritical fluid extraction. In these techniques elevated solvent temperature and/or pressure may reduce the extraction time. Microwave-assisted extraction is attracting interest because it increases the product temperature rapidly without coming into contact with it. It also provides faster start-up, reduced thermal gradient within the product,

TABLE 3.2
Comparison of Different Novel Extraction Methods

Technique	Working	Extraction Time	Sample Size	Solvent Need	Equipment Investment	Advantages	Drawbacks
Microwave assisted extraction	Sample in presence of polar solvent subjected to microwave irradiation	Comparatively less (3–30 min)	Preferably small as penetration is low (1–10 g)	High polar solvent (10–40 mL)	Moderated	• Rapid heating • No chemical residues • Water is best solvent	• Solvent should have higher dipolar moment • Temperature during extraction cannot be controlled
Supercritical fluid extraction	Sample was kept under high pressure and supercritical fluid was passed continuously	More (10–60 min)	Very small (1–5 g)	For solid trap (2–5 mL) For liquid trap (30–60 mL)	High	• Higher selectivity • Fewer impurities • Low solvent consumption	• Expensive • Requires more knowledge • Need of parameters to optimize
Accelerated solvent extraction	Solvent was maintained at high temperature and pressure and forced to pass the sample	Moderate (10–20 min)	High (1–30 g)	Conventional solvent (10–60 mL)	High	• Rapid • Fewer impurities • Low solvent consumption	• Possibly degradation of thermolabile analytes
Ultrasound assisted extraction	Sample immersed in solvent and subjected to ultrasound	Probe type: less (1–16 min) Bath type: more (10–60 min)	High (1–30 g)	As per sample need (50–200 mL)	Low	• No chemical requirement • Easy to use • Highly efficient	• Large amount of solvent • Separation of target material from solvent is necessary

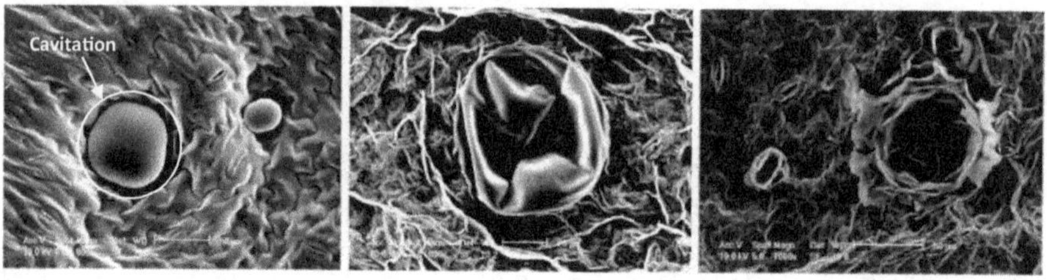

FIGURE 3.6 Collapse of cavitation bubble and releasing of plant materials in extraction of essential oil from basil.
Source: Adapted from Chemat and Khan (2011)

and selective heating. The major drawback of microwave heating is the need to use a solvent with a high dipole moment. Solvents of low dipole moment cause inefficient heating of the food. Supercritical fluid extraction requires more expensive equipment and extracts polar molecules less effectively. Accelerated solvent extraction uses modified extraction conditions (high temperature and/or pressure) to give a better extraction yield, but at the cost of degradation of thermolabile analytes. Ultrasonication of food products causes disintegration of cell wall structure which makes it a promising technology for the extraction of essential compounds (Chemat et al. 2017a). It can also significantly improve the extraction efficiency when used in combination with the above-mentioned techniques.

3.5 FREEZING

Freezing of foodstuffs is a well-known method of shelf-life extension. Of recent interest is the use of ultrasound to assist freezing. Ultrasonication creates cavitation bubbles throughout the product, promoting the nucleation of more and smaller ice crystals, which ensures the least damage to cell wall structure (Zheng and Sun, 2006).

Ice nucleation in food products is generally affected by the solvent used, impurities, surface tension, asperities, etc., and it cannot be monitored and controlled during the freezing process.

There are two major steps in the freezing process: nucleation followed by crystallization. As the temperature decreases nucleation already is the formation of tiny ice crystals throughout the product. Food morphology and ice crystal size distribution are the critical factors in nucleation. Nucleation plays a major role during industrial processing of food products as it helps to determine the ice crystal growth time and temperature. Nucleation is a probabilistic occurrence which can only be predicted by performing numerous freezing trials. Ultrasound irradiation can trigger ice nucleation, giving a linear relationship between irradiation time and temperature.

Ice crystals are a major component of frozen foods. These particularly dense and incompressible materials fracture when subjected to ultrasound waves. Cavitation induced during freezing creates bubbles which expand and contract with the changes in acoustic pressure. During the rarefaction phase, bubbles grow and dissolved gases

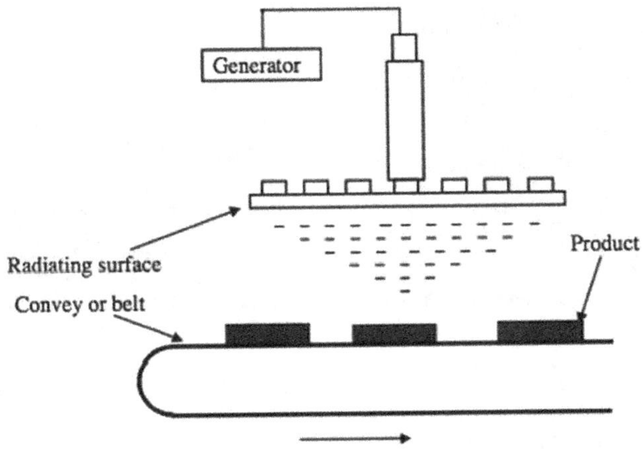

FIGURE 3.7 Ultrasound-assisted air-blast freezing system.
Source: Adapted from Zheng and Sun (2006)

within the surrounding liquid diffuse into them (Cheng et al. 2017). However, during the compression phase there is reduced bubble surface area available for diffusion of gas back into the liquid. Thus less gas is expelled than was taken in during rarefaction, and bubbles continue to grow and collapse during the overall ultrasound cycle. When they reach a critical size cavitation bubbles can serve as ice nucleation foci. Figure 3.7 shows a hybrid system having ultrasound-assisted air-blast freezing facilities.

In a study of nucleation during ultrasound-assisted freezing, Suslick (1988) investigated the effect of power ultrasound treatment on the freezing properties of liquid-based food products. The results showed that ultrasound treatment significantly increasedthe number of nuclei formed during freezing of concentrated sugar solution.

Kiani et al. (2011) studied the ultrasound-assisted nucleation characteristics of different water-based foods during freezing. The authors studied changes in nucleation time of deionized water, agar gel, and sucrose solution with and without ultrasound treatment. As nucleation in liquid and solid foods is a stochastic phenomenon, difficult to control during freezing, the study aimed to establish relationships between the different parameters which control nucleation. 25 kHz sonication was applied using an ultrasonic bath for 1, 3, 5, 10 and 15 s at temperatures ranging from 0 to –5°C. The results suggested that increased duration of ultrasound treatment raises the temperature of the product, which delays nucleation.

Sun and Li (2003) investigated the effect of immersion freezing and ultrasound-assisted immersion freezing on the properties of potato. Cryo-SEM images of frozen potato using both methods showed significant changes in the cell structure (Figure 3.8). It was found that ultrasound-assisted immersion frozen product retained better cell wall structure than the immersion frozen product. This could be because ultrasound application promotes faster freezing in the food product. The ice crystals formed had a uniform distribution, and a better cell structure was retained due to less extracellular void formation and minimal intracellular damage. It may also be that

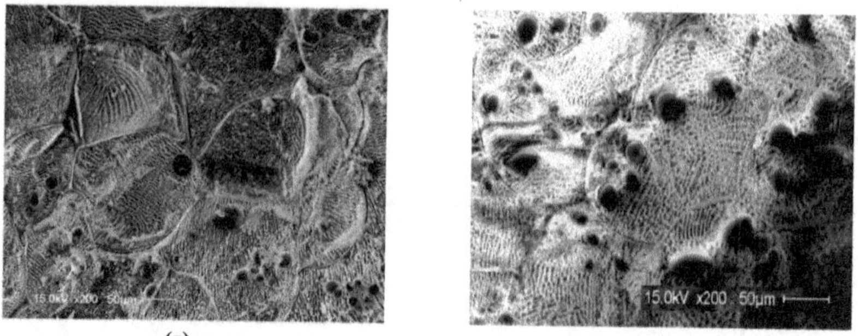

(a) (b)

FIGURE 3.8 Cryo-SEM micrographs of potato after (a) immersion freezing and (b) ultrasound-assisted immersion freezing.
Source: Adapted from Sun and Li (2003)

acoustic cavitation induces intracellular nucleation, which does not happen during immersion freezing because of insufficient supercooling. Ultrasound application also ensures the formation of smaller ice crystals which causes less cell dehydration and retains product quality.

3.6 DEHULLING

Dehulling is a major unit operation in pulse processing. It is important as the acceptance and economic value of pulses in the market depends upon the dehulling process efficiency. Dehulling is removing hull, which contains antinutritional factors, from the cotyledons of the pulses. A grain develops this protective layer to prevent insect attacks and other spoilage. Removal of hull reduces the cooking time, improves protein quality, increases digestibility and assists further processing. Dehulling is a traditional practice in the preparation of pulses for human consumption. The hull is tightly attached to the surface by gum. The conventional methods for dehulling of pulses are wet pre-treatment dehulling and dry processing, which are both time and energy consuming. Other pre-treatments to increase dehulling efficiency are gaining importance, such as chemical treatments (alcohol, sodium bicarbonate, acetic acid, etc.), conditioning with vegetable oils and urea solution, and microwave heat treatment.

Ultrasound is increasingly used for food product modification and shelf-life extension. It is also applied to the quality analysis of food products. Since ultrasound processing is a green technology, its possible use in dehulling of pulses is of growing importance. Ultrasound-induced cavitation applies temperature and pressure that help to loosen the binding gum. The simultaneous rise of temperature and pressure produces dehulled grains of sound quality. It also creates hotspots which accelerate chemical action in the extraction medium.

Sunil et al. (2018) investigated a new approach in ultrasound dehulling of black gram to improve its availability, processing, and digestibility (Figure 3.9). The authors used a bath-type ultrasound system with a tank capacity of 47 L and a maximum power input of 600 W. It was operated at room temperature. Pre-treatment

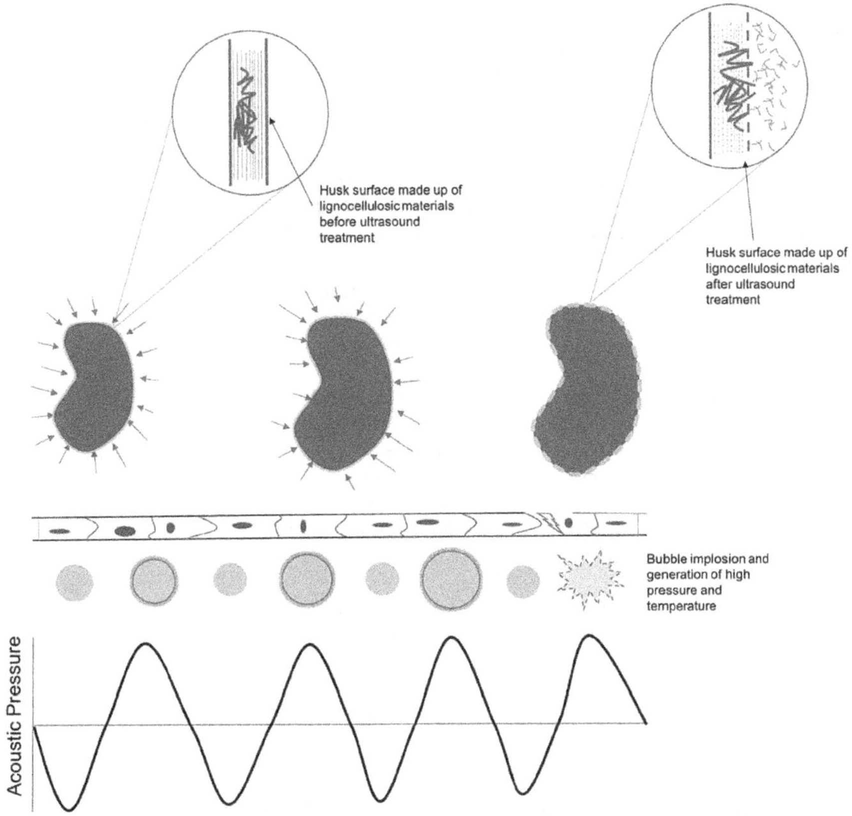

FIGURE 3.9 Changes in the hull during ultrasound-assisted dehulling of pulses.
Source: Adapted from Sunil et al. (2018)

was performed at power varying between 343 and 525 W with sonication time ranging from 1 to 3.5 h. The findings suggest that ultrasound pre-treatment increases the hydration capacity of black gram, helps to remove ant nutritional factors, diminishes cooking time, increases germination properties, and reduces hardness of the grains.

3.7 DRYING

Drying is one of the earliest methods used for food preservation. It has proven to be an effective means of retaining quality during storage. In developed countries about 25% of the total food industry energy consumption is for drying (Baeghbali et al. 2019). Drying of any food product is controlled by heat and mass transfer within and outside the product. Water movement within food is controlled by internal resistance, and movement of water vapor from the solid surface into air is controlled by external resistance.

Moisture is present in food products in three forms: free moisture, unbound moisture, and bound moisture. Free moisture evaporates during the initial stage of

TABLE 3.3
Different Ultrasound-Assisted Drying Methods

Method	Product	Findings	Reference
Ultrasound-assisted hot air drying	Passion fruit peel	• Reduced drying time • Higher rate of mass transfer • Increased diffusivity	do Nascimento et al. (2016)
Ultrasound-assisted low – temperature drying	Apple	• Drying time shortened by 80.3% • Minor quality loss	Santacatalina et al. (2016)
Microwave- and ultrasound-assisted convective drying	Strawberries	• Efficient heat and mass transfer	Szadzińska et al. (2016)
Ultrasound-assisted heat pump drying	Pea seeds	• Higher drying rate • Better germination properties	Yang et al. (2018)
Airborne ultrasound and microwave drying	Green pepper	• Less drying time • Reduced energy consumption • Better quality characteristics	Szadzińska et al. (2017)
Microwave and ultrasound-assisted convective drying	Raspberries	• Improved drying kinetics • Lower energy utilization • Better product quality	Kowalski et al. (2016)
Ultrasound-assisted atmospheric freeze drying	Eggplant	• Reduced drying time • Antioxidant content was maintained	Colucci et al. (2018)

drying. Unbound moisture includes that present inside the capillaries, which requires additional driving force for evaporation. In conventional drying methods the driving force is a thermal gradient created by the application of high temperature. However, high temperatures degrade food product appearance and nutritional value. Such limitations can be partly overcome by using novel drying technologies. Additional energy sources such as microwaves, infrared radiation and ultrasound-assisted drying, which achieve higher drying rates, are becoming popular at industrial scales.

Novel techniques reduce drying time and help to retain product quality. It is well established that products having a higher moisture content are more prone to microbial contamination. Thus novel technologies can speed up the attainment of safe conditions.

Ultrasound has mechanical effects on both the solid food product and the surrounding air, which accelerate the drying rate (Riera et al. 2011). Ultrasound dries food by heating, vibration, and synergistic effects. Absorption of ultrasound increases product temperature, which eventually raises the vapor pressure near the evaporation area. This creates a vapor pressure gradient which accelerates drying.

Ultrasound also creates vibration and turbulence in the air around the food product, causing rapid air movement and increasing the moisture uptake (Table 3.3). These two synergistic effects increase the product drying efficiency (Kowalski and Mierzwa, 2015).

3.8 ULTRASONIC OR SONO-EMULSIFICATION/ HOMOGENIZATION

In recent decades, a major trend has been the consumption of nutraceuticals containing bioactive compounds. Because most of these bioactives, such as carotenes, phytosterols etc. are oil soluble, it is difficult to deliver the compounds whilst maintaining their stability. Carriers are needed to effectively deliver the bioactives in food matrices. Couëdelo et al. (2011) reported that emulsions can provide efficient delivery of bioactives. This can give better results than conventionally used techniques such as high shear mixers (Ultraturrax) and piston homogenizers which require large amounts of emulsifiers and stabilizers (Santana et al. 2013). The efficiency of homogenization-based emulsification directly relates to the size reduction of the non-polar component in the continuous polar phase, and the storage stability of the emulsion. Recently, ultrasound has been identified as a potential method for the preparation of food emulsions. Ultrasound causes cavitation, turbulence, shear force, shock waves, and microjets, all of which promote emulsification. Little has so far been published concerning emulsification by ultrasound and more extensive studies are needed.

Li and Fogler (1978a), and later Kentish et al. (2008), recognized two stages in ultrasonic or sono-emulsification. The first stage commences with the development of instability in the oil/water interface, which leads to projection of dispersed-phase droplets into the continuous phase (Figure 3.10).

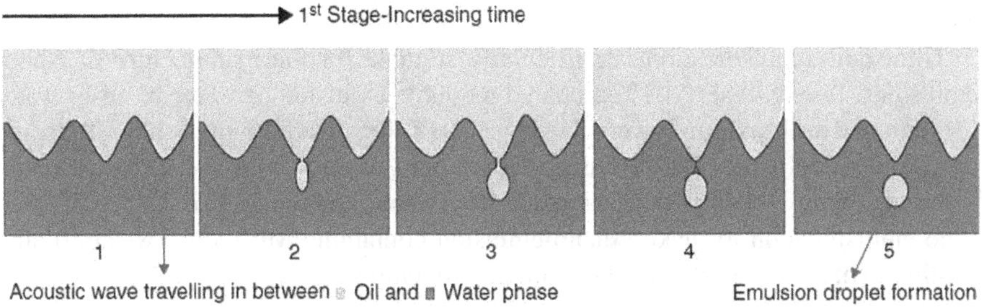

FIGURE 3.10 Instability of oil/water interface and the formation of initial droplets.
Source: Adapted from Li and Fogler (1978a)

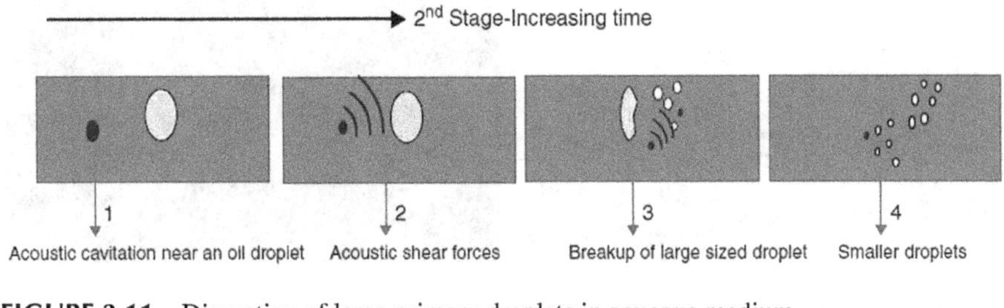

FIGURE 3.11 Disruption of large primary droplets in aqueous medium.
Source: Adapted from Li and Fogler (1978b)

This results in the intermingling of the two phases and accumulation of substantial dispersed-phase droplets. In the second stage of the acoustic emulsification, high-pressure shock waves of about 100 MPa and high local shear forces lead to the rupture of large droplets into smaller ones (Figure 3.11). The intense ultrasonic energy eventually breaks the planar interface, overcoming the Laplace pressure to produce finer droplets.

Factors controlling sono-emulsification include the sonication frequency, the power being applied, the time for which it is maintained and the operating temperature.

Bermúdez-Aguirre et al. (2011) described the use of frequencies from 20 to 40 kHz and discussed the role of shear forces at lower frequencies. These generate a strong shear force and, because of the high-power intensity, the surface capillary waves are much stronger. This reduces the time available for expansion and collapse of the bubbles, and the extent of shear is increased (Ashokkumar, 2011). The sono-chemical reactions taking place in the production of emulsions are better considered with reference to the power consumed by the equipment, or the nominal applied power (NAP), rather than the power absorbed by the liquid (Homayoonfal et al. 2014; Kaltsa et al. 2016). Several studies indicate a decrease in the droplet size of the dispersed phase with increased ultrasonic power, and the coalescence of smaller droplets into larger ones becomes prevalent (Shanmugam and Ashokkumar, 2014; Hashtjin and Abbasi, 2015). This is ascribed to the high energy density and high rate of collision under such circumstances (Gaikwad and Pandit, 2008; Kentish et al. 2008). Thus emulsification techniques can be compared on the basis of equivalent energy densities.

Ultrasonic emulsification is particularly suitable for the manufacture of Nano-emulsions. Leong et al. (2017) prepared a double emulsion of water in oil-in-water for skimmed milk and sunflower oil using 20 kHz ultrasound (Figure 3.12). Ultrasonic protocols made it possible to reduce the amount of surfactant used. The resultant water-containing oil droplets were stable for a week. Previously Li et al. (2016) used sono-emulsification to make a nano-emulsion containing MCT oil, Tween 80, and lecithin, which was used to coat curcumin with chitosan.

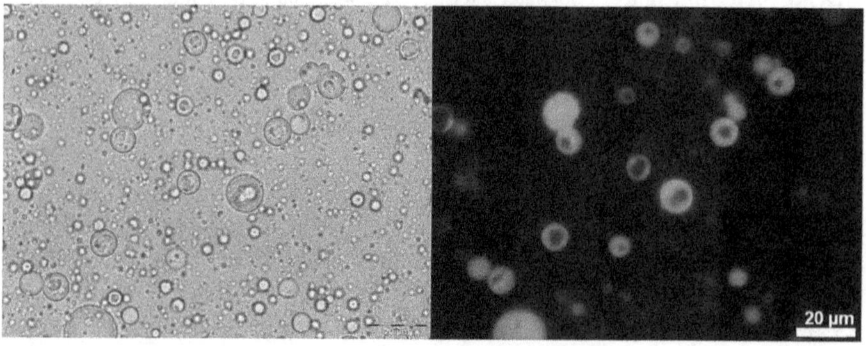

FIGURE 3.12 Double emulsion formed using ultrasonic emulsification.
Source: Adapted from Leong et al. (2017)

Rebolleda et al. (2015) formulated a stable wheat bran oil nano-emulsion, using high-intensity ultrasonication to produce small droplets. A pre-emulsification step was applied before the sono-emulsification to produce a coarse emulsion. This improves the emulsification efficiency by disrupting the planar interface between the oil and water and thus coarser droplets are developed. However, if sonication is of short duration and low ultrasonic power, instability can be observed in the final emulsion.

Some characteristics of the dispersed phase improve sono-emulsification, whilst some additives can destabilize the emulsions. D-limonene caused Ostwald ripening, and consequent destabilization of emulsions. This results from the difference in curvature radius between larger and smaller droplets. The resulting difference in chemical potential is the driving force which causes larger droplets to grow at the expense of smaller ones (Li and Lu, 2016). The zeta potential also played a role in maintaining emulsion stability. Recently sono-emulsification has been used to produce antimicrobial nano-emulsions which retained desirable attributes of the antimicrobial agents, and to illustrated-limonene's potency as an antimicrobial agent in comparison to synthetic additives (Zahi et al. 2017; Sonu et al. 2018). Table 3.4 refers to some recent studies of the inclusion of ultrasonicated emulsions in food and pharma systems.

TABLE 3.4
Effective Ultrasonic Nano-Emulsions Utilized in Food Systems

Oils	Emulsification Agents	Role	Outcome	References
Nigella sativa L.	Tween 80	Anticancer; Antioxidant; Antimicrobial	Anticancer efficacy on breast cancer cells	Periasamy et al. (2016)
Orange oil	Tween 80	Food preservation	Mitigation of yeast viability in apple juice	Sugumar et al. (2016)
D-limonene	Tween 80	Antimicrobial activity	Synergism between D-limonene and ε-polylysine	Zahi et al. (2017)
Mustard oil	Span 80	Delivery system	Process protocols	Carpenter and Saharan (2017)
Olive oil	α-tocopherol	Co-surfactant free nanoemulsion	Antioxidant and p-Anisidine activity of olive oil-based emulsion	Mehmood et al. (2017)
Flaxseed oil	Alginate–whey protein	Targeted delivery	Enrichment of broiler meat with omega 3 fatty acid	Abbasi et al. (2019)
Cinnamon oil	Tween 80	Stability and antifungal activity	Nanoemulsion with improved viscosity and particle size	Pongsumpun et al. (2019)
Orange oil	Tween 80	Delivery system	Optimized low and high energy system	Asadinezhad et al. (2019)

3.9 STRUCTURAL MODIFICATION

Structure plays an important role in the perception and palatability of a food (Moelants et al. 2014). Depending on the processing conditions, ultrasound treatment can change the food structure. Collapse of cavitation bubbles develops high pressure (up to 50 MPa) which can cleave water molecules resulting in the formation of hydroxyl and free radical groups. This changes food structure by breaking weak intermolecular bonds, such as Van der Waals interactions, leading to particle disintegration and making cellular compartments (Alarcon-Rojo et al. 2015; Jambrak et al. 2014).

The implications of ultrasound for plant-based compounds such as PPO and PME has already been covered. Ultrasound treatment proves vital for improving the structural cross section of vegetables and fruits like potatoes, pears and carrots. The treatment can also lead to secretion of response-based metabolites and oxidative bursts in the plant tissues. Peroxidase catalyzed oxidation strengthens the cell wall by the formation of cross links. Sonication treatment also enhances blanching, increasing firmness and limiting deformation and cell separation. The main observation was that PME-catalyzed demethylation of pectin facilitates the strengthening of the middle lamella through calcium mediated cross-linking of various pectin chains. Terefe et al. (2011) considered the effect of sonication treatment on potato strips. A change in the texture was observed which results from change in the surface composition of the product and even modification of the cellular structure (Figure 3.13).

Consistency in the tenderness of meat is an important quality. The sarcomere and other intermediate proteins contribute to the integrity of the meat structure. Weakening of the myofibrillar matrix through the protein structure degradation eventually enhances tenderness. The effect of ultrasound on meat depends on the species, type, and treatment conditions. It has been hypothesized that structural changes responsible for tenderization mainly result from disruption of the tissue by cavitation and activation of enzymes (Jayasooriya et al. 2004). Chang et al. (2012) observed changes in collagen using high-power low-frequency ultrasound. The results showed a decrease in hardness following 10 min of sonication (40 kHz, 1500 W). The treated samples were found to have disordered collagen fibers with a loose arrangement. Other literature suggests improvements in texture of post-rigor beef (Figure 3.14).

The changes in fat globules and protein structure during milk homogenization are other applications of ultrasonication in milk processing. Low-frequency (<100 kHz) ultrasound processing causes greater modifications in the protein structure due to unstable cavitation. Localized microstreaming has a rubbing effect on the surface that changes the structure. Low-frequency ultrasound treatment (20 kHz for 15 min) modified the case in size and eventually formed aggregates in a pH range of 6.7–8 (Liu et al. 2014). Chandrapala et al. (2013) noted an enhancement in renneting that gave better yield in cheese making. The treatment was conducted at pH 6.7 and the frequency was stable at 20 kHz. Several other references also report milk fat creaming and globule fractionation (Mohammadi et al. 2014; Nguyen and Anema, 2017; Sutariya et al. 2018). Scaling up these techniques still remains to be addressed.

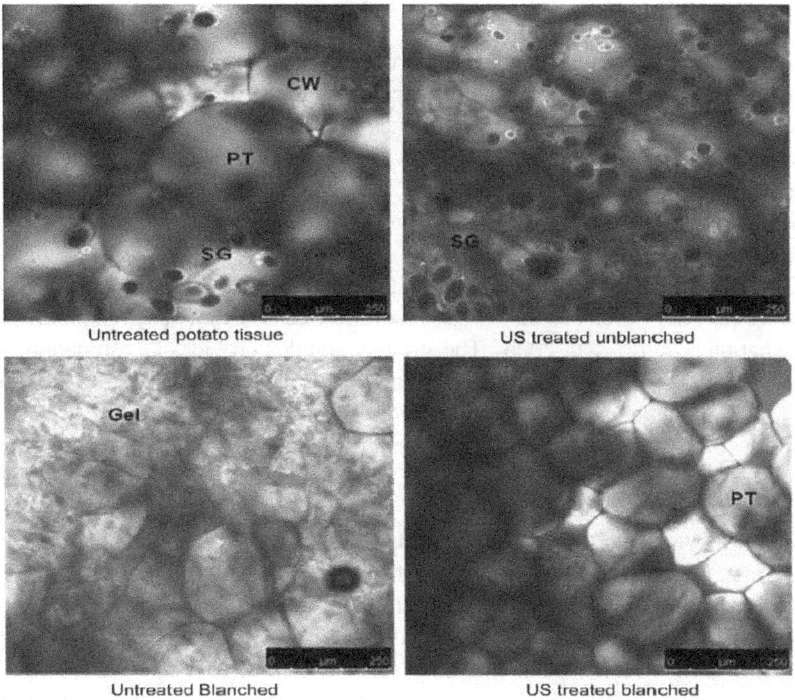

FIGURE 3.13 Confocal micrograph of potato strips (CW, Cell Wall; PT, per medullar tissue; SG, starch grain).
Source: Adapted from Terefe et al. (2011)

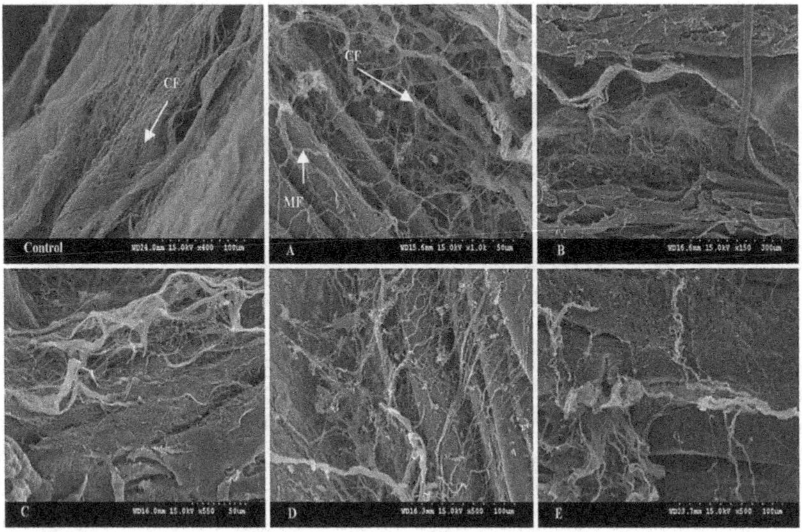

FIGURE 3.14 SEM images of beef subjected to different ultrasonic treatments.
Source: Adapted from Chang et al. (2012)

3.10 TEMPERING

Tempering is heat treatment of food to improve its structure. In fat and oil processing tempering is applied at a controlled temperature to stabilize a new plasticized fat against the polymorphic nature of fat crystals. Even though the physiochemical basis of tempering is poorly known, it plays a vital role in processing foods such as tempered shortenings.

Tempering has a great influence on the mechanical properties of whipped dairy cream. Tempering of whipped creams, warming to 15–30°C and cooling down to 4°C, significantly changed the shelf life of the product. Tempering can develop a well stiffened product which can be stored for several weeks without any major structural deformation (Drelon et al. 2006). The storage modulus provides an estimate of tempering efficiency of a stored product. Alterations in fat globule interaction during tempering might be one of the reasons for extended shelf life.

Ultrasound can be used as a pre-treatment for tempering of food. Reports indicate that researchers have already explored the use of power ultrasound for tempering of chocolate (Awad et al. 2012).

Winkelmeyer et al. (2016) used an ultrasonic spectrometric method to monitor the tempering of dark chocolate. They varied the sugar, fat, and cocoa mass content of the dark chocolate samples under different tempering protocols, and derived successful tempering combinations in which the static cooling temperature during tempering was varied from 50 to 14°C. Cocoa butter seeds were used as tempering shortenings.

In a tempered product the temperature is maintained such that a substantial proportion of the water in the product will be ice, but not all. To attain this state either the temperature of the product can be cooled below its freezing point, or it can be warmed from the temperature at which it is completely frozen. In this state the product will be rigid but not too hard to cut. If this process is combined with ultrasound treatment, micro-bubble formation and cavitation extend the efficiency of tempering.

With the assistance of ultrasound treatment, the tempering rate can be improved and better product quality achieved. In the food industry high-power ultrasound has been used to induce mechanical, physiochemical, and biochemical changes to food components (Rawson et al. 2011). Ultrasound has been applied in various unit operations such as extraction, freezing, drying, and microbial inactivation (Chemat and Khan, 2011). Ultrasound-assisted tempering can be applied to a wide variety of food products, especially meat (James et al. 2017), fruits and vegetables (Pan et al. 2003), aerated foods, honey, food enzymes, and proteins.

Tempering is one of the most crucial steps affecting the physical and chemical properties of food. It not only extends product shelf life but also provides other benefits. These include improving the stability of micro-bubbles in aerated chocolate products (Robert, 2006), and vacuum-induced bubble formation in liquid tempered chocolate (Haedelt et al. 2005). Critical points of this process are under-tempering and over-tempering. Blade breakage, decrease in yield due to shattered product, uneven slice thickness and excessive fines can result from under-tempering. Incomplete shearing of tissue and yield loss due to ragged edges result from over-tempering. These processing problems can be to some extent ameliorated with other thermal and non-thermal heat treatment techniques. Hence there may be an increased food industry

demand for the combination of tempering with cheap and easily applied techniques such as ultrasound and microwave treatments. These ultimately give a product competitive with that produced without combination treatments (James et al. 2017).

3.11 TREATMENT OF FOOD INDUSTRY WASTE WATER

Amongst all the applications explored so far, the greatest benefits are found in the remediation of food industry waste water using ultrasound.

The quality of potable water is determined by its physical, chemical, and biological properties. Because water consumption in production and processing industries is very high and in view of growing water scarcity, it is essential that wastewater should be remediated for re-use. Those industries which use the most water for production should therefore adopt waste water treatment. Suitable, reliable, and cheap waste water treatment alternatives are therefore required.

Ultrasound treatment, which consumes less energy than conventional techniques, can be used to treat waste water. It removes pollution by an advanced oxidation process and by cavitation, which decomposes complex organic compounds into simpler ones (Mahvi, 2009). Ultrasound technology has already found various applications in water treatment processes such as membrane filtration, turbidity maintenance, reduction of total suspended solids, and removal of algae. It also assists in the disinfection process, the softening of water, and removal of pollutants like halomethanes and DDT which are environmentally hazardous chemicals. Power density, frequency, treatment time and rate all affect the efficiency of this technique (Doosti et al. 2012).

Application of low-frequency ultrasound produces powerful waves which can cause chemical changes. 20–50 kHz is the most frequently used frequency range for the industrial cleaning process (Nair and Patel, 2014).

Ultrasound can cause several mechanical, acoustic, chemical, and biological changes in the process water. Bubble cavitation can generate high temperature, pressure and reactive radical species in water, leading to the thermal dissociation of water and oxygen which can oxidize dissolved organic compounds (Jiang et al. 2002).

3.12 CONCLUSION

Novel technologies can reduce the carbon footprint of food processing. Ultrasonication, in its broad spectrum of applications, can process perishable food with minimal difficulty and maintain its quality. Applications such as structure modification, increase or decrease in microbial growth rate, imaging and extraneous material detection, although still in their infancy, can employ ultrasound flexibly to save energy and time. However, the extensive studies conducted so far now need to be scaled up for industrial use.

REFERENCES

Aadil, R.M., Zeng, X.A., Zhang, Z.H., Wang, M.S., Han, Z., Jing, H. and Jabbar, S. 2015. Thermosonication: A potential technique that influences the quality of grapefruit juice. *Int J Food Sci Tech*, 50(5), 1275–1282.

Abbasi, F., Samadi, F., Jafari, S.M., Ramezanpour, S. and Shargh, M.S. 2019. Ultrasound-assisted preparation of flaxseed oil nanoemulsions coated with alginate-whey protein for targeted delivery of omega-3 fatty acids into the lower sections of gastrointestinal tract to enrich broiler meat. *Ultrason Sonochem*, 50, 208–217.

Adekunte, A., Tiwari, B., Cullen, P., Scannell, A. and O'Donnell, C. 2010. Effect of sonication on colour, ascorbic acid and yeast inactivation in tomato juice. *Food Chem*, 122(3), 500–507.

Alarcon-Rojo, A., Janacua, H., Rodriguez, J., Paniwnyk, L. and Mason, T.J. 2015. Power ultrasound in meat processing. *Meat Sci*, 107, 86–93.

Al-Hashimi, A.M., Mason, T.J. and Joyce, E.M. 2015. Combined effect of ultrasound and ozone on bacteria in water. *Environ Sci Technol*, 49(19), 11697–11702.

Arroyo, C., Cebrián, G., Condón, S. and Pagán, R. 2012a. Development of resistance in Cronobacter sakazakii ATCC 29544 to thermal and nonthermal processes after exposure to stressing environmental conditions. *J Appl Microbiol*, 112(3), 561–570.

Arroyo, C., Cebrián, G., Pagán, R. and Condón, S. 2012b. Synergistic combination of heat and ultrasonic waves under pressure for Cronobacter sakazakii inactivation in apple juice. *Food Control*, 25(1), 342–348.

Arroyo, C., Cebrián, G., Pagán, R. and Condón, S. 2011a. Inactivation of Cronobacter sakazakii by manothermosonication in buffer and milk. *Int J Food Microbiol*, 151(1), 21–28.

Arroyo, C., Cebrián, G., Pagán, R. and Condón, S. 2011b. Inactivation of Cronobacter sakazakii by ultrasonic waves under pressure in buffer and foods. *Int J Food Microbiol*, 144(3), 446–454.

Asadinezhad, S., Khodaiyan, F., Salami, M., Hosseini, H. and Ghanbarzadeh, B. 2019. Effect of different parameters on orange oil nanoemulsion particle size: Combination of low energy and high energy methods. *J Food Meas Charact*, 1–9.

Ashokkumar, M. 2011. The characterization of acoustic cavitation bubbles – An overview. *Ultrason Sonochem*, 18(4), 864–872.

Awad, T., Moharram, H., Shaltout, O., Asker, D. and Youssef, M. 2012. Applications of ultrasound in analysis, processing and quality control of food: A review. *Food Res Int*, 48(2), 410–427.

Baeghbali, V., Niakousari, M. and Ngadi, M. 2019. An update on applications of power ultrasound in drying food: A review. *J. Food Eng Technol*, 8(1), 29–38.

Balthazar, C.F., Santillo, A., Guimarães, J.T., Bevilacqua, A., Corbo, M.R., Caroprese, M., Marino, R., Esmerino, E.A., Silva, M.C. and Raices, R.S. 2019. Ultrasound processing of fresh and frozen semi-skimmed sheep milk and its effects on microbiological and physical-chemical quality. *Ultrason Sonochem*, 51, 241–248.

Bermúdez-Aguirre, D., Mobbs, T. and Barbosa-Cánovas, G.V. 2011. Ultrasound applications in food processing. In *Ultrasound technologies for food and bioprocessing*. Springer.

Bhavya, M. and Hebbar, H.U. 2019. Sono-photodynamic inactivation of Escherichia coli and Staphylococcus aureus in orange juice. *Ultrason Sonochem*, 57, 108–115.

Birmpa, A., Sfika, V. and Vantarakis, A. 2013. Ultraviolet light and ultrasound as non-thermal treatments for the inactivation of microorganisms in fresh ready-to-eat foods. *Int J Food Microbiol*, 167(1), 96–102.

Cameron, M., Mc Master, L.D. and Britz, T.J. 2009. Impact of ultrasound on dairy spoilage microbes and milk components. *Dairy Sci Technol*, 89(1), 83–98.

Carpenter, J. and Saharan, V.K. 2017. Ultrasonic assisted formation and stability of mustard oil in water nanoemulsion: Effect of process parameters and their optimization. *Ultrason Sonochem*, 35, 422–430.

Chandrapala, J., Zisu, B., Kentish, S. and Ashokkumar, M. 2013. Influence of ultrasound on chemically induced gelation of micellar casein systems. *J Dairy Res*, 80(2), 138–143.

Chang, H.-J., Xu, X.-L., Zhou, G.-H., Li, C.-B. and Huang, M. 2012. Effects of characteristics changes of collagen on meat physicochemical properties of beef semitendinosus muscle during ultrasonic processing. *Food Bioprocess Tech*, 5(1), 285–297.

Charoux, C.M., O'Donnell, C.P. and Tiwari, B.K. 2019. Effect of airborne ultrasonic technology on microbial inactivation and quality of dried food ingredients. *Ultrason Sonochem* 56: 313–317.

Chemat, F. and Khan, M.K. 2011. Applications of ultrasound in food technology: Processing, preservation and extraction. *Ultrason Sonochem*, 18(4), 813–835.

Chemat, F., Rombaut, N., Meullemiestre, A., Turk, M., Perino, S., Fabiano-Tixier, A.-S. and Abert-Vian, M. 2017a. Review of green food processing techniques. Preservation, transformation, and extraction. *Innov Food Sci Emerg*, 41, 357–377.

Chemat, F., Rombaut, N., Sicaire, A.-G., Meullemiestre, A., Fabiano-Tixier, A.-S. and Abert-Vian, M. 2017b. Ultrasound assisted extraction of food and natural products. Mechanisms, techniques, combinations, protocols and applications. A review. *Ultrason Sonochem*, 34, 540–560.

Cheng, L., Sun, D.-W., Zhu, Z. and Zhang, Z. 2017. Emerging techniques for assisting and accelerating food freezing processes: A review of recent research progresses. *Crit Rev Food Sci*, 57(4), 769–781.

Colucci, D., Fissore, D., Rossello, C. and Cárcel, J.A. 2018. On the effect of ultrasound-assisted atmospheric freeze-drying on the antioxidant properties of eggplant. *Food Res Int*, 106, 80–88.

Couëdelo, L., Boué-Vaysse, C., Fonseca, L., Montesinos, E., Djoukitch, S., Combe, N. and Cansell, M. 2011. Lymphatic absorption of α-linolenic acid in rats fed flaxseed oil-based emulsion. *Brit J Nutr*, 105(7), 1026–1035.

de Souza Soares, A., Augusto, P.E.D., Júnior, B.R.D.C.L., Nogueira, C.A., Vieira, É.N.R., de Barros, F.A.R., Stringheta, P.C. and Ramos, A.M. 2019. Ultrasound assisted enzymatic hydrolysis of sucrose catalyzed by invertase: Investigation on substrate, enzyme and kinetics parameters. *LWT - Food Sci Technol.* 107: 164–170.

Delgado-Povedano, M. and De Castro, M.L. 2015. A review on enzyme and ultrasound: A controversial but fruitful relationship. *Anal Chim Acta*, 889, 1–21.

do Nascimento, E.M., Mulet, A., Ascheri, J.L.R., de Carvalho, C.W.P. and Cárcel, J.A. 2016. Effects of high-intensity ultrasound on drying kinetics and antioxidant properties of passion fruit peel. *J Food Eng*, 170, 108–118.

Doosti, M., Kargar, R. and Sayadi, M. 2012. Water treatment using ultrasonic assistance: A review. *Proceedings of the International Academy of Ecology and Environmental Sciences*, 2(2), 96.

Drelon, N., Gravier, E., Daheron, L., Boisserie, L., Omari, A. and Leal-Calderon, F. 2006. Influence of tempering on the mechanical properties of whipped dairy creams. *Int Dairy J*, 16(12), 1454–1463.

Etzbach, L., Pfeiffer, A., Schieber, A. and Weber, F. 2019. Effects of thermal pasteurization and ultrasound treatment on the peroxidase activity, carotenoid composition, and physicochemical properties of goldenberry (Physalis peruviana L.) puree. *LWT - Food Sci Technol*, 100, 69–74.

Ferrario, M., Alzamora, S.M. and Guerrero, S. 2015. Study of the inactivation of spoilage microorganisms in apple juice by pulsed light and ultrasound. *Food Microbiol*, 46, 635–642.

Gaikwad, S.G. and Pandit, A.B. 2008. Ultrasound emulsification: Effect of ultrasonic and physicochemical properties on dispersed phase volume and droplet size. *Ultrason Sonochem*, 15(4), 554–563.

Guimarães, J.T., Silva, E.K., Alvarenga, V.O., Costa, A.L.R., Cunha, R.L., Sant'Ana, A.S., Freitas, M.Q., Meireles, M.A.A. and Cruz, A.G. 2018. Physicochemical changes and microbial inactivation after high-intensity ultrasound processing of prebiotic whey beverage applying different ultrasonic power levels. *Ultrason Sonochem*, 44, 251–260.

Haedelt, J., Pyle, D.L., Beckett, S.T. and Niranjan, K. 2005. Vacuum-induced bubble formation in liquid-tempered chocolate. *J Food Sci*, 70(2): E159–E164.

Hashtjin, A.M. and Abbasi, S. 2015. Optimization of ultrasonic emulsification conditions for the production of orange peel essential oil nanoemulsions. *J Food Sci Technol*, 52(5), 2679–2689.

Homayoonfal, M., Khodaiyan, F. and Mousavi, S.M. 2014. Walnut oil nanoemulsion: Optimization of the emulsion capacity, cloudiness, density, and surface tension. *J Disper Sci Technol*, 35(5), 725–733.

Huang, N., Cheng, X., Hu, W. and Pan, S. 2015. Inactivation, aggregation, secondary and tertiary structural changes of germin-like protein in Satsuma mandarine with high polyphenol oxidase activity induced by ultrasonic processing. *Biophys Chem*, 197, 18–24.

Islam, M.N., Zhang, M. and Adhikari, B. 2014. The inactivation of enzymes by ultrasound: A review of potential mechanisms. *Food Rev Int*, 30(1), 1–21.

Jambrak, A.R., Mason, T.J., Lelas, V., Paniwnyk, L. and Herceg, Z. 2014. Effect of ultrasound treatment on particle size and molecular weight of whey proteins. *J Food Eng*, 121, 15–23.

James, S., James, C. and Purnell, G. 2017. Microwave-assisted thawing and tempering. In *The Microwave Processing of Foods*. Elsevier.

Janghu, S., Bera, M. B., Nanda, V., and Rawson, A. 2017. Study on power ultrasound optimization and its comparison with conventional thermal processing for treatment of raw honey. *Food technology and biotechnology*, 55(4), 570–579.

Jayasooriya, S., Bhandari, B., Torley, P. and D'arcy, B. 2004. Effect of high power ultrasound waves on properties of meat: A review. *Int J Food Prop*, 7(2), 301–319.

Jiang, Y., Pétrier, C. and Waite, T.D. 2002. Effect of p H on the ultrasonic degradation of ionic aromatic compounds in aqueous solution. *Ultrason Sonochem*, 9(3), 163–168.

Juliano, P., Torkamani, A.E., Leong, T., Kolb, V., Watkins, P., Ajlouni, S. and Singh, T.K. 2014. Lipid oxidation volatiles absent in milk after selected ultrasound processing. *Ultrason Sonochem*, 21(6), 2165–2175.

Kaltsa, O., Yanniotis, S. and Mandala, I. 2016. Stability properties of different fenugreek galactomannans in emulsions prepared by high-shear and ultrasonic method. *Food Hydrocolloid*, 52, 487–496.

Kentish, S., Wooster, T., Ashokkumar, M., Balachandran, S., Mawson, R. and Simons, L. 2008. The use of ultrasonics for nanoemulsion preparation. *Innov Food Sci Emerg*, 9(2), 170–175.

Kiani, H., Zhang, Z., Delgado, A. and Sun, D.-W. 2011. Ultrasound assisted nucleation of some liquid and solid model foods during freezing. *Food Res Int*, 44(9), 2915–2921.

Kowalski, S. and Mierzwa, D. 2015. US-assisted convective drying of biological materials. *Dry Technol*, 33(13), 1601–1613.

Kowalski, S.J., Pawłowski, A., Szadzińska, J., Łechtańska, J. and Stasiak, M. 2016. High power airborne ultrasound assist in combined drying of raspberries. *Innov Food Sci Emerg*, 34, 225–233.

Krishnan, V., Kuriakose, S. and Rawson, A. 2015. Ultrasound assisted extraction of oil from rice bran: A response surface methodology approach. *J Food Process Technol*, 6(454), 2.

Leong, T.S., Zhou, M., Kukan, N., Ashokkumar, M. and Martin, G.J. 2017. Preparation of water-in-oil-in-water emulsions by low frequency ultrasound using skim milk and sunflower oil. *Food Hydrocolloid*, 63, 685–695.

Li, J., Cheng, H., Liao, X., Liu, D., Xiang, Q., Wang, J., Chen, S., Ye, X. and Ding, T. 2019. Inactivation of Bacillus subtilis and quality assurance in Chinese bayberry (Myrica rubra) juice with ultrasound and mild heat. *LWT - Food Sci Technol*. 108, 113–119.

Li, J., Hwang, I.-C., Chen, X. and Park, H.J. 2016. Effects of chitosan coating on curcumin loaded nano-emulsion: Study on stability and in vitro digestibility. *Food Hydrocolloid*, 60, 138–147.

Li, M. and Fogler, H. 1978a. Acoustic emulsification. Part 1. The instability of the oil-water interface to form the initial droplets. *J Fluid Mech*, 88(3), 499–511.

Li, M. and Fogler, H. 1978b. Acoustic emulsification. Part 2. Breakup of the large primary oil droplets in a water medium. *J Fluid Mech*, 88(3), 513–528.

Li, P.-H. and Lu, W.-C. 2016. Effects of storage conditions on the physical stability of d-limonene nanoemulsion. *Food Hydrocolloid*, 53, 218–224.

Liu, L., Jia, W., Xu, D. and Li, R. 2015. Applications of ultrasonic cutting in food processing. *J Food Process Pres*, 39(6), 1762–1769.

Liu, Z., Juliano, P., Williams, R.P., Niere, J. and Augustin, M.A. 2014. Ultrasound improves the renneting properties of milk. *Ultrason Sonochem*, 21(6), 2131–2137.

Mackersie, J.W., Timoshkin, I.V. and Mac Gregor, S.J. 2005. Generation of high-power ultrasound by spark discharges in water. *IEEE Trans Plasma Sci*, 33(5), 1715–1724.

Mahvi, A. 2009. Application of ultrasonic technology for water and wastewater treatment. *Iranian J Public Health*, 1–17.

Marchesini, G., Fasolato, L., Novelli, E., Balzan, S., Contiero, B., Montemurro, F., Andrighetto, I. and Segato, S. 2015. Ultrasonic inactivation of microorganisms: A compromise between lethal capacity and sensory quality of milk. *Innov Food Sci Emerg*, 29, 215–221.

Mehmood, T., Ahmad, A., Ahmed, A. and Ahmed, Z. 2017. Optimization of olive oil based O/W nanoemulsions prepared through ultrasonic homogenization: A response surface methodology approach. *Food Chem*, 229, 790–796.

Michelino, F., Zambon, A., Vizzotto, M.T., Cozzi, S. and Spilimbergo, S. 2018. High power ultrasound combined with supercritical carbon dioxide for the drying and microbial inactivation of coriander. *J CO2 Util*, 24, 516–521.

Moelants, K.R., Cardinaels, R., Van Buggenhout, S., Van Loey, A.M., Moldenaers, P. and Hendrickx, M.E. 2014. A review on the relationships between processing, food structure, and rheological properties of plant-tissue-based food suspensions. *Compr Rev Food Sci F*, 13(3), 241–260.

Mohammadi, V., Ghasemi-Varnamkhasti, M., Ebrahimi, R. and Abbasvali, M. 2014. Ultrasonic techniques for the milk production industry. *Measurement*, 58, 93–102.

Moody, A., Marx, G., Swanson, B.G. and Bermúdez-Aguirre, D. 2014. A comprehensive study on the inactivation of Escherichia coli under nonthermal technologies: High hydrostatic pressure, pulsed electric fields and ultrasound. *Food Control*, 37, 305–314.

Moorthy, I.G., Maran, J.P., Muneeswari, S., Naganyashree, S. and Shivamathi, C. 2015. Response surface optimization of ultrasound assisted extraction of pectin from pomegranate peel. *Int J Biol Macromol*, 72, 1323–1328.

Naik, M., Natarajan, V., Rawson, A., Rangarajan, J., and Manickam, L. 2021. Extraction kinetics and quality evaluation of oil extracted from bitter gourd (Momardica charantia L.) seeds using emergent technologies. LWT, 140, 110714.

Nair, R.R. and Patel, R.L. 2014. Treatment of dye wastewater by sonolysis process. *Int. J. Res. Mod. Eng. Emerging Technol*, 1(2), 1–6.

Nguyen, N.H. and Anema, S.G. 2017. Ultrasonication of reconstituted whole milk and its effect on acid gelation. *Food Chem*, 217, 593–601.

O'Donnell, C., Tiwari, B., Bourke, P. and Cullen, P. 2010. Effect of ultrasonic processing on food enzymes of industrial importance. *Trends Food Sci Tech*, 21(7), 358–367.

Ojha, K.S., Burgess, C.M., Duffy, G., Kerry, J.P. and Tiwari, B.K. 2018. Integrated phenotypic-genotypic approach to understand the influence of ultrasound on metabolic response of Lactobacillus sakei. *PLoS one*, 13(1): E0191053.

Palanisamy, N., Seale, B., Turner, A. and Hemar, Y. 2019. Low frequency ultrasound inactivation of thermophilic bacilli (Geobacillus spp. and Anoxybacillus flavithermus) in the presence of sodium hydroxide and hydrogen peroxide. *Ultrason Sonochem*, 51, 325–331.

Pan, Y., Zhao, L., Zhang, Y., Chen, G. and Mujumdar, A.S. 2003. Osmotic dehydration pretreatment in drying of fruits and vegetables. *Dry Technol*, 21(6), 1101–1114.

Periasamy, V.S., Athinarayanan, J. and Alshatwi, A.A. 2016. Anticancer activity of an ultrasonic nanoemulsion formulation of Nigella sativa L. essential oil on human breast cancer cells. *Ultrason Sonochem*, 31, 449–455.

Pongsumpun, P., Iwamoto, S. and Siripatrawan, U. 2019. Response surface methodology for optimization of cinnamon essential oil nanoemulsion with improved stability and antifungal activity. *Ultrason Sonochem*, 60, 104604.

Rawson, A., Tiwari, B., Tuohy, M., O'Donnell, C. and Brunton, N. 2011. Effect of ultrasound and blanching pretreatments on polyacetylene and carotenoid content of hot air and freeze dried carrot discs. *Ultrason Sonochem*, 18(5), 1172–1179.

Rebolleda, S., Sanz, M.T., Benito, J.M., Beltrán, S., Escudero, I. and San-José, M.L.G. 2015. Formulation and characterisation of wheat bran oil-in-water nanoemulsions. *Food Chem*, 167, 16–23.

Riera, E., García-Pérez, J.V., Acosta, V., Cárcel, J. and Gallego-Juárez, J.A. 2011. Computational study of ultrasound-assisted drying of food materials. *Innov Food Process Technol: Adv Multiphys Simul*, 265–301.

Robert, A. 2006. Aerated chocolate with microbubbles for improved stability. Google Patents.

Rodríguez-Calleja, J., Cebrián, G., Condón, S. and Mañas, P. 2006. Variation in resistance of natural isolates of Staphylococcus aureus to heat, pulsed electric field and ultrasound under pressure. *J Appl Microbiol*, 100(5), 1054–1062.

Saeeduddin, M., Abid, M., Jabbar, S., Wu, T., Hashim, M.M., Awad, F.N., Hu, B., Lei, S. and Zeng, X. 2015. Quality assessment of pear juice under ultrasound and commercial pasteurization processing conditions. *LWT - Food Sci Technol*, 64(1), 452–458.

Santacatalina, J., Guerrero, M., García-Pérez, J.V., Mulet, A. and Cárcel, J. 2016. Ultrasonically assisted low-temperature drying of desalted codfish. *LWT - Food Sci Technol*, 65, 444–450.

Santana, R., Perrechil, F. and Cunha, R. 2013. High-and low-energy emulsifications for food applications: A focus on process parameters. *Food Eng Rev*, 5(2), 107–122.

Sengar, A.S., Rawson, A., Muthiah, M. and Kalakandan, S.K.J.U. 2020. Comparison of different ultrasound assisted extraction techniques for pectin from tomato processing waste. *Ultrason Sonochem*, 61, 104812.

Shanmugam, A. and Ashokkumar, M. 2014. Ultrasonic preparation of stable flax seed oil emulsions in dairy systems–physicochemical characterization. *Food Hydrocolloid*, 39, 151–162.

Silva, L.C., Almeida, P.S., Rodrigues, S. and Fernandes, F.A. 2015. Inactivation of polyphenoloxidase and peroxidase in apple cubes and in apple juice subjected to high intensity power ultrasound processing. *J Food Process Pres*, 39(6), 2081–2087.

Sonu, K., Mann, B., Sharma, R., Kumar, R. and Singh, R. 2018. Physico-chemical and antimicrobial properties of d-limonene oil nanoemulsion stabilized by whey protein–maltodextrin conjugates. *J Food Sci Tech*, 55(7), 2749–2757.

Sugumar, S., Singh, S., Mukherjee, A. and Chandrasekaran, N. 2016. Nanoemulsion of orange oil with non ionic surfactant produced emulsion using ultrasonication technique: Evaluating against food spoilage yeast. *Appl Nanosci*, 6(1), 113–120.

Sun, D.-W. and Li, B. 2003. Microstructural change of potato tissues frozen by ultrasound-assisted immersion freezing. *J Food Eng*, 57(4), 337–345.

Sunil, C., Chidanand, D., Manoj, D., Choudhary, P. and Rawson, A. 2018. Effect of ultrasound treatment on dehulling efficiency of blackgram. *J Food Sci Tech*, 55(7), 2504–2513.

Suslick, K.S. 1988. *Ultrasound: Its Chemical, Physical, and Biological Effects*. VCH Publishers.

Sutariya, S., Sunkesula, V., Kumar, R. and Shah, K. 2018. Emerging applications of ultrasonication and cavitation in dairy industry: A review. *Cogent Food & Agriculture*, 4(1), 1549187.

Szadzińska, J., Kowalski, S. and Stasiak, M. 2016. Microwave and ultrasound enhancement of convective drying of strawberries: Experimental and modeling efficiency. *Int J Heat Mass Tran*, 103, 1065–1074.

Szadzińska, J., Łechtańska, J., Kowalski, S.J. and Stasiak, M. 2017. The effect of high power airborne ultrasound and microwaves on convective drying effectiveness and quality of green pepper. *Ultrason Sonochem*, 34, 531–539.

Terefe, N., Sikes, A. and Juliano, P. 2016. Ultrasound for structural modification of food products. In *Innovative Food Processing Technologies*. Elsevier.

Terefe, N.S., Pasero, C., Fernando, S., Rout, M., Woonton, B. and Mawson, R. 2011. *Application of low intensity ultrasound to improve the textural quality of processed vegetables*. In *Institute of Food Technologists (IFT) Annual Meeting*. New Orleans, LA, USA: IFT.

Tremarin, A., Canbaz, E.A., Brandão, T.R. and Silva, C.L. 2019. Modelling Alicyclobacillus acidoterrestris inactivation in apple juice using thermosonication treatments. *LWT - Food Sci Technol*, 102, 159–163.

Vercet, A., Lopez, P. and Burgos, J. 1997. Inactivation of heat-resistant lipase and protease from Pseudomonas fluorescens by manothermosonication. *J Dairy Sci*, 80(1), 29–36.

Vercet, A., Sánchez, C., Burgos, J., Montañés, L. and Buesa, P.L. 2002. The effects of manothermosonication on tomato pectic enzymes and tomato paste rheological properties. *J Food Eng*, 53(3), 273–278.

Wang, W., Chen, W., Zou, M., Lu, R., Wang, D., Hou, F., Feng, H., Ma, X., Zhong, J. and Ding, T. 2018. Applications of power ultrasound in oriented modification and degradation of pectin: A review. *J Food Eng*, 234, 98–107.

Winkelmeyer, C.B., Peyronel, F., Weiss, J. and Marangoni, A.G. 2016. Monitoring tempered dark chocolate using ultrasonic spectrometry. *Food Bioprocess Tech*, 9(10), 1692–1705.

Wu, T., Wu, C., Xiang, Y., Huang, J., Luan, L., Chen, S. and Hu, Y. 2016. Kinetics and mechanism of degradation of chitosan by combining sonolysis with H 2 O 2/ascorbic acid. *RSC Adv*, 6(80), 76280–76287.

Yang, Z., Li, X., Tao, Z., Luo, N. and Yu, F. 2018. Ultrasound-assisted heat pump drying of pea seed. *Dry Technol*, 36(16), 1958–1969.

Yildiz, G. 2019. Application of ultrasound and high-pressure homogenization against high temperature-short time in peach juice. *J Food Process Eng*, 42(3): E12997.

Zahi, M.R., El Hattab, M., Liang, H. and Yuan, Q. 2017. Enhancing the antimicrobial activity of d-limonene nanoemulsion with the inclusion of ε-polylysine. *Food Chem*, 221, 18–23.

Zheng, L. and Sun, D.-W. 2006. Innovative applications of power ultrasound during food freezing processes: A review. *Trends Food Sci Tech*, 17(1), 16–23.

4 Osmotic Dehydration in Food Processing

Santanu Malakar
National Institute of Food Technology Entrepreneurship and Management, Sonipat, Haryana, India

T. Manonmani
National Institute of Food Technology Entrepreneurship and Management, Sonipat, Haryana, India

Saptashish Deb
Sant Longowal Institute of Engineering & Technology, Longowal, Sangrur, India

Kshirod Kumar Dash
Ghani Khan Choudhury Institute of Engineering & Technology (GKCIET), Malda, West Bengal, India

CONTENTS

4.1 Introduction 68
4.2 Principles of Osmotic Dehydration 69
4.3 Factors Affecting Osmotic Dehydration 69
  4.3.1 Species and Variety 70
  4.3.2 Shape and Size 70
  4.3.3 Processing Conditions 70
  4.3.4 Types of Osmotic Agents 71
  4.3.5 Time and Temperature 71
  4.3.6 Agitation and Sample-to-Solution Ratio 72
4.4 Mass Transfer Kinetics of Osmotic Dehydration 72
4.5 Mathematical Modeling of Osmotic Dehydration 74
4.6 Application of Osmotic Dehydration in Food Processing 76
4.7 Recent Advances in Osmotic Dehydration Preparation 77
  4.7.1 Microwave Radiation 77
  4.7.2 Pulsed Vacuum Impregnation 78
  4.7.3 Ohmic Heating 78
  4.7.4 Pulsed Electric Field 79
  4.7.5 Ultrasonication 79
  4.7.6 High Hydrostatic Pressure 80
  4.7.7 Gamma Irradiation 80
  4.7.8 Osmo-Dehydrofreezing 80

4.8 Research-Oriented Problems....81
4.9 Industrial Processing and Challenges....82
4.10 Conclusion....82
References....83

4.1 INTRODUCTION

Fruits and vegetables provide lots of essential nutrients due to their vitamin, carbohydrate, mineral, and fiber content together with other nutrients beneficial for human consumption. Both production and consumption rates of fruits and vegetables are increasing worldwide. Fruits and vegetables are highly perishable, can be rapidly spoiled during storage, and have a relatively short shelf life due to their high moisture content and rapid changes in their physiological and biochemical activity such as respiration, transpiration, enzyme activity, and many other abiotic and biotic factors. Various preservation practices can enhance the shelf life and value addition of food products. Only 1.5–2% of the total production of vegetables and fruits are processed in developing countries, and a great deal of losses are incurred due to inadequate post-harvest, processing, preservation, and storage practices (Nath et al. 2018). The aim of preservation is to reduce the qualitative decay of food products during storage and handling. In this rapidly growing world, the preservation and storage of high-quality foods is a major challenge. Various preservation practices such as thermal treatment, low-temperature processing and non-thermal treatment, packaging, and cryogenic grinding are performed to prolong shelf life and handling of fruits as well as for the convenience of the consumer (Dhyani 2006). Dehydration is the most popular and effective technique for reducing moisture and increasing safe storage life. Moisture removal is accomplished by drying technology through simultaneous heat and mass transfer processes (Pragati and Preeti 2014).

Earlier, various drying techniques, such as solar drying, vacuum, microwave, freeze, spray, and fluidized-bed drying, were developed to remove moisture from food materials (Zhang and Long 2017). Recent research has mainly focused on efficient, energy-saving and economical drying techniques (Khaing Hnin et al. 2019). However, some operating conditions and other characteristics of these techniques have drawbacks in respect of energy and time savings. Food scientists and researchers are investigating various methods to improve the effective drying of food materials. Osmotic dehydration is an alternative drying technology for food preservation which decreases moisture content and water activity in food materials. This process is the most promising technique, having the advantages of low energy consumption, minimizing drying time, and maintaining food product quality (Sagar and Suresh Kumar 2010). Osmotic dehydration involves immersion of food products in a concentrated aqueous hypertonic solution of sucrose or a mixture of sugars in the case of fruits and in salt for vegetables (Sonia et al. 2015).

This method dehydrates fruit and vegetable slices in two steps: removing water through special hypertonic osmotic solvents; and then dehydrating the moisture in the dryer by air heater. The permeability of sugar in food materials is low, therefore the fruits and vegetables are infused with the osmo-active constituent in the surface layer only. Osmotic dehydration acts as a pre-drying step to increase the mass

transfer rate without changing the nutritional or functional properties and sensitivity of the food products (Yadav and Singh, 2014). Osmotic treatment of food materials is followed by further processes such as convective drying, solar drying, freeze-drying, vacuum drying, infrared drying, microwave drying, and other conventional drying methods. This chapter discusses the processes and drying kinetics of osmotic dehydration of fruits and vegetables, including research-based mathematical modeling and industrial application.

4.2 PRINCIPLES OF OSMOTIC DEHYDRATION

Osmotic dehydration is a pretreatment process that removes the desired amount of moisture from food materials to achieve shelf-life stability for further consumption. For removal of water the food product is placed in a hypertonic solution of high osmotic pressure such as common salt solution (NaCl), or solutions of $CaCl_2$, KCl, sugars (fructose, glucose, sucrose Arabic, etc.) or combinations thereof (Keerthana and Srijaya 2018). Several recent studies have shown that solvent is only transferred from an osmotic solution to foodstuffs because plant layers are semi-permeable in nature. Food materials contains a semi-permeable membrane which creates resistance to diffusion mass transfer throughout osmotic dehydration operations (Ramya and Jain 2017). Based on the different operating conditions, the plant cell membrane is partially permeable, causing significant variation in the tissue structure of food materials. Transfer of micronutrients from the food particles to the hypertonic solution occurs during osmotic dehydration (Chiralt and Fito 2003). The dehydration front moves inwards from the food product surface towards its center, which results in the spread of the food particles due to osmotic pressure. Effective dispersion of both moisture loss and solvent gain can be estimated by different techniques that give the diffusion rate of the solvent as well as water content of the food materials (Phisut 2012). The flux of osmo-active constituent penetrating the tissue of food materials changes their biochemical composition. Water content is removed by the process of osmosis at different concentrations without any phase transition of the solvent. This method is a very beneficial pretreatment for dehydration in comparison with the other methods due to positive and reversible changes both in physical and nutritional quality attributes of the food product. Water content is reduced by spreading the osmotic solution where the cell membrane only has very limited sugar content which is permitted in the tissue as a result of equalizing the concentration of the dissolved substance inside and outside fruits and vegetables which can take place by the movement of water (Maftoonazad 2012).

4.3 FACTORS AFFECTING OSMOTIC DEHYDRATION

Osmotic dehydration is affected by several process variables, including concentration and types of osmo-active solution, pretreatments, temperature, time, agitation, sample-to-solution ratio, and shape and size of the food materials (Azoubel and Murr 2004). These process variables have a further impact on the rate of mass transfer and the food product quality attributes throughout the operation of osmotic dehydration (Kaymak-Ertekin and Sultanoglu 2000), governing the specific mass transport

mechanisms that influence the food tissue and having a large impact on the total solids gain (SG), water loss (WL), rate of mass transfer, and structural changes.

4.3.1 Species and Variety

The variation in species, product, and maturity level of natural plant tissue can significantly affect the mass transfer rate between the food product and the osmotic solution. Different varieties of similar species at different levels of maturity also affect the rate of mass transfer with different responses throughout the osmotic dehydration operation (Mavroudis et al. 1998). Biochemical composition, including carbohydrate, protein, salt, and fat, and physical orientation, such as porosity, fiber orientation, plant cell arrangement and surface skin can also influence the drying kinetics of food products (Shi and Xue 2008). Indeed, the porosity of the food affects both texture and firmness. Changes in porosity caused by the osmotic process favor the action of non-entraining forces diffusion such as pressure gradients (Nieto et al. 2004).

4.3.2 Shape and Size

Size and shape are important factors in the process of mass exchange due to variation in surface area or surface-to-thickness ratio of food products. Most plant products are cut into cubes or spheres before treatment with osmotic dehydration, which facilitates the transfer of material through direct contact between cells and solution (Kowalska et al. 2008). Many researchers have studied the effect of shape and size on the mass transfer kinetics during osmo-dehydration of food products (Phisut 2012). The shape and size of solid food is also important because it determines the interfacial area and diffusion distance. The area decreases and the diffusion length increases with an increase in the size of the solid food, causing smaller food pieces to become dehydrated faster than larger counterparts of similar foods.

Lazarides et al. (1995) investigated the effect of size and shape on final solute concentrations, particularly during short-duration dehydration. Contreras and Smyrl (1981) observed that mass reduction increased approximately 1.3 times when the thickness of apple slices was reduced from 10 to 5 mm. Lerici et al. (1985) reported that a higher specific surface area gave a higher amount of SG and WL than lower specific surface area of the same food product.

4.3.3 Processing Conditions

Different pretreatments with osmotic solution and different operating conditions affect the integrity of fruit and vegetable tissue and have an intense impact on the mass transfer rate during osmotic dehydration. Many researchers report that processing conditions affect the interruption of tissue and structural change, which enhances the solute diffusivity during processing. Different pretreatments such as blanching, sulfating, freezing/thawing, and acidification, as well as process temperature, may affect mass transfer as a result of solute gain (Mavroudis et al. 1998; Azoubel and Murr 2004; Phisut 2012). Chwastek (2014) investigated the effect of steam blanching prior to osmo-dehydration and observed that WL decreased and SG increased during processing.

4.3.4 Types of Osmotic Agents

Several osmotic solutes and their combinations have been applied for the preparation of hypertonic solutions used for osmotic dehydration. The molecular weight and ionic characteristics of osmo-active solution strongly affects the rate of WL and solid uptake. The composition of the solutions (type, molecular mass of the solute) used in osmotic dehydration is a key factor in the process (Corrêa et al. 2010). The solutions are prepared from soluble crystalline solutes or water-miscible solvents, used alone or as a mixture. The constituents must be free from all toxicity, have sufficient solubility and, ideally, be inexpensive (Rastogi et al. 2004). Using different solutes in the mixture enables advantage to be taken of the respective effect of each (molar mass, diffusion properties, etc.), but also specific interactions (solute–solute and solute–food) to be developed for better control of the levels of dehydration and impregnation (Giraldo et al. 2003). Low molecular mass saccharides like sorbitol, fructose, and glucose are favorable for sugar uptake since high velocity ensures penetration of the molecule (Chavan and Amarowicz 2012). Sugar and salt solutions are mostly used for reasons of convenience and effectiveness. The osmo-active substances most frequently applied are sucrose and sodium chloride for osmo-dehydration of fruits and vegetables. Various osmotic agents are also used for dehydration, such as fructose, glucose, maltose, maltodextrin, dextrose, corn starch syrup, and polysaccharides. Numerous researchers have used other osmotic agents including ethanol, calcium chloride, invert sugar, malt dextrin, and corn syrup (Rastogi et al. 2014). Moraga et al. (2009) studied the influence of calcium lactate (2%) on the kinetics of osmo-dehydration and the effect of vacuum-impregnated and pulsed vacuum on grapefruit.

4.3.5 Time and Temperature

Time and temperature strongly affect the osmotic dehydration process (Ade-Omowaye et al. 2003). Mass transfer is reduced and weight loss increased with an increase in treatment time (Fasina et al. 2002). During the first hour of treatment, the SG and moisture loss rates are maximum, decreasing as osmotic dehydration progresses. Initially the rate of moisture loss and SG are about 20%, reducing thereafter to 10%. The rapidly enhanced rate of WL and SG during the first hour is due to the large driving force between the food products and the hypertonic solution (Chandra and Kumari 2015).

The mass transfer rate increases with increasing temperature and sugar syrup concentration (Biswal and Bozorgmehr 1992). The main effect of high temperature is faster water diffusion and solid diffusion within the product, which decreases the viscosity of the solution (Lazarides et al. 1995). Although increasing the temperature results in higher water removal in fruits and vegetables during drying, temperatures above 60°C are not recommended because of their negative impact on various quality parameters of the dried food materials (Keerthana and Srijaya 2018). Li and Ramaswamy (2006) reported that the osmotic diffusion rate is temperature and time dependent, but that a higher processing temperature favored faster moisture loss and solute uptake. During osmotic dehydration of apple slices, the amount of WL and SG increased with increase in temperature, but the water removal rate was higher than the SG rate. Though temperature enhances the osmotic process, the cell

membrane is highly sensitive and is damaged when the temperature is above 60°C (Kaymak-Ertekin and Sultanoglu 2000).

4.3.6 Agitation and Sample-to-Solution Ratio

Agitation during the osmotic process has a dynamic effect on the diffusion rate and increases the water removal rate by improving contacts between the food sample and the osmo-active solution (Phisut 2012). However, the dehydration rate only increased with the degree of agitation to a certain extent. It was observed that an adequate degree of agitation is essential to ensure removal of liquid mass transfer resistance and to provide a practically steady driving force (Arvanitoyannis et al. 2012). During the agitation of the hypertonic solution surrounding fruit products, the osmosis rate is increased due to reduction of the liquid mass transfer resistance at the surface. When the syrup surrounding the fruits is agitated, the rate of osmosis is faster because mass transfer resistance is reduced on the outer periphery of the food materials. Product-to-solution ratio significantly affects osmo-dehydration methods: the higher the product-to-solution ratio during processing, the higher the mass transfer and water removal rates (Rastogi et al. 2004). As osmotic dehydration takes place, the hypertonic solution is gradually diluted and the driving force for further mass transfer gradually decreases. A high sample-to-solution ratio is therefore essential to ensure a practically constant driving force (Singh et al. 2010). High sample-to-solution ratios maintain steady concentration of the solution and inhibit dilution (Ispir and Toğrul 2009). At industrial scale, the sample-to-solution ratio should be kept at the lowest level possible to limit the plant size and the regeneration of the hypertonic solutions. Various researchers have studied changes in solution concentration during osmotic dehydration, and the concentration change found most useful is a factor of between 4 and 10 (Chiralt and Fito 2003; Sagar and Suresh Kumar 2010; Sutar and Sutar 2013; Chwastek 2014).

4.4 MASS TRANSFER KINETICS OF OSMOTIC DEHYDRATION

Osmotic dehydration of fruits and vegetables is a complex technique of simultaneous heat and mass transfer between the product and the hypertonic solution. The combination of impregnation and dehydration processes is a consequence of changes in the biochemical composition and plant tissue of the osmo-dehydrated product. The phenomenon of osmosis involves diffusion or movement of water-soluble molecules from a higher-concentration to a lower-concentration solution through a semi-permeable membrane until equilibrium condition is achieved. Osmosis provides the movement of solvent which results in stretching and expansion of a cell due to the generation of turgor pressure inside the cell membrane as it is immersed in various hypertonic solutions. In general, all food materials contain vacuoles of solute such as sugars, salts, minerals, and organic acids within the cell membrane. The semi-permeable cell membrane of plant tissue forms the main resistance to mass transfer during osmotic dehydration (Khin et al. 2006). The driving force is the concentration gradient between the external solution and the intracellular fluid (Khin et al. 2007; Yadav and Singh 2014). Due to osmotic pressure differences within and outside the cells,

mass transfer through the cell membrane is part of the process. However, other large-scale transfer events occur, such as convection motion and dispersion of materials on the soaking hypertonic solution, resulting in the evacuation of fluid-filled intercellular capillary forces which are transported between cells (Ispir and Toğrul 2009).

In the initial stages, the cell wall layer of food materials stores a higher amount of driving force and the maximum contraction rate is observed during the process. In the second stage, the foremost driving force works for the deformation and contraction of that tissue, producing significant structural change in the food products. The last step exhibits an internal relaxation of the food material tissue, which improves mechanical transport performance (Phisut 2012; Rastogi et al. 2014). Mass transfer phenomena during the osmo-dehydration process are represented in Figure 4.1.

Mass transport during osmotic dehydration depends on different parameters (time, temperature, agitation, pressure, composition and concentration of solutions), physical properties (filtrate, shape, size, and microstructure), and multi-component mass properties (flow and their interactions). An inadequate semi-permeable membrane is established due to the complex structure and improper diffusion of solute products during treatment of osmo-active substances. Water is removed by diffusion and capillary flow, while leaching or solute uptake takes place through diffusion only. Osmotic dehydration of food materials leads to the following mass transfers:

i. Substantial water content flows out from food materials into the osmo-active solution.
ii. Solute is transferred from the hypertonic solution to the food product.
iii. Natural soluble solids existing in food materials leach into the hypertonic solution.
iv. Gases present on surfaces and in intercellular spaces are also eliminated.

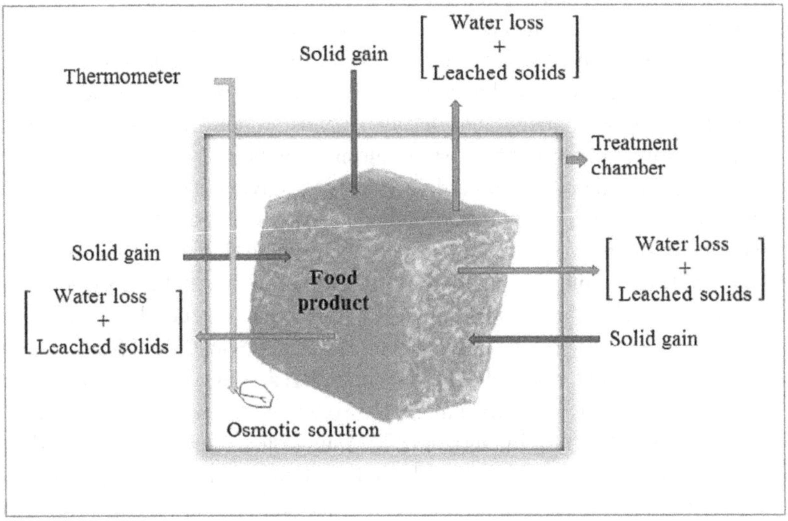

FIGURE 4.1 Mass transfer mechanisms during osmotic dehydration.
Source: Rastogi et al. (2015)

4.5 MATHEMATICAL MODELING OF OSMOTIC DEHYDRATION

Mathematical modeling and mass transfer during osmotic dehydration requires process variables and further drying processes to be optimized to attain the best-quality attributes to reduce energy consumption (Assis et al. 2017). The mass transfer kinetics of fruits and vegetables during the osmo-dehydration process are predicted by various mathematical models (Sutar and Sutar 2013). Many researchers have used the model developed by Crank (1979) for calculation of effective diffusivity for flat-plate geometry. The validity of the Page model, Azuara power law model, Lewis model, exponential model, Peleg model, penetration model, and Magee model for the calculation of WL and solute uptake has been examined by many researchers (Magee and Murphy 1983; Peleg 1988; Azuara et al. 1992). The Azuara model has been used extensively by many researchers to predict dehydration kinetics and values at equilibrium (Corzo and Gomez 2004; Singh et al. 2007). The Peleg model has also been used extensively to calculate the values at equilibrium and the flow rates of the moisture loss and solute uptake in the initial stage of the osmo-dehydration process (Sacchetti et al. 2001).

The Azuara mathematical model has been extensively applied by various researchers to prediction of the osmotic kinetics of fruits (Yadav and Singh 2014). Azuara et al. (1992) suggested a mathematical model based on liquid–solid mass transfer which is used for estimation of equilibrium point and mass transfer coefficients and to predict SG and WL. A phenomenological empirical model has been developed including different process variables and polynomial equations representing Fick's law. Empirical data were fitted to different models to obtain the best fit of coefficients. Ochoa-Martinez et al. (2007) suggested that the Azuara model equation was a better fit than Page's, Crank's, and Magee's models for prediction of the mass transfer rate during osmotic dehydration.

Bahadur and Hathan (2015) investigated the effect of different parameters on the moisture loss and solute uptake that can be characterized by the different model equations, and found that the Azuara and Magee models were the best fit for water loss and solid uptake respectively. The appropriate model for water estimation and solubility coefficient can be selected according to the size of the food. Abbasi Souraki et al. (2012) investigated mass transfer phenomena during the of osmo-dehydration of chopped green beans. The effective diffusion of moisture during osmotic dehydration is estimated by applying Fick's second law, as well as by extending a solvent to cylindrical coordinates. Care must be taken with mass transfer methods during osmotic dehydration not to dissolve the food tissue matrix and interfere with rehydration (Rastogi et al. 2014).

Peleg (1988) recommended a sorption isotherm equation and investigated its prediction using rice and milk powder. The Peleg model is generally best suited for depicting absorption of various materials while soaking (Sopade and Kaimur 1999). It has been used by various researchers to estimate WL and SG at the preliminary stage of osmotic dehydration (Sacchetti et al. 2001; Corzo and Gomez 2004; Mohebbi et al. 2011). Kaymak-Ertekin and Sultanoglu (2000) suggested a new approach to development of mathematical modeling based on liquid mass transfer in porous media. Several studies have focused on a number of empirical approaches that highlight the many factors affecting rehydration kinetics (Lewicki and Lenart 2006). Weibull analysis and Fick's second law of diffusion are widely used. The different model equations applied by different researchers to osmotic dehydration are given in Table 4.1.

TABLE 4.1
Different Model Equations Developed During Osmotic Dehydration of Food Products (Rastogi et al., 2015)

Model Name	Model Equation	Reference
Page's model	ML or $SG = \exp(-At^B)$	Page (1949)
First order kinetics	$\frac{M_t - M_e}{M_0 - M_e} = \exp(-P_5 t)$	Maskan (2001), Krokida and Marinos-Kouris (2003)
Azura's model	$ML = ML_\infty - M_m^w$ $SG = \frac{S_2 t SG_\infty}{1 + s_2 t}$	Waliszewski et al. (2002)
Exponential mathematical model	$\frac{M_t - M_e}{M_0 - M_e} = \exp(-P_1 t^{P_2})$	Misra and Brooker (1980)
Peleg's model	$M = M_0 + \frac{t}{(P_3 + P_4 \times t)}$	Ruíz Díaz et al. (2003), Peleg (1988), Ruíz Díaz et al. (2003), Bilbao-Sáinz et al. (2005)
Crank model:		Crank (1979), Rastogi et al. (1997), Waliszewski et al. (2002), Giraldo et al. (2003)
Semi-infinite plan	$\frac{ML}{ML_\infty} = 1 - \sum_{n=0}^{\infty} \frac{8}{(2n+1)^2 \pi^2} \exp \left\{ -(2n+1)^2 \frac{2\pi^2 F_0}{4} \right\}$	
Long cylinder	$\frac{ML}{ML_\infty} = 1 - \sum_{n=1}^{\infty} \frac{4}{(a^2 \alpha_n^2)} \exp(-D_e \alpha_n^2 t)$	
Spheres	$\frac{ML}{ML_\infty} = 1 - \frac{6}{\pi^2} \sum_{n=1}^{\infty} \frac{1}{n^2} \exp\left(-D_e n^2 \frac{\pi^2 t}{a^2} \right)$	
Magee's model	ML or $SG = kt^{0.5} + k_0$	Parjoko et al. (1996), Giraldo et al. (2003)
Becker model	$M_t - M_0 = M_0 + \frac{2}{\sqrt{\pi}}(M_s - M_0) \left(\frac{S}{V} \right) \sqrt{D_{eff}} \sqrt{t}$	Becker (1960), Lu et al. (1993)
Weibull model	$\frac{M_t}{M_e} = 1 - \exp\left[-\left(\frac{t}{\alpha} \right)^\beta \right]$	Sacchetti et al. (2001), Ruíz Díaz et al. (2003)
Normalized Weibull distribution function	$\frac{M_t - M_0}{M_e - M_0} = 1 - \exp\left[-\left(\frac{t \times D_{eff} \times R_g}{L^2} \right)^\beta \right]$	Marabi et al. (2003), Marabi and Saguy (2005)

(*Continued*)

TABLE 4.1
Continued

Terminologies of Above Model Equations

D_{eff}	**Effective Diffusion Coefficient (m^2/s)**	t	**Time (s)**
P_0	Water vapor pressure (Pa)	L	Distance covered by the liquid meniscus at time *t* (m)
P_1, P_2	Parameters in thin layer exponential model	V	Volume (m^3)
P_3	Constant associated to the initial rate of sorption	ρ	Density (kg/m^3)
P_4	Constant associated to the equilibrium moisture content	γ	Surface tension of the liquid (N/m)
P_5	Rehydration rate (min^{-1})	α	Scale parameter(s)
D	Diffusion coefficient (m^2/s)	β	Weibull shape parameter
M_0, M_t, M_e	Moisture content at time 0, *t*, and at equilibrium respectively (kg H_2O/kg D.S.)	k_0, k_t	Constant
M_s	Surface moisture content (kg H_2O/kg D.S.)	S	Surface area (m^2)
r	Pore radius (m)	R_g	Geometry factor (–)

4.6 APPLICATION OF OSMOTIC DEHYDRATION IN FOOD PROCESSING

The osmotic dehydration process removes about 30–50% of water from ripened fruits. The nutritional values, functional and organoleptic properties of the products are improved by osmotic dehydration. Damage to color and flavor caused by high thermal treatment can be minimized by using a mild heat treatment, and discoloration is prevented by the high sugar concentration around the pieces of fruit and vegetable (Chandra and Kumari 2015). Industrially, osmotic dehydration is applied to dried foods, candies, and dehydrated vegetables. Some researchers have already reported the effect of osmotic dehydration of food products like pineapple, banana, mango, jackfruit, guava, sapota, papaya (fruit), pineapple, carrot, ginger, garlic, and seafood.

Most research work is on osmo-dehydration of fruits and vegetables. The list includes tomato (Cataldo et al. 2011), cherry tomato (Heredia et al. 2009), red pepper (Raji Abdul Ganiy et al. 2010), potato (Eren and Kaymak-Ertekin 2007), sweet potato (Antonio et al. 2008), ivy gourd (Kulkarni and Vijayanand 2012), melon (Aminzadeh et al. 2012), watermelon (Falade et al. 2007), cantaloupe (Corzo and Gomez 2004), pumpkin (Mayor et al. 2011), sugar beet (Jokić et al. 2007), beetroot and radish, carrot (Manivannan and Rajasimman 2009; Mohebbi et al. 2011), onion, and mushroom (Sutar and Gupta 2007; Abud-Archila et al. 2008).

A great deal of work has been reported on tropical fruits such as mango (Bernardi et al. 2009), banana (Atares et al. 2011), guava, papaya (Jain et al. 2011; Ganjloo et al. 2012), jackfruit, pineapple, pear (Marani et al. 2007; Lombard et al. 2008; Saxena et al. 2009), peach, strawberries, lychee (Castelló et al. 2010; Yadav et al. 2012; Kumar 2016), kiwifruit, and apricot (Manafi et al. 2010; Tylewicz et al. 2011). A typical flow sheet for the osmotic dehydration processing of fruits is shown in Figure 4.2.

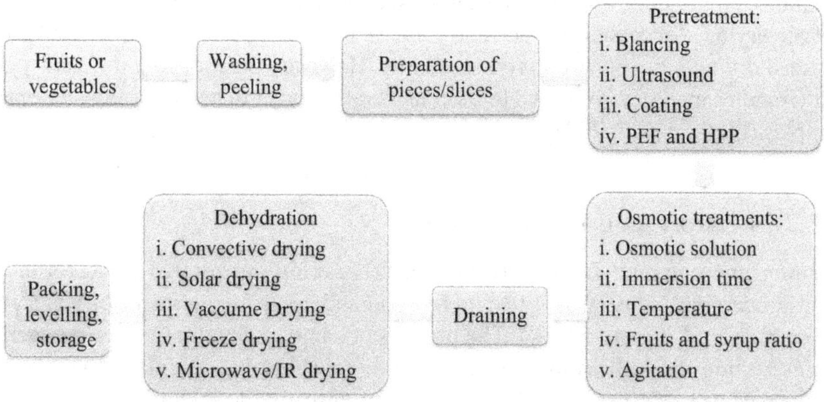

FIGURE 4.2 Process flow chart of osmotic dehydration.
Source: Chandra and Kumari (2015)

4.7 RECENT ADVANCES IN OSMOTIC DEHYDRATION PREPARATION

Advanced methods are being developed for the purpose of increasing the mass transfer rate by maximizing the osmotic pressure of high-concentration hypertonic solutions. The different techniques include ohmic heating, agitation through centrifugal force, ultrasound, pulse vacuum impregnation, pulsed electric fields, high hydrostatic pressure, gamma radiation, and extensive mass transfer before osmotic treatment for food materials.

4.7.1 Microwave Radiation

The combined effect of osmotic dehydration and microwave processing of food materials has been studied for enhancing mass transfer rate and reducing drying time. Topics included the effect on product quality of vacuum microwave-assisted osmotic dehydration of blueberries, apples, potatoes, strawberries, mushrooms, and tomatoes (Prothon et al. 2001; Torringa et al. 2001; Heredia et al. 2007; Sutar et al. 2012). Torringa et al. (2001) applied microwave and vacuum treatment throughout the osmotic dehydration progression of mushroom and found that this process reduced shrinkage and time taken, yet there was increased solute uptake and porosity due to changes in dielectric properties. Pereira et al. (2007) reported the impact of air-flow velocity, temperature and microwave power on the drying characteristics behavior of osmo-dried bananas. This study concluded that increase in microwave power resulted in reduced drying time and boosted the total quality attributes of the final dried product. Heredia et al. (2007) investigated the impact of hypertonic solutions on cherry tomatoes with a composition of 10% salt water, 27.5% sucrose and 2% calcium lactates in concurrence with microwave hot-air drying. It was found that the dried tomato product had good quality attributes and longer storage life than untreated

tomatoes. Apple cubes were treated with sucrose and dried by microwave-assisted hot-air drying. Since sucrose became infused into the cell structure, it resulted in reduced drying and improved product quality. However, the moisture diffusivity and rehydration capacity were found to be considerably lower than those of the untreated samples (Prothon et al. 2001).

4.7.2 Pulsed Vacuum Impregnation

Vacuum impregnation is extensively used to develop dehydrated products in the food-processing industry as it results in rapid and controlled transfer of water into the cell tissue (Chwastek 2014). The process of osmotic dehydration can be carried out at vacuum as well as at atmospheric pressure, leading to enlargement of the food tissue at low pressure and obstruction of gases in the integument of a tissue. When the treatment pressure is released, the osmotic solution fills the pores, resulting in enhanced mass transfer. Application of pulse vacuum to food materials in osmotic solution results in initiation of the hydrodynamic mechanism followed by an increase in solid intake and loss of water, reducing drying time and improving the quality of the food product (Gras et al. 2003). The type of hypertonic solution used and the different characteristics of the food materials must be considered during the application of combined vacuum impregnation and osmotic dehydration (Chiralt and Fito 2003). When the pulse vacuum is released after 10–20 minutes as a result of the development of a series of pulse cycles, mass transfer rate is increased when food materials are osmotically dehydrated (Santacruz-Vazquez et al. 2008). During pulsed vacuum impregnation, the product returns to atmospheric pressure and the concentrated solution penetrates massively into the pores of the food product, which has the consequence of increasing the contact surface between the product and the solution, accelerating material transfer (Fito, 1994). Corrêa et al. (2010) obtained higher water loss using vacuum only for 15 min than with ordinary osmotic drying.

4.7.3 Ohmic Heating

Ohmic heating is a method by which food products are treated with electric current. This technique passes alternating electrical current through a food sample placed between two electrodes. The food acts as an electrical resistance which generates heat within the food materials. Storage life of osmotically dehydrated products can be prolonged by combining ohmic heat treatment with vacuum dehydration (Sutar and Sutar, 2013). Moreno et al. (2011) reported that osmotically treated pears can have a shelf life of up to 25 days under pulse vacuum and ohmic heat treatment at 13 V/cm. They also confirmed that the use of ohmic heat treatment boosted the mass transfer mechanism during the process of osmo-dehydration. Xin et al. (2014) used ultrasound as pretreatment to reduce the drying time for osmo-dehydration of broccoli from 120 to 30 minutes. Moreover, its L-ascorbic acid content increased from 79.7 to 84.4%, compared with an increase from 63.4 to 72.3% in an unspecified frozen broccoli product.

4.7.4 Pulsed Electric Field

Pulsed electric field (PEF) also called electro-compression, is a non-thermal method that uses very short pulses in the range 0.5–80 kV/cm, which when applied to a food sample kept between two electrodes for microseconds to milliseconds causes changes in the structure of the tissue of the food materials (Chwastek 2014). Pulse electric field technology is mild electrical treatment which alters membrane permeability when a high voltage is applied as a series of short-duration pulses (Barba et al. 2015). Urszula Tylewicz et al. (2017) investigated and concluded that PEF treatment combined with osmotic dehydration results in water declination in the tissues of organic strawberries, thus affecting their physio-chemical properties. They reported that PEF treatment at the lowest-strength electric field (100 V/cm) positively affected mass transfer, partially saving the tissue structure and maintaining the freshness of osmotically dehydrated strawberries. Traffano-Schiffo et al. (2017) reported the effect of pulse electric fields at different voltages (100–400 V and 60 pulses) during osmotic dehydration of kiwi fruit, and concluded that the water activity of the product declined and dehydration time reduced, causing an inclination in solid and water gain (Yu et al. 2018). Yu et al. (2018) reported that PEF treatments increased the usefulness of osmotic dehydration, and that this pretreatment decreased the chemical composition and enhanced the rate of mass transfer.

4.7.5 Ultrasonication

The application of ultrasound produces a cavitation phenomenon, consisting of the formation in the liquid of gas bubbles which generate pressure fluctuations on bursting (Fernandes et al., 2009). Mass transfer during the process of diffusion is enhanced using sound waves, which stimulates a quick succession of alternate expansion and compression (Panades et al. 2008). When used at elevated power and low frequency (20–100 kHz) at nominal temperature, ultrasound removes moisture from the food product. The use of an ultrasonic field results in a high increase in the diffusion rate between the solid and liquid, accelerating degassing of the product during osmotic dehydration (Bozkir et al. 2018). Combined pulsed vacuum and ultrasound treatments are applied during osmotic dehydration, resulting in a high amount of WL and reduced firmness in food samples (Cheng et al. 2014). The reason is the development of microscopic channels that allow hypertonic solution to flow into the cellular spores of partially dried food materials (Assis et al. 2017). There is a gradual increase in the mass transfer rate when the rate of intensity is increased, since either intense cavitation or vapor occlusion occurs at the cell boundary (Rastogi et al. 2005). Ultrasound at high frequency used simultaneously with increase in sugar concentration increases the WL rate, resulting in reduced dehydration time. The dipping of food materials in osmotic solution followed by ultrasonic treatment does not cause any thermal degradation in the food product as it involves no heat (Stojanovic and Silva 2007). Natural flavor and nutritional components can be preserved by obtaining increased SG and WL, when the process of ultrasound-assisted osmo-dehydration is conducted at low temperature. Structural changes in pineapple and melon were induced when they were subjected to ultrasound-assisted osmotic dehydration,

which also led to increased water diffusivity when treated for more than 30 mins. The high soluble solid content of the diffusion solution caused a decrease in water diffusivity (Fernandes et al., 2008; Radziejewska-Kubzdela et al., 2014).

4.7.6 High Hydrostatic Pressure

High hydrostatic pressure (HHP) treatment inhibits enzymes and microbes present in food products, resulting in increased product stability. This treatment modifies the cell wall structure, increasing cell permeability. HHP is employed as a pretreatment prior to osmotic dehydration of food material so that the mass transfer rate increases throughout the drying process (Nuñez-Mancilla et al. 2014). This treatment is used to change the functional, rheological, and textural properties of the food material, and when combined with osmotic dehydration it gives increased SG. When Nuñez-Mancilla et al. (2013) subjected strawberry to HHP and osmotic dehydration in the range of 100–500 MPa, it increased the total phenolic content as the pressure increased. Some studies (Taiwo et al. 2002; Rastogi et al. 2004) highlight the fact that high-pressure pretreatment creates compaction of the cellular structure accompanied by a release of cellular components. This phenomenon results in the formation of a gel by binding of divalent ions with the esterified pectin, which limits the diffusion coefficient of solids.

4.7.7 Gamma Irradiation

The internal structure of the food products can be reformed by gamma irradiation, leading to greater permeability of cells, and resulting in improved mass transfer during air drying (Wang and Sastry 2000). Gamma irradiation is extensively applied to enhance the shelf life and retain quality attributes of a fresh food product. Application of gamma irradiation leads to retardation of microorganism growth and the formation of hyphae and fruiting bodies in vegetables, kills insects, and sterilizes food materials. Gamma radiation alters the internal structures of food products by recompensing their tissue structure as well as increasing permeability, leading to improvements in mass transfer during the osmo-dehydration operation (Rastogi et al. 2006). Wang and Chao (2003) reported that the higher the gamma radiation dose, the greater the dehydration rate and the lower the rehydration rate of potato. Rastogi et al. (2005) observed the effect of gamma irradiation on the osmotic dehydration process, and found the moisture diffusivity rate was higher than the solution diffusivity rate, but at 10 Brix there was less water intake and increased solute loss. When gamma-irradiated potato (3.0–12.0 kGy) was osmotically dehydrated, diffusion rates increased as the firmness reduced, resulting in an increase in cell permeability (Rastogi et al. 2006).

4.7.8 Osmo-Dehydrofreezing

Osmo-dehydrofreezing is the combined process of osmo-dehydration and freezing (Lowithun and Charoenrein 2009). It shrinks the energy requirement during freezing as well as the cost of distribution and packaging. Products resulting from this process

have a low drip loss rate and measurable structural changes (Dermesonlouoglou et al. 2008). The moisture content is reduced due to treatment with osmotic solution as a result of less moisture being available during freeze-drying, so food quality can eventually be maintained after thawing. Many researchers have found that improved food products with maximum textural attributes are obtained by osmo-dehydrofreezing. Increased textural quality in terms of firmness in apples as a result of osmotic dehydration followed by freezing was reported by Tregunno and Goff (1996).

4.8 RESEARCH-ORIENTED PROBLEMS

Osmotic dehydration is the drying technique with the greatest potential for preservation of fruits and vegetables and this has led to many studies by different researchers on mass transfer mechanisms, modeling, and microstructural changes in osmotically dehydrated food products. Mass transfer phenomena are complex mechanisms involving changes to cell walls and structures. These tissue cells can be influenced by transport mechanisms; therefore microstructural analysis is desirable. Researchers are constantly trying to improve the micro-pores in epithelial cells in order to improve mass transfer rates prior to osmotic dehydration. Identification of the end point and reuse of the concentration hypertonic solution for further processing is an important area of research to ensure the energy efficiency and economic viability of this process. The osmotic process is less expensive than conventional infusion or vacuum drying (freeze-drying). Improving the potential energy savings and the overall quality of food products are essential steps in the osmo-dehydration process. Some researchers have recently tried a new approach to the rehydration of dried food particles that cannot be fully described by the osmotic treatment mechanism. Water imbalance, capillary flow, and porous media are the suggested mechanisms considered relevant for explaining water entry into dry food cells. Innovative development of the dehydration process is required to infuse and trap bioactive compounds in the sample for the addition of nutritional value.

Osmotic dehydration has the following advantages:

i. Higher retention of flavor and nutritional characteristics is maintained when sugar syrup is used as an osmo-active agent.
ii. Prevention of enzymatic reaction and oxidative browning of osmo-dried food materials.
iii. Reduction of drying time and energy consumption due to the reduced water removal load in the dryer.
iv. Increase in the density of the product, and improved textural quality after reconstitution.

It also has the following limitations.

i. Osmotic dehydration is constrained by the availability of fruits of the desired quality, hence activity is highly seasonal.
ii. Acidity and pH level are reduced by osmotic treatment, which reduces the sensory characteristic of some food products.

iii. Sugar uptake may not be required in certain products, hence rapid rinsing of water drained in fruits may be necessary.
iv. The cost of the concentrated solution is key, such that the management of the total amount of concentrated solution employed may be a major factor in the viability of the process.
v. The optimization of different factors affecting performance during the osmotic dehydration process is complicated by two main components i) syrup recycling; and ii) reduction of the syrups volume involved.

4.9 INDUSTRIAL PROCESSING AND CHALLENGES

Osmotic dehydration has shown great potential for impregnation/concentration at a reasonable technical and economic cost. Nevertheless, there is abundant room for further advanced research in this area, and for development and implementation of novel techniques for more widespread industrial application that can overcome difficulties associated with the design of equipment for continuous large-scale movement and operations. The use of highly concentrated sucrose solutions causes two major difficulties: (i) the need to pay careful attention to microbiological safety; and (ii) the viscosity of the syrup is so high that mass transfer resistance to the solvent needs to be reduced. Although fresh syrup is not added, sucrose syrup can be recycled at least five times without spoiling fruit quality. To make this process economically viable, syrup must be recovered and reused. However, this problem can be solved by mixing osmotic solution with an antioxidant or by adopting better packaging material for industrial application. Equipment management, especially when handling perishable foods, requires special attention when planning for automated control and online monitoring during operation.

4.10 CONCLUSION

Osmotic dehydration is an energy-efficient technique for preserving fruit and vegetable products without affecting their quality. Osmotic dehydration causes low thermal deterioration of functional nutrients due to the water removal being performed at lower temperature. It offers several advantages, including reduction of color and taste degradation, prevention of enzyme breakdown, and reduced energy costs. It has been used effectively for 30–50% weight reduction in food products but further drying and packaging of food materials is required for extension of safe storage life and consumer convenience. The low energy consumption and quality retention advantages offered by osmotic dehydration suggests it has a vital potential role to play in the food-processing industry. Today's consumers care deeply about nutritional quality, so this dehydration process is concerned with quality improvement as well as reduction in microbial attacks and enzymatic reactions. Recent developments in osmotic dehydration have reduced osmosis time and increased moisture loss with minimal SG. This technique is highly recommended for fruits, vegetables, and other food crops as it reduces volume, prolongs storage life and makes transport more convenient. Small-scale entrepreneurs and industries can use this process for the drying of food products at the level of pilot-plant operations.

REFERENCES

Abbasi Souraki, B., Ghaffari, A., and Bayat, Y. 2012. Mathematical modeling of moisture and solute diffusion in the cylindrical green bean during osmotic dehydration in salt solution. *Food and Bioproducts Processing*, *90*(1), 64–71.

Abud-Archila, M., Vázquez-Mandujano, D.G., Ruiz-Cabrera, M.A., Grajales-Lagunes, A., Moscosa-Santillán, M., Ventura-Canseco, L.M.C., and Dendooven, L. 2008. Optimization of osmotic dehydration of yam bean (Pachyrhizuserosus) using an orthogonal experimental design. *Journal of Food Engineering*, *84*(3), 413–419.

Ade-Omowaye, B.I.O., Rastogi, N.K., Angersbach, A., and Knorr, D. 2003. Combined effects of pulsed electric field pre-treatment and partial osmotic dehydration on air drying behavior of red bell pepper. *Journal of Food Engineering*, *60*(1), 89–98.

Aminzadeh, R., Sargolzaei, J., and Abarzani, M. 2012. preserving melons by osmotic dehydration in a ternary system followed by air-drying. *Food and Bioprocess Technology*, *5*(4), 1305–1316.

Antonio, G.C., Azoubel, P.M., Elizabeth, F., Murr, X., and Park, K.J. 2008. Osmotic dehydration of sweet potato in ternary solutions. *Ciência e Tecnologia de Alimentos*, 28(3), 696–701.

Arvanitoyannis, I.S., Veikou, A., and Panagiotaki, P. 2012. Osmotic dehydration: Theory, methodologies, and applications in fish, seafood, and meat products. *Progress in Food Preservation*, 161–189.

Assis, F.R., Morais, R.M.S.C., and Morais, A.M.M.B. 2017. Mathematical modelling of osmotic dehydration kinetics of apple cubes. *Journal of Food Processing and Preservation*, *41*(3): 183–192.

Atares, L., Sousa Gallagher, M.J., and Oliveira, F.A.R. 2011. Process conditions effect on the quality of banana osmotically dehydrated. *Journal of Food Engineering*, *103*(4), 401–408.

Azoubel, P.M., and Murr, F.E.X. 2004. Mass transfer kinetics of osmotic dehydration of cherry tomato. *Journal of Food Engineering*, *61*(3), 291–295.

Azuara, E., Cortes, R., Garcia, H.S., and Beristain, C.I. 1992. Kinetic model for osmotic dehydration and its relationship with Fick's second law. *International Journal of Food Science & Technology*, *27*(4), 409–418.

Bahadur, S., and Hathan, S. 2015. Elephant foot yam (amorphophalluspaeoniifolius): Osmotic dehydration and modelling. *Journal of Food Processing & Technology*, *6*(10), 210–219.

Barba, F.J., Parniakov, O., Pereira, S.A., Wiktor, A., Grimi, N., Boussetta, N., Vorobiev, E. 2015. Current applications and new opportunities for the use of pulsed electric fields in food science and industry. *Food Research International*, *77*, 773–798.

Becker, H.A. 1960. On the absorption of liquid water by the wheat kernel. *Cereal Chemistry*, *37*, 309–323.

Bernardi, S., Bodini, R.B., Marcatti, B., Petrus, R.R., and Favaro-Trindade, C.S. 2009. Quality and sensorial characteristics of osmotically dehydrated mango with syrups of inverted sugar and sucrose. *Scientia Agricola*, *66*(1), 40–43.

Bilbao-Sáinz, C., Andrés, A., and Fito, P. 2005. Hydration kinetics of dried apple as affected by drying conditions. *Journal of Food Engineering*, *68*(3), 369–376.

Biswal, R.N., and Bozorgmehr, K. 1992. Mass transfer in mixed solute osmotic dehydration of apple rings. *Transactions of the American Society of Agricultural Engineers*, *35*(1), 257–262.

Bozkir, H., Rayman Ergün, A., Tekgül, Y., and Baysal, T. 2018. Ultrasound as pretreatment for drying garlic slices in microwave and convective dryer. *Food Science and Biotechnology*, *28*(2), 347–354.

Castelló, M.L., Fito, P.J., and Chiralt, A. 2010. Changes in respiration rate and physical properties of strawberries due to osmotic dehydration and storage. *Journal of Food Engineering*, *97*(1), 64–71.

Cataldo, A., Cannazza, G., De Benedetto, E., Severini, C., and Derossi, A. 2011. An alternative method for the industrial monitoring of osmotic solution during dehydration of fruit and vegetables: A test-case for tomatoes. *Journal of Food Engineering*, *105*(1), 186–192.

Chandra, S., and Kumari, D. 2015. Recent development in osmotic dehydration of fruit and vegetables: A review. *Critical Reviews in Food Science and Nutrition*, *55*(4), 552–561.

Chavan, U.D., and Amarowicz, R. 2012. Osmotic dehydration process for preservation of fruits and vegetables. *Journal of Food Research*, *1*(2): 202–211.

Cheng, X.F., Zhang, M., Adhikari, B., and Islam, M.N. 2014. Effect of power ultrasound and pulsed vacuum treatments on the dehydration kinetics, distribution, and status of water in osmotically dehydrated strawberry: A combined nmr and dsc study. *Food and Bioprocess Technology*, *7*(10), 2782–2792.

Chiralt, A., and Fito, P. 2003. Food science and technology international transport mechanisms in osmotic dehydration: The role of structure. *Food Science and Technology International*, *9*(3), 179–186.

Chwastek, A. 2014. Methods to increase the rate of mass transfer during osmotic dehydration of foods. *Acta Scientiarum Polonorum, Technologia Alimentaria*, *13*(4), 341–350.

Contreras, J.E., and Smyrl, T.G. 1981. An evaluation of osmotic concentration of apple rings using corn syrup solids solutions. *Canadian Institute of Food Science and Technology Journal*, *14*(4), 310–314.

Corrêa, J., Pereira, L., Vieira, G., and Hubinger, M. 2010. Mass transfer kinetics of pulsed vacuum osmotic dehydration of guavas. *Journal of Food Process Engineering*, 96(5), 498–504.

Corzo, O., and Gomez, E.R. 2004. Optimization of osmotic dehydration of cantaloupe using desired function methodology. *Journal of Food Engineering*, *64*(2), 213–219.

Crank, J. 1979. *The Mathematics of Diffusion*. Oxford University Press.

Dermesonlouoglou, E.K., Pourgouri, S., and Taoukis, P.S. 2008. Kinetic study of the effect of the osmotic dehydration pre-treatment to the shelf life of frozen cucumber. *Innovative Food Science and Emerging Technologies*, *9*(4), 542–549.

Dhyani, A. 2006. Post-harvest technology of fruits and vegetables. In *Innovative Production Systems in Horticulture*. Greater Noida, U.P: Institute of Horticulture Technology, pp. 36–40.

Eren, I., and Kaymak-Ertekin, F. 2007. Optimization of osmotic dehydration of potato using response surface methodology. *Journal of Food Engineering*, *79*(1), 344–352.

Falade, K.O., Igbeka, J.C., and Ayanwuyi, F.A. 2007. Kinetics of mass transfer, and colour changes during osmotic dehydration of watermelon. *Journal of Food Engineering*, *80*(3), 979–985.

Fasina, O., Fleming, H., and Thompson, R. 2002. Mass transfer and solute diffusion in brined cucumbers. *Journal of Food Science*, *67*(1), 181–187.

Fernandes, F., Gallão, M.I., and Rodrigues, S. 2009. Effect of osmosis and ultrasound on pineapple cell tissues structure during dehydration. *Journal of Food Engineering*, 90(3), 186–190.

Fernandes, F.A.N., Gallão, M.I., and Rodrigues, S. 2008. Effect of osmotic dehydration and ultrasound pre-treatment on cell structure: Melon dehydration. *LWT - Food Science and Technology*, *41*(4), 604–610.

Fito, P. 1994. Modelling of vacuum osmotic dehydration of food. *Journal of Food Engineering*, 22(2), 313–328.

Ganjloo, A., Rahman, R.A., Bakar, J., Osman, A., and Bimakr, M. 2012. Kinetics modeling of mass transfer using Peleg's equation during osmotic dehydration of seedless guava (psidiumguajava l.): Effect of process parameters. *Food and Bioprocess Technology*, *5*(6), 2151–2159.

Giraldo, G., Talens, P., Fito, P., and Chiralt, A. 2003. Influence of sucrose solution concentration on kinetics and yield during osmotic dehydration of mango. *Journal of Food Engineering*, *58*(1), 33–43.

Gras, M.L., Vidal, D., Betoret, N., Chiralt, A., and Fito, P. 2003. Calcium fortification of vegetables by vacuum impregnation: Interactions with cellular matrix. *Journal of Food Engineering*, *56*(2–3), 279–284.

Heredia, A., Barrera, C., and Andrés, A. 2007. Drying of cherry tomato by a combination of different dehydration techniques: Comparison of kinetics and other related properties. *Journal of Food Engineering*, *80*(1), 111–118.

Heredia, A., Peinado, I., Barrera, C., and Grau, A.A. 2009. Influence of process variables on colour changes, carotenoids retention and cellular tissue alteration of cherry tomato during osmotic dehydration. *Journal of Food Composition and Analysis*, *22*(4), 285–294.

Ispir, A., and Toğrul, I.T. 2009. Osmotic dehydration of apricot: Kinetics and the effect of process parameters. *Chemical Engineering Research and Design*, *87*(2), 166–180.

Jain, S.K., Verma, R.C., Murdia, L.K., Jain, H.K., and Sharma, G.P. 2011. Optimization of process parameters for osmotic dehydration of papaya cubes. *Journal of Food Science and Technology*, *48*(2), 211–217.

Jokić, A., Gyura, J., Lević, L., and Zavargó, Z. 2007. Osmotic dehydration of sugar beet in combined aqueous solutions of sucrose and sodium chloride. *Journal of Food Engineering*, *78*(1), 47–51.

Kaymak-Ertekin, F., and Sultanoglu, M. 2000. Modelling of mass transfer during osmotic dehydration of apples. *Journal of Food Engineering*, *46*(4), 243–250.

Keerthana, H., and Srijaya, M. 2018. Recent developments in osmotic dehydration technique for improving the post-harvest quality of fruits and vegetables. *Research Journal of Pharmaceutical, Biological and Chemical Sciences*, *9*(26), 26–33.

Khin, M.M., Zhou, W., and Perera, C.O. 2006. A study of the mass transfer in osmotic dehydration of coated potato cubes. *Journal of Food Engineering*, *77*(1), 84–95.

Khin, M.M., Zhou, W., and Yeo, S.Y. 2007. Mass transfer in the osmotic dehydration of coated apple cubes by using maltodextrin as the coating material and their textural properties. *Journal of Food Engineering*, *81*(3), 514–522.

Kowalska, H., Lenart, A., and Leszczyk, D. 2008. The effect of blanching and freezing on osmotic dehydration of pumpkin. *Journal of Food Engineering*, *86*(2), 30–38.

Krokida, M.K., and Marinos-Kouris, D. 2003. Rehydration kinetics of dehydrated products. *Journal of Food Engineering*, *57*(1), 1–7.

Kulkarni, S.G., and Vijayanand, P. 2012. Effect of pretreatments on quality characteristics of dehydrated Ivy Gourd (Cocciniaindica L.). *Food and Bioprocess Technology*, *5*(2), 593–600.

Kumar, S. 2016. A review article: Modified atmosphere packaging of fruits and vegetables: A promising concept. *Journal of Bio-Technology, Food Technology, Agriculture and Innovative Research*, *1*(1), 26–29.

Lazarides, H.N., Katsanidis, E., and Nickolaidis, A. 1995. Mass transfer kinetics during osmotic preconcentration aiming at minimal solid uptake. *Journal of Food Engineering*, *25*(2), 151–166.

Lee, K.T., Farid, M., and Nguang, S.K. 2006. The mathematical modelling of the rehydration characteristics of fruits. *Journal of Food Engineering*, *72*(1), 16–23.

Lerici, C.R., Pinnavaia, G., Rosa, M.D., and Bartolucci, L. 1985. Osmotic dehydration of fruit: Influence of osmotic agents on drying behavior and product quality. *Journal of Food Science*, *50*(5), 1217–1219.

Lewicki, P.P., and Lenart, A. 2006. Osmotic dehydration of fruits and vegetables. In *Handbook of Industrial Drying*, Taylor & Francis, pp. 375–400.

Li, H., and Ramaswamy, H. 2006. Osmotic dehydration of apple cylinders: I. Conventional batch processing conditions. *Drying Technology*, *24*(5), 619–630.

Lombard, G.E., Oliveira, J.C., Fito, P., and Andrés, A. 2008. Osmotic dehydration of pineapple as a pre-treatment for further drying. *Journal of Food Engineering*, *85*(2), 277–284.

Lowithun, N., and Charoenrein, S. 2009. Influence of osmo-dehydrofreezing with different sugars on the quality of frozen rambutan. *International Journal of Food Science and Technology*, *44*(11), 2183–2188.

Lu, R., Siebenmorgen, T.J., and Archer, T.R. 1993. Absorption of water in long grain rice during soaking. *Journal of Food Process Engineering*, *17*(4), 141–154.

Maftoonazad, N. 2012. Use of osmotic dehydration to improve fruits and vegetables quality during processing. *Recent Patents on Food, Nutrition & Agriculture*, *2*(3), 233–242.

Magee, T.R., and Murphy, W.R. 1983. Internal mass transfer during osmotic dehydration of apple slices in sugar solutions. *Irish Journal of Food Science and Technology*, *7*(2), 147–155.

Manafi, M., Hesari, J., Peighambardoust, H., and Khoyi, M.R. 2010. Osmotic dehydration of apricot using salt-sucrose solutions. *World Academy of Science Engineering and Technology*, *44*(3), 1098–1101.

Manivannan, P., and Rajasimman, M. 2009. Osmotic dehydration of beetroot in salt solution: Optimization of parameters through statistical experimental design. *International Journal of Chemical Biomolecular Engineering*, *1*(4), 215–222.

Marabi, A., Livings, S., Jacobson, M., and Saguy, I.S. 2003. Normalized Weibull distribution for modeling rehydration of food particulates. *European Food Research and Technology*, *217*(4), 311–318.

Marabi, A., and Saguy, I.S. 2005. Viscosity and starch particle size effects on rehydration of freeze-dried carrots. *Journal of the Science of Food and Agriculture*, *85*(4), 700–706.

Marani, C.M., Agnelli, M.E., and Mascheroni, R.H. 2007. Osmo-frozen fruits: Mass transfer and quality evaluation. *Journal of Food Engineering*, *79*(4), 1122–1130.

Maskan, M. 2001. Drying, shrinkage and rehydration characteristics of kiwifruits during hot air and microwave drying. *Journal of Food Engineering*, *48*(2), 177–182.

Mavroudis, N.E., Gekas, V., and Sjöholm, I. 1998. Osmotic dehydration of apples. Shrinkage phenomena and the significance of initial structure on mass transfer rates. *Journal of Food Engineering*, *38*(1–4), 101–123.

Mayor, L., Moreira, R., and Sereno, A.M. 2011. Shrinkage, density, porosity and shape changes during dehydration of pumpkin (Cucurbitapepo L.) fruits. *Journal of Food Engineering*, *103*(1), 29–37.

Misra, M.K., and D.B. Brooker. 1980. Thin-layer drying and rewetting equations for shelled yellow corn. *Transactions of the ASAE*, 23(5), 1254–1260.

Mohebbi, M., Akbarzadeh-T, M.R., Shahidi, F., and Zabihi, S.M. 2011. Modeling and optimization of mass transfer during osmosis dehydration of carrot slices by neural networks and genetic algorithms. *International Journal of Food Engineering*, *7*(2).

Moraga, M.J., Moraga, G., Fito, P.J., and Martínez-Navarrete, N. 2009. Effect of vacuum impregnation with calcium lactate on the osmotic dehydration kinetics and quality of osmo-dehydrated grapefruit. *Journal of Food Engineering*, *90*(3), 372–379.

Moreno, J., Simpson, R., Sayas, M., Segura, I., Aldana, O., and Almonacid, S. 2011. Influence of ohmic heating and vacuum impregnation on the osmotic dehydration kinetics and microstructure of pears (cv. Packham's Triumph). *Journal of Food Engineering*, *104*(4), 621–627.

Nath, A., Meena, L.R., Kumar, V., and Panwar, A.S. 2018. Postharvest management of horticultural crops for doubling farmer's income. *Journal of Pharmacognosy and Phytochemistry*, *1*, 2682–2690.

Nieto, A.B., Salvatori, D.M., Castro, M.A., and Alzamora, S.M. 2004. Structural changes in apple tissue during glucose and sucrose osmotic dehydration: Shrinkage, porosity, density and microscopic features. *Journal of Food Engineering*, *61*(3), 269–278.

Nuñez-Mancilla, Y., Pérez-Won, M., Uribe, E., Vega-Gálvez, A., and Di Scala, K. 2013. Osmotic dehydration under high hydrostatic pressure: Effects on antioxidant activity, total phenolics compounds, vitamin C and colour of strawberry (Fragariavesca). *LWT - Food Science and Technology*, *52*(2), 151–156.

Nuñez-Mancilla, Y., Vega-Gálvez, A., Pérez-Won, M., Zura, L., García-Segovia, P., and Di Scala, K. 2014. Effect of osmotic dehydration under high hydrostatic pressure on microstructure, functional properties and bioactive compounds of strawberry (fragariavesca). *Food and Bioprocess Technology*, *7*(2), 516–524.

Ochoa-Martinez, C.I., Ramaswamy, H.S., and Ayala-Aponte, A.A. 2007. A comparison of some mathematical models used for the prediction of mass transfer kinetics in osmotic dehydration of fruits. *Drying Technology*, *25*(10), 1613–1620.

Page, G.E. 1949. *Factors Influencing the Maximum of Air-Drying Shelled Corn in Thin Layer*. Purdue, Indiana: Purdue University.

Panades, G., Castro, D., Chiralt, A., Fito, P., Nuñez, M., and Jimenez, R. 2008. Mass transfer mechanisms occurring in osmotic dehydration of guava. *Journal of Food Engineering*, *87*(3), 386–390.

Parjoko, R.M.S., Buckle, K.A., and Perera, C.O. 1996. Osmotic dehydration kinetics of pineapple wedges using palm sugar. *LWT - Food Science and Technology*, *29*(5–6), 452–459.

Peleg, M. 1988. An empirical model for the description of moisture sorption curves. *Journal of Food Science*, *53*(4), 1216–1217.

Pereira, N.R., Marsaioli, A., and Ahrné, L.M. 2007. Effect of microwave power, air velocity and temperature on the final drying of osmotically dehydrated bananas. *Journal of Food Engineering*, *81*(1), 79–87.

Phisut, N. 2012. Factors affecting mass transfer during osmotic dehydration of fruits. *International Food Research Journal*, *19*(1), 7–18.

Pragati, S., and Preeti, B. 2014. Technological revolution in drying of fruit and vegetables. *International Journal of Science and Research*, *3*(10), 705–711.

Prothon, F., Ahrné, L.M., Funebo, T., Kidman, S., Langton, M., and Sjöholm, I. 2001. Effects of combined osmotic and microwave dehydration of apple on texture, microstructure and rehydration characteristics. *LWT - Food Science and Technology*, *34*(2), 95–101.

Radziejewska-Kubzdela, E., Biegańska-Marecik, R., and Kidoń, M. 2014. Applicability of vacuum impregnation to modify physico-chemical, sensory and nutritive characteristics of plant origin products – A review. *International Journal of Molecular Sciences*, *15*(9), 16577–16610.

Raji Abdul Ganiy, O., Falade Kolawole, O., and Abimbolu Fadeke, W. 2010. Effect of sucrose and binary solution on osmotic dehydration of bell pepper (chilli) (Capsicum spp.) varieties. *Journal of Food Science and Technology*, *47*(3), 305–309.

Ramya, V., and Jain, N.K. 2017. A review on osmotic dehydration of fruits and vegetables: An integrated approach. *Journal of Food Process Engineering*, *40*(3), 1–22.

Rastogi, N., Adsare, S., Karley, D., Raghavarao, K.S.M.S., and Niranjan, K. 2015. Osmotic dehydration: Applications and recent advances. In: *Drying Technologies for Foods: Fundamentals and Applications*, 1st ed. New Delhi: New India Publishing Agency.

Rastogi, N.K., Nayak, C.A., and Raghavarao, K.S.M.S. 2004. Influence of osmotic pre-treatments on rehydration characteristics of carrots. *Journal of Food Engineering*, *65*(2), 287–292.

Rastogi, N.K., Raghavarao, K.S.M.S., and Niranjan, K. 1997. Mass transfer during osmotic dehydration of banana: Fickian diffusion in cylindrical configuration. *Journal of Food Engineering*, *31*(4), 423–432.

Rastogi, N.K., Raghavarao, K.S.M.S., and Niranjan, K. 2005. Developments in osmotic dehydration. In *Emerging Technologies for Food Processing*, Academic Press (pp. 221–249).

Rastogi, N.K., Raghavarao, K.S.M.S., and Niranjan, K. 2014. Recent developments in osmotic dehydration. *Emerging Technologies for Food Processing*, *8*(6), 181–212.

Rastogi, N.K., Raghavarao, K.S.M.S., Niranjan, K., and Knorr, D. 2002. Recent developments in osmotic dehydration: Methods to enhance mass transfer. *Trends in Food Science and Technology*, *13*(2), 48–59.

Rastogi, N.K., Suguna, K., Nayak, C.A., and Raghavarao, K.S.M.S. 2006. Combined effect of γ-irradiation and osmotic pretreatment on mass transfer during dehydration. *Journal of Food Engineering*, *77*(4), 1059–1063.

Ruíz Díaz, G., Martínez-Monzó, J., Fito, P., and Chiralt, A. 2003. Modelling of dehydration-rehydration of orange slices in combined microwave/air drying. *Innovative Food Science and Emerging Technologies*, *4*(2), 203–209.

Sacchetti, G., Gianotti, A., and Dalla Rosa, M. 2001. Sucrose-salt combined effects on mass transfer kinetics and product acceptability. *Study on apple osmotic treatments. Journal of Food Engineering*, *49*(2–3), 163–173.

Sagar, V.R., and Suresh Kumar, P. 2010. Recent advances in drying and dehydration of fruits and vegetables: A review. *Journal of Food Science and Technology*, *47*(1), 15–26.

Santacruz-Vazquez, C., Santacruz-Vazquez, V., Jaramillo-Flores, M.E., Chanona-Perez, J., Welti-Chanes, J., and Gutierrez-Lopez, G.F. 2008. Application of osmotic dehydration processes to produce apple slices enriched with β-carotene. *Drying Technology*, *26*(10), 1265–1271.

Saxena, A., Bawa, A.S., and Raju, P.S. 2009. Optimization of a multitarget preservation technique for jackfruit (Artocarpusheterophyllus L.) bulbs. *Journal of Food Engineering*, *91*(1), 18–28.

Shi, J., and Xue, S.J. 2008. Application development of osmotic dehydration technology in food processing. In *Preservation*, Taylor & Francis, pp. 187–188).

Singh, B., Kumar, A., and Gupta, A.K. 2007. Study of mass transfer kinetics and effective diffusivity during osmotic dehydration of carrot cubes. *Journal of Food Engineering*, *79*(2), 471–480.

Singh, B., Panesar, P.S., Nanda, V., and Kennedy, J.F. 2010. Optimization of osmotic dehydration process of carrot cubes in mixtures of sucrose and sodium chloride solutions. *Food Chemistry*, *123*(3), 590–600.

Sonia, N.S., Mini, C., and Geethalekshmi, P.R. 2015. Osmotic dehydration: A novel drying technique of fruits and vegetables – A review. *Journal of Agricultural Engineering and Food Technology*, *2*(2), 80–85.

Stojanovic, J., and Silva, J.L. 2007. Influence of osmotic concentration, continuous high frequency ultrasound and dehydration on antioxidants, colour and chemical properties of rabbiteye blueberries. *Food Chemistry*, *101*(3), 898–906.

Sutar, N., and Sutar, P.P. 2013. Developments in osmotic dehydration of fruits and vegetable – A review. *Trends in Post-Harvest Technology*, *1*(1), 20–36.

Sutar, P.P., and Gupta, D.K. 2007. Mathematical modeling of mass transfer in osmotic dehydration of onion slices. *Journal of Food Engineering*, *78*(1), 90–97.

Sutar, P.P., Raghavan, G.V.S., Gariepy, Y., Prasad, S., and Trivedi, A. 2012. Optimization of osmotic dehydration of potato cubes under pulsed microwave vacuum environment in ternary solution. *Drying Technology*, *30*(13), 1449–1456.

Taiwo, K.A., Angersbach, A., and Knorr, D. 2002. Influence of high intensity electric field pulses and osmotic dehydration on the rehydration characteristics of apple slices at different temperatures. *Journal of Food Engineering*, *52*(3), 185–192.

Torringa, E., Esveld, E., Scheewe, I., Van Den Berg, R., and Bartels, P. 2001. Osmotic dehydration as a pre-treatment before combined microwave-hot-air drying of mushrooms. *Journal of Food Engineering*, *49*(2–3), 185–191.

Traffano-Schiffo, M.V., Laghi, L., Castro-Giráldez, M., Tylewicz, U., Romani, S., Ragni, L., and Fito, P.J. 2017. Osmotic dehydration of organic kiwifruit pre-treated by pulsed electric fields: Internal transport and transformations analyzed by NMR. *Innovative Food Science and Emerging Technologies*, *41*, 259–266.

Tregunno, N.B., and Goff, H.D. 1996. Osmodehydro-freezing of apples: Structural and textural effects. *Food Research International*, *29*(5–6), 471–479.

Tylewicz, U., Fito, P.J., Castro-Giráldez, M., Fito, P., and Dalla Rosa, M. 2011. Analysis of kiwifruit osmo-dehydration process by systematic approach systems. *Journal of Food Engineering*, *104*(3), 438–444.

Tylewicz, U., Tappi, S., Mannozzi, C., Romani, S., Dellarosa, N., Laghi, L., and Dalla Rosa, M. 2017. Effect of pulsed electric field (PEF) pre-treatment coupled with osmotic dehydration on physico-chemical characteristics of organic strawberries. *Journal of Food Engineering*, *213*, 2–9.

Waliszewski, K.N., Delgado, J.L., and Garcia, M.A. 2002. Equilibrium concentration and water and sucrose diffusivity in osmotic dehydration of pineapple slabs. *Drying Technology*, *20*(2), 527–538.

Wang, J., and Chao, Y. 2003. Effect of gamma irradiation on quality of dried potato. *Radiation Physics and Chemistry*, *66*(4), 293–297.

Wang, W.C., and Sastry, S.K. 2000. Effects of thermal and electrothermal pretreatments on hot air-drying rate of vegetable tissue. *Journal of Food Process Engineering*, *23*(1), 299–319.

Xin, Y., Zhang, M., and Adhikari, B. 2014. Freezing characteristics and storage stability of broccoli (brassica oleracea l. var. botrytis l.) underosmo-dehydrofreezing and ultrasound-assisted osmo-dehydrofreezing treatments. *Food and Bioprocess Technology*, 7(6), 1736–1744. doi:10.1007/s11947-013-1231-4

Yadav, A.K., and Singh, S.V. 2014. Osmotic dehydration of fruits and vegetables: A review. *Journal of Food Science and Technology*, *51*(9), 1654–1673.

Yadav, B.S., Yadav, R.B., and Jatain, M. 2012. Optimization of osmotic dehydration conditions of peach. *Journal of Food Science and Technology*, *49*(5), 547–555.

Yu, Y., Jin, T.Z., Fan, X., and Wu, J. 2018. Biochemical degradation and physical migration of polyphenolic compounds in osmotic dehydrated blueberries with pulsed electric field and thermal pretreatments. *Food Chemistry*, *239*, 1219–1225.

5 Pulsed Light Technology Applied in Food Processing

Prasanna Bhalerao, Rishab Dhar and Snehasis Chakraborty
Institute of Chemical Technology, Matunga, Maharashtra, India

CONTENTS

5.1 Introduction 91
5.2 Mechanism and Critical Process Parameters of Pulsed Light 92
5.2.1 Design and Working of the PL System 92
5.2.2 Critical Process Parameters 94
5.2.2.1 Properties of a Sample 95
5.2.2.2 Target Microorganisms or Enzymes 95
5.2.2.3 Fluence and Associated Parameters 96
5.3 Literature Review on Photochemical, Photothermal, Photophysical, and Photo-Reactivation Mechanisms 100
5.3.1 Photochemical, Photothermal, and Photophysical 100
5.3.2 Photo-Reactivation 101
5.4 Summary of PL-Related Studies of Liquid Foods Using Various Quality Parameters 102
5.4.1 Microbial Inactivation in Liquid Foods 102
5.4.2 Effect of PL Treatment on Enzymes 107
5.4.3 Effect of Pulsed Light on Color, Sensory, and Biochemical Attributes 108
5.4.4 PL in Combination with Other Technologies 109
5.5 Conclusion 114
Acknowledgment 114
References 115

5.1 INTRODUCTION

The growing needs and evolving lifestyle of society have increased demand for processed foods. Consumers want these food products to be not only safe and stable but also nutritious and minimally processed. The conventional thermal processing of food products is known to be a reliable pasteurization technology providing adequate microbial safety and long shelf life. In addition to its benefits, high-temperature

processing brings undesirable changes in the product such as degradation of nutrients, flavor components, color pigments, and overall sensory profile. Rattanathanalerk et al. (2005) observed quality loss in pineapple juice due to thermal processing. Many researchers have highlighted the advantages of alternative technologies over thermal processing (Pereira and Vicente, 2010; Zhang et al., 2019). Therefore, non-thermal technologies have attracted many researchers to ultra-sonication, ultraviolet (UV), infrared, pulse electric field, pulsed light (PL), ionizing radiation, cold plasma, high-pressure processing, and ozone processing technologies (Rawson et al., 2011; Pan et al., 2017; Zhang et al., 2019). Pulsed light technology also referred to as pulsed UV technology has been explored significantly and is regarded as a very hot topic for research. It uses light over a broad spectrum from near-infrared (700–1100 nm) and visible (380–700 nm) through to the ultraviolet (200–380 nm) (Elmnasser et al., 2007; Hwang et al., 2015). Pulsed light is recognized as a safe processing technology for food products with a maximum fluence limit of 12 J/cm^2 (FDA 2019). To ensure consumer safety, the US Food and Drug Administration (2004) recommends a 5-log cycle reduction of resistive pathogens. Although equipment costs are high, operating costs are lower, reduction of microorganisms is faster and more flexible than continuous UV, making it a promising processing technology. Also, it does not promote the production of organic volatile compounds, and may only produce reduced solid waste (Pereira and Vicente, 2010). The inert xenon-based flash lamps used for PL are more environmentally friendly than mercury lamps (Gómez-López et al., 2007). PL has shown rapid microbial reduction (Krishnamurthy et al., 2007; Pataro et al., 2011; Zhu et al., 2019) and can inactivate enzymes (Manzocco et al., 2013; Pellicer and Gómez-López, 2017; Pellicer et al., 2019) to ensure stability and overall shelf life with minimal loss of other heat-sensitive nutrients (Valdivia-Nájar et al., 2018; Kwaw et al., 2018). It also acts as an excellent surface disinfector (Rowan 2019; Pedrós-Garrido, 2018).

This chapter explores the use of PL for liquid food products. Although it is a non-thermal technique, prolonged use can also induce heating effects (Huang and Chen, 2005).

5.2 MECHANISM AND CRITICAL PROCESS PARAMETERS OF PULSED LIGHT

Pulsed inactivation of microbes or enzymes is achieved through a set of optimized process parameters, including understanding the technology and its design, which will help to build new lab-scale equipment or modify existing equipment, and to utilize it to its fullest extent. It is also important to know the impacts of all the critical process parameters on the final efficiency of PL inactivation of the target microorganisms and enzymes.

5.2.1 Design and Working of the PL System

The pulsed light system consists of five basic components:

1) High-voltage power supply: provides electrical power to the storage capacitor
2) Storage capacitor: a temporary powerhouse for electrical energy and flash lamp

3) Pulse forming network (PFN): decides wavelength and pulse width
4) Gas flash lamp: the heart of the system

Trigger signal: releases electrical energy to the flash lamp other accessories include an air blower, optical sensor, and collimator.

The flash lamp consists of a light electrode surrounded by a transparent jacket filled with an inert gas such as xenon or krypton. Light sources equipped with xenon lamps have better efficiency and higher inactivation power. These flash lamps convert 45–50% of the input electrical energy to pulsed radiant energy (Xenon Corp., 2005). The jackets used for flash lamps are made of 1 mm-thick quartz. The metallic electrodes are connected to the high-voltage capacitor. The electrodes provide electric current to the gas. The functionality of the lamp is affected by the performance of the anode and cathode. The anode has a longer lifespan than the cathode, as it has more surface area to withstand the electron bombardment. The entire flash lamp assembly is sealed with solder, rod, and ribbon seals. The continuous bombardment of high-voltage high-current pulses excites the electrons surrounding the gas atoms, causing them to emit photons corresponding to the wavelength ranges given above. The release of pulsating light, for example in square wave-form, is achieved by integrating a PFN made up of an array of capacitors. The formation of highly charged particles triggers temperature rises, hence cooling devices need to be provided to ensure functionality and extend the lifespan of the flash lamps. Higher pulse transfer operations are performed by connecting one or more flash lamp units to the same power and high voltage units. The electrical energy supplied to the gas chamber is emitted in the form of short intense bursts. There are numerous patented product-specific items of equipment designed to control individual treatments (Green et al., 2005).

The high current discharge through gas-filled flash lamps generates broad-spectrum white flashes that are 20,000 times more intense in a few milliseconds (Dunn et al., 1995). The machine can convert 50% of the total electricity into light. The spectral distribution of flash lamps is 25% ultraviolet, 45% visible light, and 30% infrared. A pulse light system can deliver high-energy pulses in a short time; the rate of light flashes varies from 1–20 flashes/sec and is sufficient to produce pasteurization of sterilization effect. The severity of the process parameters depends on the end application. To date, PL processing has been adopted for continuous sterilization of packaging film, aseptic processing, surface treatments of solid and liquid foods, and plastic packaging materials. It is difficult to achieve the desired results for meat and fish with PL processing, as certain parts of the samples cannot be covered under flashlight sources (Ohlsson and Bengtsson, 2002).

Parameters to be considered in the design of food preservation equipment include ozone build-up, the surface area of food product to be exposed to the light source, the dimensions of each treatment unit, and the desired degree of decontamination. A cooling unit facility may be required where the food under treatment is temperature sensitive (Green et al., 2005). Pulsed light systems can be of either batch or continuous type, depending on the usage. Process parameters such as pulse width, frequency, and the number of pulses are managed through a control panel in both types of system.

The batch process device developed by Xenon Corp. (Waltham, MA) consists of a control panel, a quartz chamber loaded with gas, and a flash lamp. The light source

is mounted in a horizontal position and an adjustable tray is provided to hold the samples. The PL design can be modified to cover the full 360° area of the sample by mounting more than one flash lamp. A single control panel can operate up to 8 flash lamps (Proctor, 2011).

A pulsed light system has been operated in continuous mode for milk decontamination (Krishnamurthy et al., 2007) and various fruit juices (Palgan et al., 2011a; Pataro et al., 2011).

The lamp housing has a quartz window to allow maximum transmission of light and is connected to an air blower to protect the lamp from overheating and remove any ozone formed. To capture the amount of fluence received by the sample, optical sensors or radiometers are needed at the same elevation. The optical sensor may be connected to a computer. A collimator may be placed between the sample and the light source to deliver more uniform and parallel rays to the sample. Since this is not available ready-made, it needs to be inserted manually. This will help to improve the accuracy, uniformity, and repeatability of experiments, but at the same time will absorb a significant amount of the emitted radiation which will increase treatment time and reduce PL efficiency (Gómez-López and Bolton, 2016). There are three principal companies manufacturing PL systems at the commercial and international level: Xenon Corporation (USA), Claranor (France), and SteriBeam System (Germany). While batch-type PL systems are widely available from manufacturers, many researchers have also used the continuous type by customizing it jointly with the manufacturers or by modifying the batch type to a continuous one themselves. Batch and continuous PL systems are illustrated in Figures 5.1 and 5.2.

5.2.2 Critical Process Parameters

1. Properties of a sample
2. Target microorganism or enzyme
3. Fluence and associated parameters

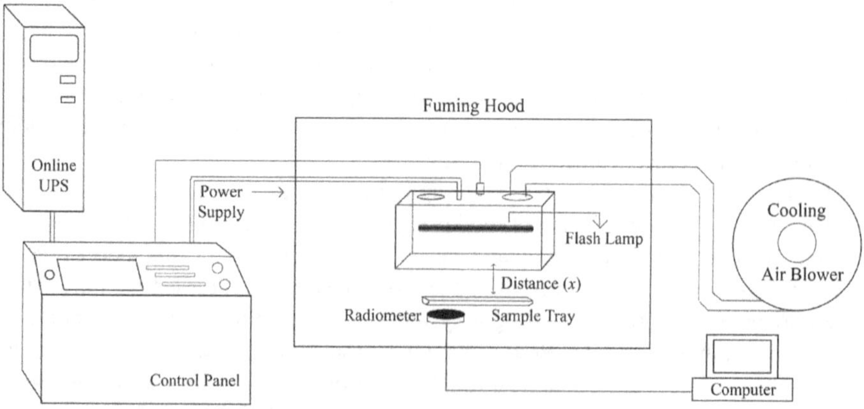

FIGURE 5.1 Batch type of pulsed light processing unit

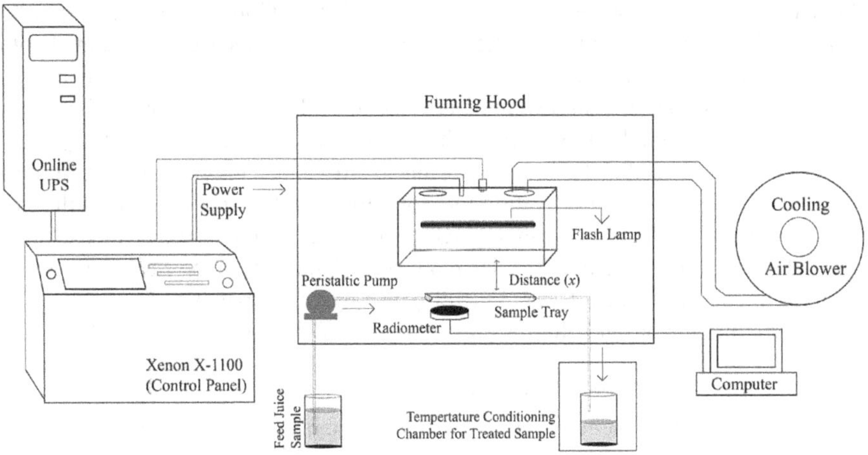

FIGURE 5.2 Continuous type of pulsed light processing unit

5.2.2.1 Properties of a Sample

PL has limited penetration power, so solid and opaque liquid samples will only receive surface treatment. Light undergoes scattering, refraction, reflection, and absorption as it enters the sample medium from the air. An opaque sample will only absorb on the surface and the rest will be reflected; no radiation will penetrate the deeper layers. Optically transparent liquid will be fully treated, but translucent liquid samples will be partially treated, depending on the turbidity and presence of suspended solids. The dependence of the efficacy of PL-based inactivation of a microorganism (*Pseudomonas aeruginosa*) on the optical characteristics of the sample medium was discussed by Hwang et al. (2015). It is obvious that microbial inactivation will be greatly reduced as the medium transparency is decreased (Palgan et al., 2011a; Ferrario et al., 2013). Similarly, Aguirre et al. (2005) report poorer inactivation of *Listeria innocua* in a culture medium with more coloration than with less.

The volume or thickness of the sample under irradiation also affects the effectiveness of PL because the penetration of light decreases with increasing path length. Under PL treatment of clove honey sample, a thickness of about 2 mm resulted in a higher reduction in the count of *Clostridium sporogenes* than a layer 8 mm thick (Hillegas and Demirci, 2003).

5.2.2.2 Target Microorganisms or Enzymes

The target microorganisms or enzymes may have variable resistance to PL inactivation. For example, *E. coli* was found to be more susceptible to PL irradiation than *L. innocua* artificially inoculated in apple and orange juices (Pataro et al., 2011). It was observed that the susceptibility of microorganisms towards PL is highest in the case of gram-negative bacteria, medium for gram-positive bacteria, and lowest for fungal spores (Rowan et al., 1999; Levy et al., 2012).

Microbes inoculated on the upper surface of the agar medium are easily inactivated. The efficacy of PL for microbial inactivation treatment on the agar medium is

affected by the size of the inoculum (Uesugi et al. 2007). Gómez-López et al. (2005) found no significant log reduction after PL treatment of agar surface inoculated with higher counts of *L. monocytogenes*. The authors found that the microorganisms on the surface of the agar medium were inactivated after the initial few pulses but shadowed the rest from the light, which resulted in inadequate disinfection. Thus, initial microbial load affects the efficiency of PL inactivation, just as Ferrario and Guerrero (2018) observed a higher inactivation rate for a lower initial size of the inoculum of *A. acidoterrestris* in PL-treated apple juice. Uniform exposure of the surface of food material is necessary to achieve the desired results. Similarly, in the case of enzymes, at a fixed fluence, polyphenol oxidase (PPO) inactivation also reduced when the PPO concentration in the buffer solution was above 10 U (Manzocco et al., 2013).

5.2.2.3 Fluence and Associated Parameters

Fluence is one of the most important process parameters in PL treatment. It is the light energy received on the per unit surface area of the sample during experimentation, also known as dose, total fluence, radiant energy fluence, or spherical radiant exposure (energy per unit spherical cross-sectional area). The SI unit is $J{\cdot}m^{-2}$. The IUPAC-recommended symbols are used to denote all the associated photochemical nomenclature (Braslavsky, 2006). Fluence and spherical radiant exposure are denoted by F_o and H_o, respectively. Fluence is a process parameter that is also associated with other terminologies: fluence rate, treatment time, pulse period, frequency, and peak power.

- *Fluence rate* is the light energy per unit area and per unit time, in other words, fluence per unit time (Eq. 5.1). It is also known as irradiance, intensity (only for qualitative description) or average radiant power. It is denoted by E_o ($J{\cdot}m^{-2}{\cdot}s^{-1}$ or $W{\cdot}m^{-2}$).
- *Treatment time* is overall exposure time (t) for the sample under PL radiation, expressed in seconds (s).
- *Pulse period.* As energy is delivered in pulsating form, which includes ON time and OFF time in one wave cycle, pulse period or width is the ON time interval during which energy is delivered.
- *Frequency* is the number of pulses per second, also referred to as pulse repetition rate (PRR), denoted by υ and expressed in hertz ($Hz = s^{-1}$). It can also be calculated as a reciprocal of the pulse period (Eq. 5.2).
- *Peak power* is the measure of the energy of pulse per unit pulse duration (period) and can also be written as a product of fluence rate and respective surface area, which can be denoted as total radiant power, P, and expressed in watt (W) or $J{\cdot}s^{-1}$ (Eq. 5.3).

$$F_o = E_o \times t = (Fluence\ per\ pulse) \times (Number\ of\ pulses) \tag{5.1}$$

$$v = \frac{Number\ of\ pulses}{Treatment\ time} = \frac{1}{Pulse\ period} \tag{5.2}$$

$$P = E_o \times (Surface\ area) \tag{5.3}$$

The light energy emitted by the lamp is generally different from what is received by the sample. Thus, it is important to quantify it using optical sensors or radiometers. The actual fluence received by the sample can be controlled by changing the treatment time (number of pulses), voltage, pulse frequency, relative position of the sample, and distance from the lamp. Controlling fluence will, in turn, affect sample heating, the peak power of the light pulses, and the spectrum distribution of radiation.

5.2.2.3.1 Treatment Time (Number of Pulses) and Sample Heating

Fluence continuously increases with the increase in treatment time or the number of pulses and higher fluence results in more inactivation of microbes and enzymes. Better inactivation of *S. aureus* was observed in milk at lower flow rates, as longer holding time in a continuous processing unit causes higher absorption of pulsed UV energy (Krishnamurthy et al. 2007). Higher levels of microbial inactivation were observed at longer treatment times (Levy et al., 2012; Karaoglan et al., 2017). The same trend was visible in the case of inactivation of enzymes like lipase and lipoxygenase (Alhendi et al., 2018; Jeon et al., 2019).

Longer treatment time will induce surface heat due to the presence of infrared and visible range in the PL spectrum. It also possible that the heat from the lamp may reach out to the sample as well. Prolonged exposure to PL (50–100 pulses) at a higher fluence rate (5.6 J/cm^2/pulse) caused the burning of the honey surface and saw the temperature rise from 20 to 80–100°C (Hillegas and Demirci, 2003). Therefore, to avoid unnecessary temperature rise, pulsed light equipment needs to include a cooling system. Ferrario et al. (2013) used an external cooling method to keep the treated sample temperature below 20°C by keeping the petri dish carrying the sample in a secondary container with ice flakes. The heat dependency and photothermal effects of PL have been explored by many researchers (Gómez-López et al., 2005; Pataro et al. 2011; Heinrich et al., 2016). Other researchers, however (Luksiene et al., 2007; Krishnamurthy et al., 2007), have observed no significant heating during PL processing.

5.2.2.3.2 Relative Position of the Sample and Distance from the Lamp

The intensity of the light source and fluence rate is a function of distance. According to the inverse square law, the fluence is inversely proportional to the squared distance (Gaertner 2012). With an increase in distance, penetration depth decreases significantly compared with the initial fluence emitted by the light source, which results in weaker lethality of the PL radiation and lower inactivation of both microorganisms and enzymes (Gómez-López et al., 2005; Karaoglan et al. 2017; Alhendi et al., 2018).

The energy received by the samples is also affected by the sample's position and orientation with respect to the light source. Gómez-López et al. (2005) found a significant reduction in microbial inactivation for a group of food pieces placed very close to the side shelf and not directly below the lamp.

5.2.2.3.3 Voltage and Spectrum Distribution

Higher voltage supply directly increases the lamp's output fluence. Caution should be observed at very high levels of fluence which can cause surface burning. Many of the

available PL system models have a variable input voltage option to control output fluence. The literature shows that with higher voltage levels, higher inactivation of microorganisms (Luksiene et al., 2007; Levy et al., 2012; Hwang et al., 2019) and of enzymes (Jeon et al., 2019) occurs. It was also observed that longer treatment time at lower voltage was not as effective as shorter treatment at higher voltage, for equivalent fluence (Kramer et al., 2017). The reason is that higher lamp discharge voltage can shift the spectrum towards the lower wavelength range (blue shift), i.e., at a higher voltage or fluence range, the increased proportion of UV helps enhance inactivation in a comparatively shorter time. Kramer et al. (2017) also report that the proportion of UV in the multi-spectrum PL overall varied from 16.3 to 25.5%, depending on the voltage range of 1–3 kV. Similar conclusions regarding the voltage-triggered blue shift were drawn by Levy et al. (2012), and Gómez-López and Bolton (2016).

Previous research comparing inactivation by continuous UV and by PL for equivalent fluence found that PL was better, but when only the UV-C dose coming from PL and continuous UV-C were compared, they showed identical results. In addition, PL was completely ineffective when the UV-C range was cutoff (Levy et al., 2012; Kramer et al., 2017).

5.2.2.3.4 Frequency and Peak Power

Peak power is the pulse energy per unit pulse period. PL energy is delivered in the form of flashes or pulses and the peak power (proportional to fluence per pulse) carried by each of the light pulses may or may not have a different impact to continuous UV radiation, which carries a fixed power during sample illumination (Figure 5.3). Controlling the pulse width (reciprocal of pulse frequency) also may or may not affect inactivation of the target microorganism or enzyme. Luksiene et al. (2007) and Kramer et al. (2017) found no significant influence of frequency (1–5 Hz) on microbial reduction. However, at a higher frequency, treatment time was reduced.

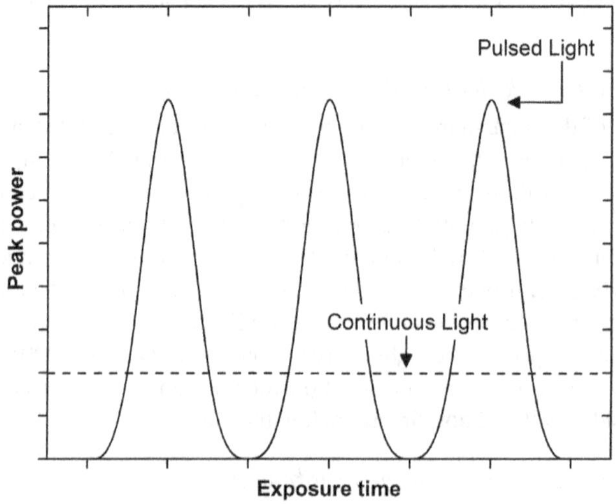

FIGURE 5.3 Comparison between the peak power of pulsed light and that from continuous UV

In photochemistry, the Bunsen–Roscoe reciprocity (BRR) law states that the biological effects due to photons only depend on the total photons absorbed or total fluence absorbed, irrespective of the rate of absorption at a given time. As fluence is a product of fluence rate (average peak power per unit area) and time, the law states that it does not matter whether the path of higher fluence rate and shorter time or lower fluence rate and longer time is selected to deliver a fixed fluence; only the total delivered fluence matters. Researchers have explored the microbial inactivation efficiency of PL and continuous UV, and compared the effects of the delivery of power in pulsed and in continuous form for equivalent fluence value. Rice and Ewell (2001) investigated the dependence of peak power on UV inactivation of bacterial spores. The authors reported that there was no significant difference in the survival count of *B. subtilis* spores after being treated with UV laser having high peak power (248 nm) and continuous UV having low peak power (254 nm) at similar delivered photon energies. They concluded that neither peak power while delivering photon energy nor the rate of delivery is important, and the only advantage of a higher peak power delivering source is shorter treatment time. This result agrees with the BRR law. Similar conclusions were drawn by Otaki et al. (2003) who also observed no noticeable difference between the efficiencies in the inactivation of two coliphages and three *E. coli* strains by low-pressure UV lamp (254 nm) and broad-spectrum xenon lamp at same germicidal doses. On the other hand, Bowker et al. (2011) observed deviation from the BRR law in the case of *E. coli* but found the law holds true for coliphages. The authors assumed that this difference was due to the action of bacterial repair mechanisms for *E. coli*. In another study of *A. niger*, it was reported that at the same fluence, 250 mJ/cm^2 delivered by a low-pressure UV-C lamp, a reduction of 1 and 2 log cycles was observed due to two different fluence rates having values of 0.022 mW/cm^2 (low) and 0.11 mW/cm^2 (high) (Taylor-Edmonds et al., 2015). This clearly violated the BRR law. Takeshita et al. (2003) also report a violation of the BRR law in the case of *S. cerevisiae*. Unlike viruses (coliphage) and spores, vegetative bacteria have repair mechanisms that may cause deviation from the BRR law. Kramer et al. (2017) also reported higher inactivation in the *E. coli* and *L. innocua* by a higher fluence rate in a shorter time at equivalent total fluence. The authors stated that compliance with the BRR law is insignificantly affected by the damage repair mechanisms. Furthermore, while studying solar disinfection of *S. flexneri*, Bosshard et al. (2009) discovered that the BRR principle failed when solar irradiation was increased 2- to 3-fold. This suggests that, depending on the target microorganism or enzyme, the BRR law will hold only within a certain range of fluence rate, and will fail on exceeding the limit. For the same fluence, Levy et al. (2012) compared the impact of the UV-C range delivered by continuous UV-C lamp and PL and observed no difference. The authors stated that particularly in the case of *A. niger*, the contribution of the photothermal effect cannot be neglected. Thus, the peak power effects of xenon lamp-based PL, mercury vapor-based UV, and LED-based UV may not be directly comparable because of their dissimilar emission spectra and different kinds of effects. More focused research on this topic is required before any firm conclusions can be drawn regarding the impact of fluence rate for equivalent fluence, spectrum range, and other process parameters.

5.3 LITERATURE REVIEW ON PHOTOCHEMICAL, PHOTOTHERMAL, PHOTOPHYSICAL, AND PHOTO-REACTIVATION MECHANISMS

Pulsed light (PL) affects microbial inactivation through photochemical, photothermal, and photophysical pathways. Microorganisms may also possess partial healing capabilities through photo-reactivation mechanisms.

5.3.1 Photochemical, Photothermal, and Photophysical

Microbial inactivation by PL is due to structural changes caused by photochemical and photothermal effects. The changes caused by the photochemical effect are cell disruption (Wekhof, 2000; Wekhof et al., 2001; Fine and Gervais, 2004), protein elution (Takeshita et al., 2003), vacuole expansion (Takeshita et al., 2003), and cell membrane destruction (Fine and Gervais, 2004). The photothermal effect causes changes such as cell wall damage, cytoplasmic membrane shrinkage, cellular content leakage, and mesosome disintegration (Krishnamurthy et al., 2007). Continuous exposure to UV light has a germicidal effect on microbes by altering their DNA and RNA structures. The inactivation mechanism in living microorganisms is frequently explained by photochemical and photothermal effects. However, considerable research has been focused on the photochemical effect, which is more lethal than the photothermal effect (Rowan et al., 1999). The photochemical effect occurs as a consequence of DNA damage. Exposure to UV light induces changes in carbon–carbon double bonds and affects the DNA and RNA structure of living microbes. The light initiates the formation of pyrimidine dimers such as thymine and cytosine. The dimers result in clonogenic death of microorganisms, as they damage single-and double-stranded DNA, and restrict the cell replication process (Gómez-López et al., 2007). Several studies have highlighted the importance of processing in the UV-C range because of its higher energy output (Rowan et al., 1999). UV-C processing is reported to produce maximum germicidal efficiency as it contributes significantly to total inactivation (Rowan et al., 1999; Levy et al., 2012).

Rowan et al. (1999) observed a photochemical microbial inactivation (2 and 6 log CFU/mL) in six strains (*Listeria monocytogenes*, *Escherichia coli*, *Salmonella enteritidis*, *Pseudomonas aeruginosa*, *Bacillus cereus*, and *Staphylococcus aureus*) with the use of low-and high-energy UV sources. The combined effect of UV-C with other light sources showed higher inactivation for *Botrytis cinera* and *Monilia fructigena* (Marquenie et al., 2003). Wang et al. (2005) confirmed the inactivation of *E. coli* with a photochemical effect and achieved the best results with short pulse width and higher energy doses. In addition to DNA and RNA damage, PL can induce the double-stranded generation of hydroxyl radicals, resulting in an increased amount of intracellular reactive oxygen species, accelerating bacterial death both from outside and inside (Zhu et al., 2019).

The photothermal effect occurs due to a temporary rise in temperature or overheating of samples when exposed under pulsed or UV light. Wekhof (2000) hypothesized that when microbial cells are exposed at higher energy levels (>0.5 J/cm^2), structural changes such as cell deformation and rupture result in a reduction in viable

counts. Temporary overheating in *Aspergillus niger* produced an internal explosion and resulted in the evacuation of the light pulse (Wekhof et al., 2001). Takeshita et al. (2003) confirmed that the photothermal effect in bacterial cells is a function of water vaporization and membrane disruption. Photothermal effects also produce changes in cell structure and enlargement of vacuoles (Proctor, 2011). Excluding the primary photochemical effects of UV, the light pulses also simply transfer a significant amount of heat energy onto the product's surface (Mertens and Knorr, 1992), which may eventually be spread to the rest of the sample body to increase the overall sample temperature.

The rise in temperature after PL treatment depends on the intensity of the pulse and dwell time, so only high-power PL doses and longer irradiation time will contribute to the photothermal and photochemical effects in killing or inactivating the microorganisms. Pulsed UV treatment in the infrared region can achieve temperatures similar to mild heat treatments (Demirci and Krishnamurthy, 2011).

Several authors report structural changes in microbial cultures (photophysical) with photochemical and photothermal effects. This effect is attributed to the disturbing effect of the intermittent high-energy supply on cells. PL processing brings about significant changes in the structure of *S. cerevisiae*. PL-treated samples had damaged membrane structure, expanded vacuoles, and uneven shape (Takeshita et al., 2003). The severity of cell wall damage is measured by concentration of eluted protein. Increased eluted protein concentration was observed in yeasts after processing at higher fluence rates (Takeshita et al., 2003). Cheigh et al. (2012) and Cheigh et al. (2013) studied the effect of UV-C and intense pulse light (IPL) treatment on the structure of the foodborne pathogen *L. monocytogenes*, and *Escherichia coli* O157: H7. Owing to its high energy density and broad spectrum, IPL processing produced a milder photochemical effect that was more destructive than that of UV-C. Light bursts exhibited several internal changes, such as breakdown of the cell wall and DNA strands, and cytoplasmic shrinkage. Krishnamurthy et al. (2010) reported several changes in the internal structure of *S. aureus* (cell wall damage, disintegration, cellular content leakage, cytoplasmic membrane shrinkage) with the use of pulsed UV light. The superiority of PL over continuous UV was discussed by Levy et al. (2012) and Orlowska et al. (2013) for *Aspergillus niger* and *E. coli*, respectively. Nicorescu et al. (2013) studied the combined photochemical and photothermal effect on the structural deformation of spices inoculated with *Bacillus subtilis*. The authors concluded that cell walls were disrupted by the photothermal effect and the DNA structure by the photochemical effect. Xu and Wu (2016) confirmed that the structural disruption of *E. coli* after PL processing was due to overheating, intracellular water vaporization, and membrane disruption.

5.3.2 Photo-Reactivation

Microorganisms damaged by PL or UV irradiation due to photochemical, photothermal, and photophysical mechanisms may heal if exposed to daylight (310–700 nm). This happens because of the microbial DNA repair process catalyzed by DNA photolyase which needs daylight to act (Salcedo et al., 2007). The photolyase enzyme utilizes light energy to split the UV- or PL-induced cyclobutane dimers present in the

damaged DNA via radical mechanism (Gómez-López et al., 2007). While this mechanism can repair a fraction of the sublethal damage inflicted by photochemical effects, it is ineffective against the irreversible damage done by photothermal and photophysical effects (Lasagabaster and de Marañón, 2017). A certain degree of photo-reactivation can occur depending on the severity of the PL treatment, length of exposure to the reactivating daylight, and time spent in the dark between treatment and daylight. Lasagabaster and de Marañón (2012, 2017) observed significant levels of photo-reactivation in *L. innocua* and *L. monocytogeenes*. After PL treatment of *L. innocua* at a fluence of 0.2 J/cm^2 and storing the culture under daylight for 180 min, the value of relative log reduction (Log N/N$_0$) decreased from 4.5 to 2.3. Although no significant difference was reported in photo-reactivation following daylight illumination time ranging from 30 to 180 min, maximum reactivation of 2.2 log was obtained by an initial 30 min. At the same time, however, no photo-reactivation was visible when treatment fluence was 0.4 J/cm^2 after achieving a log reduction of 6–6.4. In the case of *L. monocytogenes*, Gómez-López et al. (2005) discovered lower photo-reactivation of about 0.3 log after samples were illuminated with sunlight for 4 h. In contrast, no photo-reactivation was observed in the case of *B. cereus* and *L. monocytogenes* at both low (0.078 J/cm^2) and high (0.195 J/cm^2) fluence treatments. Similarly, Otaki et al. (2003) observed no photo-repair mechanisms for *E. coli*. This indicates that not every microorganism has photo-repair mechanisms, and even those that do have different fluence-based inactivation thresholds and sensitivity to sunlight.

5.4 SUMMARY OF PL-RELATED STUDIES OF LIQUID FOODS USING VARIOUS QUALITY PARAMETERS

5.4.1 Microbial Inactivation in Liquid Foods

PL is useful for inactivation of microorganisms that are thermally resistant, and cause spoilage and diseases, especially in liquid food. PL-based inactivation of microorganisms in liquid foods is summarized in Table 5.1. Pulse light treatments in liquid foods are strongly affected by the transparency and shielding effect of microbes (Sauer and Moraru, 2009). Of all fluids, water has maximum transparency. Other fluids such as fruit juices, sugar solution, and wines exhibit limited transparency as they have a significant amount of solid content. The efficacy of pulse light flashes is compromised with less transparent samples. Similarly, the penetration of UV radiation is affected by excess solid content (Bintsis et al., 2000). Maftei et al. (2005) observed higher log reduction (3.21) for *Penicillium expansum*-inoculated apple juice with thin-layer (6 mm) pulse light treatment. In a similar way, higher penetration of UV light was achieved in apple cider (Wright et al., 2000).

The efficacy of PL treatment for inactivating different microbial cultures follows the or dergram-negative bacteria > gram-positive bacteria > bacterial spores > fungal spores (Rowan et al., 1999; Anderson et al., 2000; Levy et al., 2012). The variation in the sensitivity of microorganisms towards light can be attributed to their different structures and to composition of the membranes and cell walls. Gram-positive microorganisms have a thicker peptidoglycan layer (20–80 nm) in their cell wall than

TABLE 5.1
Microbial Inactivation in Liquid Foods after Pulsed Light Processing

FOOD PRODUCT: Microorganism	Treatment Condition	Maximum log Reduction	References
I. MILK: *S. aureus*	Continuous flow (20–40 mL/min), 1–3 no. of passes, 3 Hz pulse frequency, 3800 V, 5–11 cm distance, and 1.07–0.80 W/cm^2 fluence rate respective to distance	7.29	Krishnamurthy et al. (2007)
II. (1) APPLE JUICE (2) APPLE CIDER: (a) *E. coli* ATCC 25922 (b) *E. coli* O157: H7	Frequency 3 Hz, pulse width 360 µs, fluence 12.6 J/cm^2, with static and turbulence (orbital shaking)	(1) Under static (a) 2.66 (b) 2.52 Under turbulence: 7.15 (approximately same for both surrogates) (2) Under static (a) 2.32 (b) 3.22 Under turbulence: 5.76 (approximately same for both surrogates)	Sauer and Moraru (2009)
III. INFANT BEVERAGE: *L. monocytogenes*	Frequency 15 Hz, voltage 10–25 kV, pulse width 1.5 µs, distance 60 mm, treatment time 3.5 s	5	Choi et al. (2010)
IV. (1) APPLE JUICE: (a) *E. coli* (b) *L. innocua* (2) ORANGE JUICE: (c) *E. coli* (d) *L. innocua* (3) MILK: (e) *E. coli* (f) *L. innocua*	Fluence 7–28 J/cm^2	(a) 4.7 (b) 1.93 (c) 1.0 (d) 1.0 (e) 1.06 (f) 0.84	Palgan et al. (2011a)
V. (1) APPLE JUICE: (a) *E. coli* (b) *L. innocua* (2) ORANGE JUICE: (c) *E. coli* (d) *L. innocua*	Continuous flow 17 mL/min, fluence 4 J/cm^2	(a) 4.0 (b) 2.9 (c) 2.98 (d) 0.93	Pataro et al. (2011)
VI. SUGAR SYRUP: (a) *S. cerevisiae* (b) *B. subtilis* spores (c) *G. stearothermophilus* spores (d) *A. acidoterrestris* spores, (e) *A. niger*	Fluence (a) 1.23 J/cm^2 (b) 1.5 J/cm^2 (c) 1.86 J/cm^2 (d) 1.86 J/cm^2 (e) 1.2 J/cm^2	(a) 5.4 (b) 4.2 (c) >4 (d) 3 (e) 1.3	Chaine et al. (2012)

(Continued)

TABLE 5.1
Continued

FOOD PRODUCT: Microorganism	Treatment Condition	Maximum log Reduction	References
VII. APPLE JUICE: *P. expansum*	Fluence 32 J/cm^2	3.76	Maftei et al. (2005)
VIII. (1) DISTILLED WATER (2) WHEY (3) DILUTED WHEY (4) SKIMMED WHEY: *L. innocua*	Fluence 11 J/cm^2	(1) 5.0 (2 & 4)<0.5 (3 with dilutions; 75%, 50%, 25%, and 12.5%) 0.5, 1.4, 2.3, and 5.4, respectively	Artíguez and de Marañón (2015)
IX. (1) APPLE JUICE (2) ORANGE JUICE (3) STRAWBERRY JUICE: (a) Total aerobic mesophiles (b) Total yeast and mold (c) *E. coli* (d) *L. innocua* (e) *S. enteritidis* (f) *S. cerevisiae*	Fluence 71.6 J/cm^2	(1a) 0.1–0.7 (1b) 0.5–0.6 (1c) 2.1 (1d) 1.6 (1e) 2.4 (1f) 1.0 (2a) 0.1–0.7 (2b) 0.5–0.6 (2c–2f) 0.3–0.8 (for all inoculated strains) (3a) 0.1–0.7 (3b) 0.5–0.6 (3c–3f) 0.3–0.8 (for all inoculated strains)	Ferrario et al. (2015b)
X. (1) MINERAL WATER (2) ISOTONIC BEVERAGE (3) APPLE JUICE (4) ORANGE JUICE (5) GRAPE JUICE (6) PLUM JUICE (7) CARBONATED BEVERAGE (8) MILK (9) COFFEE WITHOUT MILK BEVERAGE: *P. aeruginosa*	Fluence: 0.97 J/cm^2 (for W and I), 12.7–24.35 J/cm^2 (for A, C, and P), 29.21 J/cm^2 (for M, B, O, and G)	7.0 (for 1, 2, 3, 6, and 7) 0.5–2.0 (for 4, 5, 8 and 9)	Hwang et al. (2015)
XI. GOAT'S MILK: *E. coli*	Fluence 10 J/cm^2	6	Kasahara et al. (2015)
XII. WATER: (a) M. norovirus (b) *E. coli* (c) Total aerobic and facultative anaerobic heterotrophs	Continuous flow 12–40 L/min, fluence (a) 4.3 J/cm^2, (b) 14.02 J/cm^2, (c) 13.05 J/cm^2	(a) 3.35 (b) 4.79 (c) 2.91	Yi et al. (2016)

TABLE 5.1
Continued

FOOD PRODUCT: Microorganism	Treatment Condition	Maximum log Reduction	References
XIII. TURNIP JUICE: *C. inconspicua*	Fluence 19.71 J/cm^2	2.8	Karaoglan et al. (2017)
XIV. APPLE JUICE: *A. acidoterrestris*	Fluence 71.6 J/cm^2	3.0–3.5	Ferrario and Guerrero (2018)
XV. PINEAPPLE JUICE: (a) Total aerobic count (b) Total yeast and mold count	Fluence 160–1479 J/cm^2	Below detection limit, count for 'a' & 'b' at 1479 J/cm^2. At 757 J/cm^2 (a) approximately 5 (b) approximately 5	Vollmer et al. (2020)

gram-negative (1–2 nm) ones (Alcamo, 1997). Apple juice, when inoculated with *S. cerevisiae*, requires more severe PL treatment than applies to *L. innocua*, *E. coli*, and *Salmonella enteritidis* (Ferrario et al., 2015b). Some authors reported a different trend in the susceptibility of microbes with PL treatment. Bacteria are less resistant than fungal spores (Anderson et al., 2000) and viruses are more resistant than bacteria (Huang et al., 2017). Each bacterium has its unique structural composition and the time required to achieve target inactivation is different for each bacterial species. For instance, *E. coli* (gram-negative bacteria) are easily inactivated with PL treatment, while inactivation of *L. innocua* (gram-positive bacteria) requires higher energy and a larger number of pulses (MacGregor et al., 1998; Otaki et al., 2003; Ramos-Villarroel et al., 2011; Ramos-Villarroel et al., 2012).

The inactivation trend of different microbial species based on their resistance to PL processing at lower fluence is *Listeria innocua* > *Pseudomonas fluorescens* > mesophilic bacteria > *Bacillus* species (Luksiene et al., 2012; Hilton et al., 2017). Hilton et al. (2017) studied the combined effect of mild heating (<50°C) and PL processing on inactivation of three microbial strains (*E. coli*, *P. fluorescens*, and *L. innocua*). The authors reported that PL treatments were sufficient to achieve the inactivation of *E. coli* and *P. fluorescens*, while *L. innocua* required a combination of mild heating (50°C) and PL treatments. Inactivation of *Listeria* species isolated from fish products required more severe PL treatment than other fish spoilage bacteria. *Listeria* species (*L. innocua* and *L. monocytogenes*) showed higher resistance to PL treatment; its complete inactivation required treatment of higher fluence (0.312 J/cm^2) and a larger number of pulses (6) (Lasagabaster and de Marañón, 2012).

The adequacy of PL treatment for microbial inactivation of liquid foods depends on certain physicochemical parameters, such as color, thickness, viscosity, pH, total soluble solids, and flow conditions of the product (Choi et al., 2010; Artíguez et al., 2011; Ferrario and Guerrero, 2016). The clearest liquid food is drinking water; Yi et al. (2016) explored the inactivation capability of PL (4.3–14.02 J/cm^2 fluence) in case of inoculated *E.coli*, *Murine norovirus*, and total aerobic and facultative anaerobic microorganisms in water, and observed log reductions of 4.79, 3.35, and 2.91,

respectively. The effect of PL on total aerobic and total yeast and mold count in pineapple was recently studied by Vollmer et al. (2020), who observed an approximate reduction of 5 log cycle in both total aerobic and yeast-mold count after irradiating with 757 J/cm^2 fluence. Subsequently, all the microbial counts fell below the detection limit after treatment with 1479 J/cm^2 fluence. In continuous pulsed UV treatment of milk inoculated with *Staphylococcus aureus*, a reduction of 7.23–7.26 log cycle was observed at two processing conditions: (1) 8 cm sample distance, 20 mL/min flow rate, 1 pass; and (2) 11 cm sample distance, 20 mL/min flow rate, 2 passes (Krishnamurthy et al., 2007). The efficacy of PL treatment in reducing the *E. coli* count in milk samples was significantly affected by increasing levels of total solid content (Palgan et al., 2011a; Miller et al., 2012; Kasahara et al., 2015). Microbial inactivation in fruit juices is influenced by transparency (Palgan et al., 2011a; Pataro et al., 2011; Hwang et al., 2015). PL treatment of 12.17–24.35 J/cm^2 was enough to ensure 7-log reduction for transparent beverages such as apple juice, carbonated drink, and plum juice, while a lower reduction (1–1.9 log) was observed in food samples with higher solid content (grape juice, milk, and coffee) even after treatment with higher fluence levels (12.17–24.35 J/cm^2) (Hwang et al., 2015). Karaoglan et al. (2017) reported a reduction of 2.80 log cycles of *Candida inconspicua* under PL fluence of 19.71 J/cm^2. The study by Ferrario et al. (2015b) suggested that PL treatments are more effective in reducing microbial counts of clearer juice (apple) rather than pulpy fruit juices (orange and strawberry). The efficacy of PL to achieve microbial inactivation differs with the composition of the inoculation medium. For instance, the counts were reduced by 4.6 log cycles when distilled water was used as an inoculation medium. Conversely, lower inactivation (3-log) was attained at 1.8 J/cm^2 when sugar syrup acted as the treatment medium (Chaine et al. 2012). In case of inactivation of *Listeria innocua* by continuous PL (11 J/cm^2) processing, better inactivation was achieved compared to whey samples, and higher inactivation was observed with higher dilutions, when treating different dilutions of whey (Artíguez and de Marañón, 2015). This again highlights the transparency factor in PL penetration.

The optical properties of liquid food samples have a significant impact on the efficiency of PL processing. For instance, colored fruit juices have higher absorbance values for UV-C, thus more severe treatment is required for microbial inactivation of such samples (Ferrario et al., 2013; Kramer et al., 2017). PL treatments are less effective for milk processing because milk fat causes scattering of light (Miller et al., 2012). PL treatments are more efficient for treating foods with higher carbohydrate content (fruits and vegetables) than those rich in fats and proteins (Gómez-López et al., 2005). Several researchers reported higher microbial inactivation in apple juice treated with continuous PL processing (Sauer and Moraru, 2009; Palgan et al., 2011a; Pataro et al., 2011; Muñoz et al., 2012; Ferrario et al., 2013; Ferrario et al., 2015b; Hwang et al., 2015; Ferrario and Guerrero, 2018). The flow properties of liquid foods have a significant impact on PL efficiency. Chaine et al. (2012) found that shorter residence time is sufficient to ensure microbial inactivation in liquid foods, and higher flows of liquid foods could be treated by slight modification to design parameters. The rotational movement of liquid samples under a pulsed or UV light source facilitates higher energy transfer and disintegrates clusters/clumps of microbial cells, enhancing the lethality of the treatment. Sauer and Moraru (2009) observed differences in *E. coli*

count for apple juice and cider after PL treatment under static and rotatory conditions (3000 rpm). PL treatments under static conditions managed to reduce microbial counts by only 2.66 log cycles. In contrast, the turbulence created with rotatory movement imparted a higher log reduction in both apple juice (7.29) and apple cider (5.49). Miller et al. (2012) also supported the use of rotatory movements for PL inactivation of *E. coli* in reconstituted milk. Innocente et al. (2005) observed a combined effect of photochemical and photothermal damage in milk samples treated at higher fluence levels (26.25 J/cm^2). PL treatment (28 J/cm^2) of apple juice showed a higher log reduction for *E. coli* (≥ 4.7) than *L. innocua* (1.93). Further, there was no cell recovery even after 48 h (Palgan et al., 2011a). Several authors have reported that PL treatments did not affect physical characteristics (pH, Brix, and color) of apple juice (Muñoz et al., 2012; Maftei et al., 2005; and Palgan et al., 2011a).

5.4.2 Effect of PL Treatment on Enzymes

Enzymes are proteins composed of several linkages between amino acids. The inactivation of enzymes with PL treatment is associated with the structural deformation caused by the photochemical effect. Bombardment with UV light produces major changes in the sequencing of amino acids. The photolysis and photoionization phenomena cause electrical excitation of several amino acid molecules (cysteine, tryptophan, and tyrosine) and create thiol groups. The newly created thiol group breaks the disulfide and peptide linkages and eventually damages the secondary (β sheet and α helix) structure of the protein. Several authors have used PL treatment for inactivation of enzymes (polyphenoloxidase (PPO), peroxidase (POD), and polygalacturonase (PG)) in different model solutions.

In their patent, Dunn et al. (1989) claimed that the browning caused by PPO enzymes in potato slices can be inhibited by PL treatment (2–5 flashes at a fluence of 3 J/cm^2). Similar observations were later made for banana and apple slices. Faster inactivation of enzymes can be achieved by treatments with a higher fluence level. PPO was completely inactivated after exposure of 5 pulses at a fluence of 8.75 J/cm^2. The inactivation phenomena were associated with several structural changes: cross-linking of amino acids (Trp, Tyr, Phe, Met, and Cys), breakdown of disulfide bonds, and unfolding of protein structure (Manzocco et al., 2013). Treatment of 10 pulses at an intensity of 500 J/pulse showed complete inactivation of POD due to the breakdown of the secondary and tertiary structure of a protein (Wang et al., 2017). However, Pellicer et al. (2017) and Pellicer et al. (2018) state that under PL treatment the inactivation of POD and PPO was due to the all-or-none pathway of protein unfolding with no observed aggregation of the protein.

PL has a broad spectrum; light in the UV range was found to be the main cause of enzymein activation because proteins absorb significant amounts of UV from the incident light (Hollósy, 2002). During conventional continuous UV, inactivation of enzymes was achieved mainly by absorption of UV-C with a peak wavelength of 254 nm. This indicates that PL is a promising non-thermal technology that has the potential to inactivate enzymes.

Exposure of short-wave light pulses (1–60) at a fixed voltage (2.5 kV), fluence rate (128 J/cm^2), and energy rate per pulse (500 J) showed more than 90%

inactivation in many spoilage-causing enzymes, including POD (Pellicer and Gómez-López, 2017), PPO (Pellicer et al., 2018), and PG (Pellicer et al., 2019). Several authors have reported that the higher the fluence level, the greater the inactivation (Manzocco et al., 2013; Pellicer and Gómez-López, 2017; Donsingha and Assatarakul, 2018; Pellicer et al., 2018). During PL treatment of pineapple juice at a fluence of 1479 J/cm^2, the enzyme activity reduced by 42% and 50% for POD and PPO, respectively (Vollmer et al., 2020). The same authors reported that the enzyme activity of bromelain remained almost unaffected in the treated (160–1479 J/cm^2) pineapple juice, which appears to be an advantage in view of its health-promoting activity of helping with protein digestion. Manzocco et al. (2013) observed complete inactivation of PPO at 8.75 J/cm^2 in a model solution (4 or 30 U activity). The authors also stated that a higher fluence is required when initial enzyme concentration was higher, reflecting the fact that higher protein content (enzymes) hinders the uniform penetration of UV or PL, which might be due to a shadowing effect. During PL treatment with higher fluence rates and longer holding times, surface heat might be generated due to the contribution of the infrared portion of the PL spectrum (Demirci and Krishnamurthy, 2011).

5.4.3 Effect of Pulsed Light on Color, Sensory, and Biochemical Attributes

UV can make a variety of changes in food products induced by the formation of free radicals, causing different kinds of photochemical reactions which may damage various biochemicals, including vitamins, and antioxidants, along with color and sensory changes (Koutchma, 2009). Ibarz et al. (2005) reported a decrease in non-enzymatic browning index (NEBI) after UV-visible irradiance of fruit juices (apple, peach, and lemon). The overall change in color and increase in browning can be attributed to light-induced Maillard reaction, photo-degradation of color pigments (Guerrero-Beltrén and Barbosa-Cénovas, 2006), and enzymatic browning (Ferrario and Guerrero, 2016) caused by possibly surviving PPO and POD in the sample after PL treatment.

A slightly noticeable change in the total color (ΔE) was observed in treated (7, 5, and 28 J/cm^2) apple juice; according to the scale of Cserhalmi et al. (2006) this means $0.5<\Delta E<1.5$. But no significant NEBI was observed in the treated samples. In addition to that, again no significant change in antioxidant capacity (AOX) was visible after an exposure of 7 and 5 J/cm^2, whereas the sample treated at 28 J/cm^2 showed a loss of 5.5% AOX. Following the same trend the sensory evaluation by consumer acceptability test also revealed that longer treatment time (8 s) or higher fluence exposure (28 J/cm^2) showed a negative impact on the juice's flavor (Palgan et al., 2011a). In a different study, no significant color change in terms of ΔE and non-enzymatic browning index (NEBI) was observed after PL treatment of 3.3 J/cm^2 as per the product (orange + carrot juice blend) flow rate of 20.8 mL/min, with temperature not rising above 30°C (Caminiti et al., 2012). The authors also reported no significant change in total phenolic content (TPC) and only a slight increase in AOX.

Ferrario and Guerrero (2016) experimented with the effect of PL on the color parameters of two kinds of apple juice; commercial apple juice (CAJ) and natural apple juice (NAJ). The authors observed that, after a continuous PL treatment of

about 0.73 J/cm^2 at a flow rate of 155 mL/min with 5 passes, in case of CAJ, the a^* value decreased and the b^* value increased, and in case of NAJ the L^* value decreased and the b^* value increased. PL exhibited an insignificant impact on the sensory attributes – color, flavor, taste and overall acceptability – of PL-treated (500 J per pulses, 0.5 Hz frequency, and 10–25 pulses) carrot juice (Zhu et al., 2019). Kwaw et al. (2018) studied the effect of PL (7.26 J/cm^2) on AOX and the entire phenolic profile, which includes 19 polyphenolic compounds, in fermented mulberry juice. There was a notable increase in the overall TPC and AOX. The authors state that the increase can be attributed to the photochemical effects of PL and improved extraction which helps in increasing the availability of the component for the reaction. This rise in TPC and AOX can also happen due to depolymerization of the polyphenolic compounds (Bhat et al., 2007). However, Vollmer et al. (2020) reported that after PL treatment of pineapple juice at about 579 J/cm^2, the authors discovered 5% loss of AOX, 29% loss of vit-C (ascorbic acid), 47% loss of TPC, and total carotenoid content varied in the range of 183–240 μg/mL (over the entire PL fluence range of 160–579 J/cm^2). Subsequently, in case of color parameters of the treated pineapple juice, a minor change of about 0.8–1.3 in total color (ΔE^*) and no significant change in the browning index was observed. It can be seen that, depending on the treatment conditions, sample type and compositions of TPC and AOX, the PL irradiation may increase or decrease the TPC or AOX level, or affect their extractability positively or negatively.

Abuagela et al. (2018) explored the impact of PL on quality parameters of peanut oil and the reduction of the level of aflatoxin present in it. It was observed that PL is capable of reduction of aflatoxin in peanut oil, for about 84.8% reduction after a treatment time of 600 s with a corresponding rise in the temperature to approximately 200°C, whereas a reduction of 55.6% in the peanut oil surrounded with ice bath was visible after a PL treatment of 600 s. This indicates that the degradation of aflatoxin was due to both photochemical and photothermal effects. The authors also reported that for longer treated samples when temperature increased up to more than 100°C, temperature-induced primary oxidation of the peanut oil was prominent but overall, no significant changes were observed in the color parameters of the oil with respect to the untreated samples. Comprehensively, it can be understood that shorter PL treatment processes or provision of sample cooling (by using cold water or ice baths) can lead to lower loss of color, flavor, and nutrients like polyphenols and antioxidants, but at the same time, depending on the sample, may also help decrease toxins like aflatoxin (in peanut oil) and increase phenolic content and antioxidants.

5.4.4 PL in Combination with Other Technologies

Pulsed light has great potential as a non-thermal processing technology, but it has some limitations regarding penetration, non-uniform exposure, and temperature rise when used for a longer time. To overcome its few shortcomings and obtain better efficiencies in shorter treatment times, to avoid sample heating, and to inactivate resistant microorganisms and enzymes, many researchers have explored PL treatment in combination with other technologies for processing liquid food. Table 5.2 summarizes research exploring the combined effects of PL with thermal or non-thermal processing techniques in the liquid food domain.

TABLE 5.2
Literature Covering the Effects on Liquid Foods of Processing Technologies in Combination with Pulsed Light

Combined Processing Technology	FOOD PRODUCT: Target Attribute	Treatment Conditions	Inferences	References
I. (1) PEF+PL (2) PL+PEF	APPLE JUICE: (a) *E. coli* (b) pH, TSS, TA, ΔE, and NEBI (c) TPC and AOX (d) Sensory	Continuous flow rate: 13.4–17 mL/min PEF: 24 and 34 kV/cm and supply of total specific energy 130.5 and 261.9 J/mL PL: 4–5.1 J/cm^2	(1a) better synergism. (1a, 2a) 5 log cycle reduction achieved. (1b, 2b) No significant change. (2c) No significant change as per control, most acceptable	Caminiti et al. (2011)
II. (1) PL+TS (2) TS+PL	ORANGE JUICE: *E. coli*	PL: 4.03–5.1 J/cm^2 TS: '2.8 min at 40 °C' to '5 min at 50°C'	No effect of sequence (1 & 2), stronger impact PL than TS PL (5.1 J/cm^2)+TS (2.8 min, 40°C): maximum inactivation of 3.93 log CFU/mL	Muñoz et al. (2011)
III. (1) PL+PEF (2) PL+MTS	APPLE + CRANBERRY JUICE BLEND: (a) *E. coli* (b) *P. fermentans*	PL: 360 μs pulse width, 3 Hz frequency, 1.21 J/cm^2/pulse, 1.9 cm distance, 40 cm of the pipe carrying fluid underexposure PEF: 18 Hz pulse frequency, 1 μs pulse duration, 34 kV/cm, 93 μs time MTS: 20 kHz frequency, 23 μm amplitude, 750 W power output, 400 kPa pressure (with sample cooling facility)	Effect of sequence not explored, (1–2 & a–b) all combination and for each microbe achieved 6 or close to 6 log CFU/mL reduction	Palgan et al. (2011b)
IV. PL+MTS	ORANGE + CARROT JUICE BLEND: (a) pH, TSS, ΔE and NEBI (b) TPC and AOX (c) PME (d) Sensory	PL: 3.3 J/cm^2 (as per flow rate 20.8 mL/min) MTS: 20 kHz frequency, 31 μm amplitude, 1 kW power output, 400 kPa pressure, 35–63°C temperature, 2.2 min time (with assisted cooling)	(a) No significant change in pH, slight variation in TSS (around mean 9 brix), ΔE = 4.51, NEBI = 0.91 (b) 9.5% loss in TPC and non-significant loss in AOX. (c) 77% inactivation (d) No significant changes compared to the thermal treated control	Caminiti et al. (2012)

TABLE 5.2
Continued

Combined Processing Technology	FOOD PRODUCT: Target Attribute	Treatment Conditions	Inferences	References
V. (1) PL+TS (2) TS+PL	APPLE JUICE: (a) *E. coli* (b) pH, TSS, ΔE and NEBI (c) AOX	PL: 4.03–5.1 J/cm^2 or 51.5–65.4 J/mL TS: 24 kHz frequency, 100 μs amplitude, at '40°C, 2.9 min' to '50°C, 5 min'	(1a & 2a) All combinations reached approximately 6 log CFU/mL reduction, showed additive effect. (b) No significant changes in pH, TSS, and NEBI but change in color was visible having ΔE = 2.4 and 1.3 for treatment (1) and treatment (2), respectively. (c) No significant variation in AOX	Muñoz et al. (2012)
VI. US + PL	APPLE JUICE, COMMERCIAL (CAJ) AND NATURAL (NAJ): (a) *A. acidoterrestris* spores (b) *S. cerevisiae*	PL: 2.4–71.6 J/cm^2 US: 20 kHz frequency, 95.2 μm amplitude, 600 W power output, 10 and 30 min time, 20, 30 and 44°C temperature	(CAJ-a) 3.0 (CAJ-b) 6.4 (NAJ-a) 2.0 (NAJ-b) 5.8	Ferrario et al. (2015a)
VII. US + PL	APPLE JUICE, COMMERCIAL (CAJ) AND NATURAL (NAJ): (a) *E. coli* (b) *S. enteritidis* (c) *S. cerevisiae* (d) TMA (e) YM (f) Residual BI (g) Sensory	PL: 0.73 J/cm^2 with corresponding flow rate 155 mL/min US: 20 kHz frequency, 95.2 μm amplitude, 600 W net output capacity, 30 min time, 25°C temperature maintained	No significant difference between CAJ and NAJ for all attributes (a) 5.9 (b) 6.3 (c) 3.7 (d) 2.2 and (e) 2.0 log CFU/mL after 10 days in 4°C. (f) No residual BI observed (g) Overall acceptability score 6.1 (like slightly)	Ferrario and Guerrero (2016)
VIII. US + PL	APPLE JUICE, COMMERCIAL (CAJ) AND NATURAL (NAJ): *S. cerevisiae*	PL: 71.6 J/cm^2 US: 20 kHz frequency, 95.2 μm amplitude, 600 W power output, 10 and 30 min time, 20 and 44°C temperature	(CAJ at 20°C) 4.9 (NAJ at 20 C) 3.9 (CAJ at 40°C) 6.4 (NAJ at 40 C) 5.8	Ferrario and Guerrero (2017)

(*Continued*)

TABLE 5.2
Continued

Combined Processing Technology	FOOD PRODUCT: Target Attribute	Treatment Conditions	Inferences	References
IX. (1) US + PL (2) PL+US	FERMENTED MULBERRY JUICE: (a) AOX (b) TPC (c) TPA (d) TAC (e) TF	PL: 7.26 J/cm^2 US: 28 kHz frequency, 60 W power output, 15 min time, 5°C temperature maintained	(1a) 4.4–23.2% rise (1b) 8.4% rise (1c) 12% rise (1d) 13.1% rise (1e) 3.4% rise (2a) 4.1–22.3% rise (2b) 7.1% rise (2c) 6% rise (2d) 19.2% rise (2e) 4.1% rise	Kwaw et al. (2018)

CFU: colony-forming unit, PL: pulsed light, PEF: pulsed electric field, TSS: total soluble solids, TA: titratable acidity, ΔE: total change in color, NEBI: non-enzymatic browning, TPC: total phenolic content, AOX: antioxidant capacity, TS: thermo-sonication, MTS: manothermosonication, PME: pectin methyl-esterase, US: ultrasound, TMA: total mesophilic aerobic population, YM: yeast and mold population, BI: browning index, TPA: total phenolic acids, TAC: total anthocyanins, TF: total flavanols.

Caminiti et al. (2011) studied the effect of the combination of pulsed electric field (PEF) and pulsed light (PL) on inactivation of *E. coli* K12, physicochemical, biochemical and sensory parameters. Results showed that combination treatments starting with PEF and followed by PL caused better inactivation of *E. coli* compared to PL and PEF. The authors assumed that this kind of synergism can happen because of higher damage to the cell membrane by PEF, and enhanced susceptibility to UV coming from PL. The physico-chemical parameters were not much affected by any of the combination treatments, and in the case of sensory parameters, the PL+PEF treatment was the most acceptable and caused less difference compared to untreated. In a different study, Palgan et al. (2011b) observed approximately 6 log reduction of *E. coli* K12 and *Pichiafermentans* inoculated in a blend of apple and cranberry juice after combined treatments of PL+PEF and PL+MTS (manothermosonication). MTS, in general, is a technology that involves pressure application in addition to heating and sonication. The order of processing was not explored. It was also observed that the sample treated with PL+PEF reached a shelf life of 21 days. Very little literature is available regarding the effects of combination treatments with PL on enzymes. Caminiti et al. (2012) experimented the effect of many non-thermal processing techniques and their combinations, such as PL and MTS, in comparison with thermal pasteurization (72°C for 26 s), on various parameters, including physico-chemical properties (pH, total soluble solids (TSS), color in terms of total color change (ΔE), and non-enzymatic browning (NEBI)), biochemical content (total phenolic content (TPC) and antioxidant capacity (AOX)), pectin methylesterase (PME) enzyme activity, and sensory evaluation. In the case of PL+MTS, higher change in color ΔE = 4.51

and lower NEBI = 0.19 compared to the thermal control having ΔE = 2.91 and NEBI = 0.197 was observed. Compared to an insignificant loss of 1% in TPC by thermal control, PL+MTS resulted in a significant loss of 9.5%. The combined non-thermal treatment achieved 77% inactivation of PME activity, whereas the thermal control achieved 92% inactivation. In sensory evaluation, the pasteurized sample was found to be the best in terms of overall acceptability. Here, we can see that overall, compared to thermal control, non-thermal combination processing may not lead to better results in terms of parameters other than microbial inactivation which also depends on different samples and their interaction with different processing technologies.

The effect of combined hurdle processing of PL and ultrasound (US) and thermal (usually mild) assisted ultrasound/thermo-sonication (TS) has been explored by many researchers in cases of liquid food products such as apple, orange, and fermented mulberry juices (Muñoz et al., 2011; Muñoz et al., 2012; Ferrario et al., 2015a; Ferrario and Guerrero, 2016, 2017; Kwaw et al., 2018). In a study by Muñoz et al. (2011) processing of orange juice with PL and TS was individually executed continuously, with the experimental conditions for PL treatment the same as those followed by Caminiti et al. (2011); the processing conditions for TS were constant wave frequency of 24 kHz, 100 µm as maximum amplitude, and maximum deliverable acoustic power density by the probe was 85 W/cm^2. The TS unit was utilized at its full power output capacity. No significant impact of the sequence of treatment on inactivation of *E. coli* was visible, but PL (2.42 log reduction) showed higher inactivation than TS (1.10 log reduction) when used experimentally as standalone techniques. However, when PL and TS are combined, inactivation increased a maximum of up to 3.93 log CFU/mL. Subsequently, Muñoz et al. (2012) reported 6-log cycle reduction of *E. coli* after PL and TS combination treatment of apple juice, regardless of the treatment sequence. Except for the total change in color ΔE, no notable variation was observed in pH, TSS, NEBI, and AOX after processing. The authors also observed higher inactivation of *E. coli* by PL than TS, when experimented separately.

Microbial spores may prove to be quite resistant even under hurdle processing with PL. When apple juice (commercial (CAJ) and natural (NAJ)) inoculated with *Alicyclobacillus acidoterrestris* (ATCC 49025) spores and *Saccharomyces cerevisiae* was treated with US, no inactivation happened in *A. acidoterrestris* for both CAJ and NAJ (Ferrario et al., 2015a). Subsequently, after single PL treatment, 3.0 and 2.0 log cycle reductions were achieved in CAJ and NAJ, respectively. Even combined treatment of US + PL resulted in similar inactivation of *A. acidoterrestris* compared to single PL treatment. In the case of *S. cerevisiae*, both US and PL individually showed a significant reduction of 2.5 log cycle (for both CAJ and NAJ) and 2.0 (NAJ) to 3.7 (CAJ) log cycle, respectively. However, combined treatment of US then PL resulted in 6.4 and 5.8 log reductions in CAJ and NAJ, respectively, which indicates an additive effect. After 15 days' storage, in the CAJ treated with '10 min US + 20 s PL', the *A. acidoterrestris* spores remained almost the same, but the *S. cerevisiae* recovered about 1.1 log cycles. By contrast, in the CAJ treated solely with PL, *S. cerevisiae* recovered by 2.4 log cycles. Ferrario and Guerrero (2016) discovered a higher contribution to the inactivation by PL in combined treatment of apple juice (commercial and natural) with US + PL, and observed an additive effect, keeping in mind their individual effects.

Little literature is currently focused on the effect of PL with other technologies on the biochemical and entire categorical profile of specific products. Kwaw et al. (2018) studied the effect of PL, US, and their combination on the phenolic components (including 19 polyphenolic components) of fermented mulberry juice. It was reported that PL and US treatments can enhance the phenolic components by the action of isomerization, depolymerization, condensation, and degradation of phytonutrients into phenolic products/by-products, and by also improving extractability. The synergistic photochemical and acoustic influences induced by PL and US, respectively, may also play a role in enhancement.

5.5 CONCLUSION

Pulsed light (PL) is non-thermal technology that is known for its quick and environmentally friendly decontamination qualities. PL is an excellent non-thermal processing alternative, capable of attaining major microbial reduction in the case of surface decontamination and of processing liquid food products with adequate transparency. PL technology has the potential to achieve adequate shelflife, food safety, minimum change in sensory parameters, and retention of nutrients and health-promoting biochemical components. The effectiveness of PL depends on the type of food material, target microorganism or enzyme, their initial count or concentration, and treatment conditions (total fluence, fluence rate, voltage, treatment time/number of pulses, the distance between the sample and the light source, the spectral range of the light).

Plenty of work has been done on microbial inactivation (natural as well as artificially added). However, there is insufficient literature on the effects on indigenous enzymes and biochemicals. Individual research focusing on mechanisms of microbial and enzyme inactivation is available, but literature highlighting the biochemical loss or enhancement mechanisms remains scarce. It was observed that longer PL treatment generates heat, which then requires some cooling mechanism to control the sample temperature if we wish to use it as a non-thermal technique. Just like other light-based processing technologies, PL radiation also struggles to penetrate in case of solids, suspensions with a significant amount of particulate matter, and opaque liquid food products. Thus, to complement the limitations of PL, many researchers have studied the effects of PL in combination with other thermal or non-thermal technologies. More literature is needed on this domain, covering the effects of PL-based hurdle processing on microorganisms, enzymes, biochemical nutrients, sensory profile, and associated mechanisms. Furthermore, to validate the effectiveness of PL on various quality attributes, large-scale studies are necessary for promotion of industrial applications.

ACKNOWLEDGMENT

The authors acknowledge the funding given by the Science and Engineering Research Board (SERB), Department of Science and Technology (DST), Govt. of India (File No. ECR/2016/0055).

REFERENCES

Abuagela, M.O., B.M. Iqdiam, G.L. Baker, and A.J. MacIntosh. 2018. Temperature-controlled pulsed light treatment: Impact on aflatoxin level and quality parameters of peanut oil. *Food Bioproc. Tech.* 11: 1350–1358. doi:10.1007/s11947-018-2105-6

Aguirre, J.S., E. Hierro, M. Fernández, and G.D. García de Fernando. 2005. Modelling the effect of light penetration and matrix colour on the inactivation of *Listeria innocua* by pulsed light. *Innov. Food Sci. Emerg. Technol.* 26: 505–510. doi:10.1016/j.ifset.205.05.011

Alcamo, I.E. 1997. *Fundamentals of Microbiology*, 5th ed. Menlo Park, CA: Benjamin Cummings.

Alhendi, A., W. Yang, R. Goodrich-Schneider, and P.J. Sarnoski. 2018. Total inactivation of lipoxygenase in whole soya bean by pulsed light and the effect of pulsed light on the chemical properties of soya milk produced from the treated soya beans. *Int. J. Food Sci. Tech.* 53: 457–466. doi:10.1111/ijfs.13604

Anderson, J.G., N.J. Rowan, S.J. MacGregor, R.A. Fouracre, and O. Farish. 2000. Inactivation of food-borne enteropathogenic bacteria and spoilage fungi using pulsed-light. *IEEE Trans. Plasma Sci.* 28: 83–88. doi:10.1109/27.842870

Artíguez, M.L., A. Lasagabaster, and I.M. de Marañón. 2011. Factors affecting microbial inactivation by Pulsed Light in a continuous flow-through unit for liquid products treatment. *Procedia Food Sci.* 1: 786–791. doi:10.1016/j.profoo.2011.09.119

Artíguez, M.L., and I.M. de Marañón. 2015. Improved process for decontamination of whey by a continuous flow-through pulsed light system. *Food Control.* 47: 599–605. doi:10.1016/j.foodcont.205.08.006

Bhat, R., K.R. Sridhar, and K. Tomita-Yokotani. 2007. Effect of ionizing radiation on antinutritional features of velvet bean seeds (*Mucunapruriens*). *Food Chem.* 103: 860–866. doi:10.1016/j.foodchem.2006.09.037

Bintsis, T., E. Litopoulou-Tzanetaki, and R.K. Robinson. 2000. Existing and potential applications of ultraviolet light in the food industry – A critical review. *J. Sci. Food Agric.* 80: 637–645. doi:10.1002/(SICI)1097-0010(20000501)80:6<637::AID-JSFA603>3.0.CO;2-1

Bolton, J.R., and K.G. Linden. 2003. Standardization of Methods for Fluence (UV Dose) Determination in Bench-Scale UV Experiments. *J. Environ. Eng.* 129: 209–215. doi:10.1061/(ASCE)0733-9372(2003)129:3(209)

Bosshard, F., M. Berney, M. Scheifele, H.-U. Weilenmann, and T. Egli. 2009. Solar disinfection (SODIS) and subsequent dark storage of *Salmonella typhimurium* and *Shigellaflexneri* monitored by flow cytometry. *Microbiology* 155: 1310–1317. doi:10.1099/mic.0.024794-0

Bowker, C., A. Sain, M. Shatalov, and J. Ducoste. 2011. Microbial UV fluence-response assessment using a novel UV-LED collimated beam system. *Water Res.* 45: 2011–2019. doi:10.1016/j.watres.2010.12.005

Braslavsky, S.E. 2006. Glossary of terms used in photochemistry. *Chem Int - News Magazine for IUPAC.* 28. doi:10.1515/ci.2006.28.1.28b

Caminiti, I.M., F. Noci, D.J. Morgan, D.A. Cronin, and J.G. Lyng. 2012. The effect of pulsed electric fields, ultraviolet light or high intensity light pulses in combination with manothermosonication on selected physico-chemical and sensory attributes of an orange and carrot juice blend. *Food and Bioprod. Process.* 90: 442–448. doi:10.1016/j.fbp.2011.11.006

Caminiti, I.M., I. Palgan, F. Noci, A. Muñoz, P. Whyte, D.A. Cronin, D.J. Morgan, and J.G. Lyng. 2011. The effect of pulsed electric fields (PEF) in combination with high intensity light pulses (HILP) on *Escherichia coli* inactivation and quality attributes in apple juice. *Innov. Food Sci. Emerg. Technol.* 12: 118–123. doi:10.1016/j.ifset.2011.01.003

Chaine, A., C. Levy, B. Lacour, C. Riedel, and F. Carlin. 2012. Decontamination of Sugar Syrup by Pulsed Light. *J. Food Prot.* 75: 913–917. doi:10.4315/0362-028X.JFP-11-342

Cheigh, C.I., H.J. Hwang, and M.S. Chung. 2013. Intense pulsed light (IPL) and UV-C treatments for inactivating *Listeria monocytogenes* on solid medium and seafoods. *Food Res. Int.* 54: 745–752. doi:10.1016/j.foodres.2013.08.025

Cheigh, C.I., M.H. Park, M.S. Chung, J.K. Shin, and Y.S. Park. 2012. Comparison of intense pulsed light- and ultraviolet (UVC)-induced cell damage in *Listeria monocytogenes* and *Escherichia coli* O157: H7. *Food Control.* 25: 654–659. doi:10.1016/j.foodcont.2011.11.032

Choi, M.S., C.I. Cheigh, E.A. Jeong, J.K. Shin, and M.S. Chung. 2010. Nonthermal sterilization of *Listeria monocytogenes* in infant foods by intense pulsed-light treatment. *J. Food Eng.* 97: 504–509. doi:10.1016/j.jfoodeng.2009.11.008

Cserhalmi, Z., Á. Sass-Kiss, M. Tóth-Markus, and N. Lechner. 2006. Study of pulsed electric field treated citrus juices. *Innov. Food Sci. Emerg. Technol.* 7: 49–54. doi:10.1016/j.ifset.2005.07.001

Demirci, A., and K. Krishnamurthy. 2011. Pulsed ultraviolet light. In: H.Q. Zhang, G.V. Barbosa-Cánovas, V.M. Balasubramaniam, C.P. Dunne, D.F. Farkas, and J.T.C. Yuan (eds), *Nonthermal Processing Technologies for Food.* Chichester: Wiley-Blackwell, pp. 249–261.

Donsingha, S., and K. Assatarakul. 2018. Kinetics model of microbial degradation by UV radiation and shelf life of coconut water. *Food Control* 92: 162–168. doi:10.1016/j.foodcont.2018.04.030

Dunn, J., T. Ott, and W. Clark. 1995. Pulsed-light treatment of food and packaging. *Food Tech. (Chic.)* 49: 95–98.

Dunn, J.E., R.W. Clark, J.F. Asmus, J.S. Pearlman, K. Boyer, F. Painchaud, and G.A. Hofmann. 1989. *Methods for Preservation of Foodstuff.* Patent No. 4, 871, 559. Washington, DC: U.S. Patent and Trademark Office.

Elmnasser, N., S. Guillou, F. Leroi, N. Orange, A. Bakhrouf, and M. Federighi. 2007. Pulsed-light system as a novel food decontamination technology: A review. *Can. J. Microbiol.* 53: 813–821. doi:10.1139/W07-042

FDA. 2019. Code of Federal Regulations. 21CFR179.41. Pulsed light for the treatment of food. Available at https://www.accessdata.fda.gov/scripts/cdrh/cfdocs/cfcfr/CFRSearch.cfm?fr=179.41. Accessed on 4 Feb 2020.

Ferrario, M., S.M. Alzamora, and S. Guerrero. 2013. Inactivation kinetics of some microorganisms in apple, melon, orange and strawberry juices by high intensity light pulses. *J. Food Eng.* 118: 302–311. doi:10.1016/j.jfoodeng.2013.04.007

Ferrario, M., S.M. Alzamora, and S. Guerrero. 2015a. Study of the inactivation of spoilage microorganisms in apple juice by pulsed light and ultrasound. *Food Microbiol.* 46: 635–642. doi:10.1016/j.fm.205.06.017

Ferrario, M., S.M. Alzamora, and S. Guerrero. 2015b. Study of pulsed light inactivation and growth dynamics during storage of *Escherichia coli* ATCC 35218, *Listeria innocua* ATCC 33090, *Salmonella* Enteritidis MA44 and *Saccharomyces cerevisiae* KE162 and native flora in apple, orange and strawberry juices. *Int. J. Food Sci. Tech.* 50: 2498–2507. doi:10.1111/ijfs.12918

Ferrario, M., and S. Guerrero. 2016. Effect of a continuous flow-through pulsed light system combined with ultrasound on microbial survivability, color and sensory shelf life of apple juice. *Innov. Food Sci. Emerg. Technol.* 34: 25–224. doi:10.1016/j.ifset.2016.02.002

Ferrario, M., and S. Guerrero. 2017. Impact of a combined processing technology involving ultrasound and pulsed light on structural and physiological changes of *Saccharomyces cerevisiae* KE 162 in apple juice. *Food Microbiol.* 65: 83–94. doi:10.1016/j.fm.2017.01.012

Ferrario, M.I., and S.N. Guerrero. 2018. Inactivation of *Alicyclobacillusacidoterrestris* ATCC 49025 spores in apple juice by pulsed light. Influence of initial contamination and required reduction levels. *Rev. Argent. Microbiol.* 50: 3–11. doi:10.1016/j.ram.2017.04.002

Fine, F., and P. Gervais. 2004. Efficiency of Pulsed UV Light for Microbial Decontamination of Food Powders. *J. Food Prot.* 67: 787–792. doi:10.4315/0362-028X-67.4.787

Gaertner, A.A. 2012. Optical radiation measurement. In L. Cocco (ed.), *Modern Metrology Concerns*. Intech, pp. 223–262.

Gómez-López, V.M., and J.R. Bolton. 2016. An approach to standardize methods for fluence determination in bench-scale pulsed light experiments. *Food Bioproc. Tech.* 9: 1040–1048. doi:10.1007/s11947-016-1696-z

Gómez-López, V.M., F. Devlieghere, V. Bonduelle, and J. Debevere. 2005. Factors affecting the inactivation of micro-organisms by intense light pulses. *J. Appl. Microbiol.* 99: 460–470. doi:10.1111/j.1365-2672.2005.02641.x

Gómez-López, V.M., P. Ragaert, J. Debevere, and F. Devlieghere. 2007. Pulsed light for food decontamination: A review. *Trends Food Sci. Tech.* 18: 464–473. doi:10.1016/j.tifs.2007.03.010

Green, S., N. Basaran, B. Swanson, L. Bogh-Sorenson, and P. Zenthen (ed.). 2005. *Food Preservation Techniques*. Washington, DC: Woodhead Publishing House, CRC Press.

Guerrero-Beltrén, J.A., and G.V. Barbosa-Cénovas. 2006. Inactivation of Saccharomyces cerevisiae and Polyphenoloxidase in Mango Nectar Treated with UV Light. *J. Food Prot.* 69: 362–368. doi:10.4315/0362-028X-69.2.362

Heinrich, V., M. Zunabovic, A. Petschnig, H. Müller, A. Lassenberger, E. Reimhult, and W. Kneifel. 2016. Previous Homologous and Heterologous Stress Exposure Induces Tolerance Development to Pulsed Light in *Listeria monocytogenes*. *Front. Microbiol.* 7: 490. doi:10.3389/fmicb.2016.00490

Hillegas, S.L., and A. Demirci. 2003. *Inactivation of Clostridium sporogenes in Clover Honey by Pulsed UV-light Treatment*. In *ASAE Annual Meeting*. Las Vegas, NV: American Society of Agricultural and Biological Engineers, pp. 1–7.

Hilton, S.T., J.O. de Moraes, and C.I. Moraru. 2017. Effect of sublethal temperatures on pulsed light inactivation of bacteria. *Innov. Food Sci. Emerg. Technol.* 39: 49–54. doi:10.1016/j.ifset.2016.11.002

Hollósy, F. 2002. Effects of ultraviolet radiation on plant cells. *Micron.* 33: 179–197. doi:10.1016/S0968-4328(01)00011-7

Huang, Y., and H. Chen. 2005. A novel water-assisted pulsed light processing for decontamination of blueberries. *Food Microbiol.* 40: 1–8. doi:10.1016/j.fm.2013.11.017

Huang, Y., M. Ye, X. Cao, and H. Chen. 2017. Pulsed light inactivation of murine norovirus, Tulane virus, *Escherichia coli* O157: H7 and *Salmonella* in suspension and on berry surfaces. *Food Microbiol.* 61: 1–4. doi:10.1016/j.fm.2016.08.001

Hwang, H.J., C.I. Cheigh, and M.S. Chung. 2015. Relationship between optical properties of beverages and microbial inactivation by intense pulsed light. *Innov. Food Sci. Emerg. Technol.* 31: 91–96. doi:10.1016/j.ifset.2015.06.009

Hwang, H.J., J.H. Seo, C. Jeong, C.I. Cheigh, and M.S. Chung. 2019. Analysis of bacterial inactivation by intense pulsed light using a double-Weibull survival model. *Innov. Food Sci. Emerg. Technol.* 56: 102185. doi:10.1016/j.ifset.2019.102185

Ibarz, A., J. Pagán, R. Panadés, and S. Garza. 2005. Photochemical destruction of color compounds in fruit juices. *J. Food Eng.* 69: 155–160. doi:10.1016/j.jfoodeng.2004.08.006

Innocente, N., A. Segat, L. Manzocco, M. Marino, M. Maifreni, I. Bortolomeoli, A. Ignat, and M.C. Nicoli. 2005. Effect of pulsed light on total microbial count and alkaline phosphatase activity of raw milk. *Int. Dairy J.* 39: 108–112. doi:10.1016/j.idairyj.205.05.009

Jeon, M.-S., K.-M. Park, H. Yu, J.-Y. Park, and P.-S. Chang. 2019. Effect of intense pulsed light on the deactivation of lipase: Enzyme-deactivation kinetics and tertiary structural changes by fragmentation. *Enzyme Microb. Tech.* 124: 63–69. doi:10.1016/j.enzmictec.2019.02.001

Karaoglan, H.A., N.M. Keklik, and N. Develi Işikli. 2017. Modeling Inactivation of *Candida inconspicua* Isolated from Turnip Juice using Pulsed UV Light: Modeling Inactivation of *Candida inconspicua*. *J. Food Process Eng.* 40: E12418. doi:10.1111/jfpe.12418

Kasahara, I., V. Carrasco, and L. Aguilar. 2015. Inactivation of *Escherichia coli* in goat milk using pulsed ultraviolet light. *J. Food Eng.* 152: 43–49. doi:10.1016/j.jfoodeng.205.11.012

Koutchma, T. 2009. Advances in ultraviolet light technology for non-thermal processing of liquid foods. *Food Bioproc. Tech.* 2: 138–155. doi:10.1007/s11947-008-0178-3

Kramer, B., J. Wunderlich, and P. Muranyi. 2017. Impact of treatment parameters on pulsed light inactivation of microorganisms on a food simulant surface. *Innov. Food Sci. Emerg. Technol.* 42: 83–90. doi:10.1016/j.ifset.2017.05.011

Krishnamurthy, K., A. Demirci, and J.M. Irudayaraj. 2007. Inactivation of *Staphylococcus aureus* in milk using flow-through pulsed UV-light treatment system. *J. Food Sci.* 72: M233–M239. doi:10.1111/j.1750-3841.2007.00438.x

Krishnamurthy, K., J.C. Tewari, J. Irudayaraj, and A. Demirci. 2010. Microscopic and spectroscopic evaluation of inactivation of *Staphylococcus aureus* by pulsed UV light and infrared heating. *Food Bioproc. Tech.* 3: 93–104. doi:10.1007/s11947-008-0084-8

Kwaw, E., Y. Ma, W. Tchabo, M.T. Apaliya, A.S. Sackey, M. Wu, and L. Xiao. 2018. Impact of ultrasonication and pulsed light treatments on phenolics concentration and antioxidant activities of lactic-acid-fermented mulberry juice. *LWT - Food Sci. Tech.* 92: 61–66. doi:10.1016/j.lwt.2018.02.016

Lasagabaster, A., and I.M. de Marañón. 2005. Survival and growth of *Listeria innocua* treated by pulsed light technology: Impact of post-treatment temperature and illumination conditions. *Food Microbiol.* 41: 76–81. doi:10.1016/j.fm.205.02.001

Lasagabaster, A., and I.M. de Marañón. 2012. Sensitivity to pulsed light technology of several spoilage and pathogenic bacteria isolated from fish products. *J. Food Prot.* 75: 2039–2044. doi:10.4315/0362-028X.JFP-12-071

Lasagabaster, A., and I.M. de Marañón. 2017. Comparative study on the inactivation and photoreactivation response of *Listeria monocytogenes* seafood isolates and a *Listeria innocua* surrogate after pulsed light treatment. *Food Bioproc. Tech.* 10: 1931–1935. doi:10.1007/s11947-017-1972-6

Levy, C., X. Aubert, B. Lacour, and F. Carlin. 2012. Relevant factors affecting microbial surface decontamination by pulsed light. *Int. J. Food Microbiol.* 152: 168–174. doi:10.1016/j.ijfoodmicro.2011.08.022

Luksiene, Z., I. Buchovec, K. Kairyte, E. Paskeviciute, and P. Viskelis. 2012. High-power pulsed light for microbial decontamination of some fruits and vegetables with different surfaces. *J. Food Agric. Environ.* 10: 162–167.

Luksiene, Z., V. Gudelis, I. Buchovec, and J. Raudeliuniene. 2007. Advanced high-power pulsed light device to decontaminate food from pathogens: Effects on *Salmonella typhimurium* viability in vitro: In vitro decontamination of food using high-power pulsed light device. *J. Appl. Microbiol.* 103: 1545–1552. doi:10.1111/j.1365-2672.2007.03403.x

MacGregor, S.J., N.J. Rowan, L. Mc Ilvaney, J.G. Anderson, R.A. Fouracre, and O. Farish. 1998. Light inactivation of food-related pathogenic bacteria using a pulsed power source. *Lett. Appl. Microbiol.* 27: 67–70. doi:10.1046/j.572-765X.1998.00399.x

Maftei, N.A., A.Y. Ramos-Villarroel, A.I. Nicolau, O. Martín-Belloso, and R. Soliva-Fortuny. 2005. Influence of processing parameters on the pulsed-light inactivation of *Penicilliumexpansum* in apple juice. *Food Control.* 41: 27–31. doi:10.1016/j.foodcont.2013.12.023

Manzocco, L., A. Panozzo, and M.C. Nicoli. 2013. Inactivation of polyphenoloxidase by pulsed light. *J. Food Sci.* 78: E1183–E1187. doi:10.1111/1750-3841.12216

Marquenie, D., A.H. Geeraerd, J. Lammertyn, C. Soontjens, J.F. Van Impe, C.W. Michiels, and B.M. Nicolai. 2003. Combinations of pulsed white light and UV-C or mild heat treatment to inactivate conidia of *Botrytis cinerea* and *Moniliafructigena*. *Int. J. Food Microbiol.* 85: 185–196. doi:10.1016/S0168-1605(02)00538-X

Mertens, B., and D. Knorr. 1992. Developments of nonthermal processes for food preservation. *Food Tech. (Chic.).* 46: 124–133.

Miller, B.M., A. Sauer, and C.I. Moraru. 2012. Inactivation of *Escherichia coli* in milk and concentrated milk using pulsed-light treatment. *J. Dairy Sci.* 95: 5597–5603. doi:10.3168/jds.2012-575

Muñoz, A., I.M. Caminiti, I. Palgan, G. Pataro, F. Noci, D.J. Morgan, D.A. Cronin, P. Whyte, G. Ferrari, and J.G. Lyng. 2012. Effects on *Escherichia coli* inactivation and quality attributes in apple juice treated by combinations of pulsed light and thermosonication. *Food Res. Int.* 45: 299–305. doi:10.1016/j.foodres.2011.08.020

Muñoz, A., I. Palgan, F. Noci, D.J. Morgan, D.A. Cronin, P. Whyte, and J.G. Lyng. 2011. Combinations of High Intensity Light Pulses and Thermosonication for the inactivation of *Escherichia coli* in orange juice. *Food Microbiol.* 28: 1200–1204. doi:10.1016/j.fm.2011.04.005

Nicorescu, I., B. Nguyen, M. Moreau-Ferret, A. Agoulon, S. Chevalier, and N. Orange. 2013. Pulsed light inactivation of *Bacillus subtilis* vegetative cells in suspensions and spices. *Food Control.* 31: 151–157. doi:10.1016/j.foodcont.2012.09.047

Ohlsson, T., and N. Bengtsson (eds). 2002. *Minimal Processing Technologies in the Food Industry.* Cambridge: Woodhead Publishing, CRC Press.

Orlowska, M., T. Koutchma, M. Grapperhaus, J. Gallagher, R. Schaefer, and C. Defelice. 2013. Continuous and pulsed ultraviolet light for nonthermal treatment of liquid foods. Part 1: Effects on quality of fructose solution, apple juice, and milk. *Food Bioproc. Tech.* 6: 1580–1592. doi:10.1007/s11947-012-0779-8

Otaki, M., A. Okuda, K. Tajima, T. Iwasaki, S. Kinoshita, and S. Ohgaki. 2003. Inactivation differences of microorganisms by low pressure UV and pulsed xenon lamps. *Water Sci. Tech.* 47: 185–190. doi:10.2166/wst.2003.0193

Palgan, I., I.M. Caminiti, A. Muñoz, F. Noci, P. Whyte, D.J. Morgan, D.A. Cronin, and J.G. Lyng. 2011a. Effectiveness of High Intensity Light Pulses (HILP) treatments for the control of *Escherichia coli* and *Listeria innocua* in apple juice, orange juice and milk. *Food Microbiol.* 28: 5–20. doi:10.1016/j.fm.2010.07.023

Palgan, I., I.M. Caminiti, A. Muñoz, F. Noci, P. Whyte, D.J. Morgan, D.A. Cronin, and J.G. Lyng. 2011b. Combined effect of selected non-thermal technologies on *Escherichia coli* and *Pichiafermentans* inactivation in an apple and cranberry juice blend and on product shelf life. *Int J of Food Microbiol* 151: 1–6. doi:10.1016/j.ijfoodmicro.2011.07.019

Pan, Y., D.-W. Sun, and Z. Han. 2017. Applications of electromagnetic fields for nonthermal inactivation of microorganisms in foods: An overview. *Trends in Food Sci. Tech.* 64: 13–22. doi:10.1016/j.tifs.2017.02.05

Pedrós-Garrido, S., S. Condón-Abanto, I. Clemente, J.A. Beltrán, J.G. Lyng, D. Bolton, N. Brunton, and P. Whyte. 2018. Efficacy of ultraviolet light (UV-C) and pulsed light (PL) for the microbiological decontamination of raw salmon (*Salmosalar*) and food contact surface materials. *Innov. Food Sci. Emerg. Technol.* 50: 124–131. doi:10.1016/j.ifset.2018.10.001

Pellicer, J.A., and V.M. Gómez-López. 2017. Pulsed light inactivation of horseradish peroxidase and associated structural changes. *Food Chem.* 237: 632–637. doi:10.1016/j.foodchem.2017.05.151

Pellicer, J.A., P. Navarro, and V.M. Gómez-López. 2018. Pulsed Light Inactivation of Mushroom Polyphenol Oxidase: A Fluorometric and Spectrophotometric Study. *Food Bioproc. Tech.* 11: 603–609. doi:10.1007/s11947-017-2033-x

Pellicer, J.A., P. Navarro, and V.M. Gómez-López. 2019. Pulsed light inactivation of polygalacturonase. *Food Chem.* 271: 109–113. doi:10.1016/j.foodchem.2018.07.194

Pereira, R.N., and A.A. Vicente. 2010. Environmental impact of novel thermal and non-thermal technologies in food processing. *Food Res. Int.* 43: 1936–1943. doi:10.1016/j.foodres.2009.09.013

Proctor, A. (ed.). 2011. *Alternatives to Conventional Food Processing*. Cambridge: Royal Society of Chemistry (RSC).

Ramos-Villarroel, A.Y., N. Aron-Maftei, O. Martín-Belloso, and R. Soliva-Fortuny. 2012. The role of pulsed light spectral distribution in the inactivation of *Escherichia coli* and *Listeria innocua* on fresh-cut mushrooms. *Food Control.* 24: 206–213. doi:10.1016/j.foodcont.2011.09.029

Ramos-Villarroel, A.Y., O. Martín-Belloso, and R. Soliva-Fortuny. 2011. Bacterial inactivation and quality changes in fresh-cut avocado treated with intense light pulses. *Eur. Food Res. Tech.* 233: 395–402. doi:10.1007/s00217-011-1533-6

Rattanathanalerk, M., N. Chiewchan, and W. Srichumpoung. 2005. Effect of thermal processing on the quality loss of pineapple juice. *J. Food Eng.* 66: 259–265. doi:10.1016/j.jfoodeng.2004.03.016

Rawson, A., A. Patras, B.K. Tiwari, F. Noci, T. Koutchma, and N. Brunton. 2011. Effect of thermal and non thermal processing technologies on the bioactive content of exotic fruits and their products: Review of recent advances. *Food Res. Int.* 44: 1875–1887. doi:10.1016/j.foodres.2011.02.053

Rice, J.K., and M. Ewell. 2001. Examination of peak power dependence in the UV inactivation of bacterial spores. *Appl. Environ. Microbiol.* 67: 5830–5832. doi:10.1128/AEM.67.12.5830-5832.2001

Rowan, N.J. 2019. Pulsed light as an emerging technology to cause disruption for food and adjacent industries – Quo vadis? *Trends in Food Sci. Tech.* 88: 316–332. doi:10.1016/j.tifs.2019.03.027

Rowan, N.J., S.J. MacGregor, J.G. Anderson, R.A. Fouracre, L. Mc Ilvaney, and O. Farish. 1999. Pulsed-light inactivation of food-related microorganisms. *Appl. Environ. Microbiol.* 65: 1312–1315. doi:10.1128/AEM.65.3.1312-1315.1999

Salcedo, I., J.A. Andrade, J.M. Quiroga, and E. Nebot. 2007. Photoreactivation and dark repair in UV-treated microorganisms: Effect of temperature. *App. Environ. Microbiol.* 73: 1594–1600. doi:10.1128/AEM.0255-06

Sauer, A., and C.I. Moraru. 2009. Inactivation of *Escherichia coli* ATCC 25922 and *Escherichia coli* O157: H7 in apple juice and apple cider, using pulsed light treatment. *J. Food Prot.* 72: 937–944. doi:10.4315/0362-028X-72.5.937

Takeshita, K, et al. 2003. Damage of yeast cells induced by pulsed light irradiation. *Int. J. Food Microbiol.* 85: 151–158. doi:10.1016/S0168-1605(02)00509-3

Taylor-Edmonds, L., T. Lichi, A. Rotstein-Mayer, and H. Mamane. 2015. The impact of dose, irradiance and growth conditions on *Aspergillusniger* (renamed *A. brasiliensis*) spores low-pressure (LP) UV inactivation. *J. Environ. Sci. Heal. A.* 50: 341–347. doi:10.1080/10934529.2015.987519

Uesugi, A.R., S.E. Woodling, and C.I. Moraru. 2007. Inactivation kinetics and factors of variability in the pulsed light treatment of *Listeria innocua* cells. *J. Food Prot.* 70: 2518–2525. doi:10.4315/0362-028X-70.11.2518

Valdivia-Nájar, C.G., O. Martín-Belloso, and R. Soliva-Fortuny. 2018. Kinetics of the changes in the antioxidant potential of fresh-cut tomatoes as affected by pulsed light treatments and storage time. *J. Food Eng.* 237: 56–153. doi:10.1016/j.jfoodeng.2018.05.029

Vollmer, K., S. Chakraborty, P.P. Bhalerao, R. Carle, J. Frank, and C.B. Steingass. 2020. Effect of pulsed light treatment on natural microbiota, enzyme activity, and phytochemical composition of pineapple (*Ananascomosus* [L.] merr.) juice. *Food Bioproc. Tech.* doi:10.1007/s11947-020-02460-7

Wang, B., Y. Zhang, C. Venkitasamy, B. Wu, Z. Pan, and H. Ma. 2017. Effect of pulsed light on activity and structural changes of horseradish peroxidase. *Food Chem.* 234: 20–25. doi:10.1016/j.foodchem.2017.04.59

Wang, T., S.J. MacGregor, J.G. Anderson, and G.A. Woolsey. 2005. Pulsed ultra-violet inactivation spectrum of *Escherichia coli*. *Water Res.* 39: 2921–2925. doi:10.1016/j.watres.2005.04.067

Wekhof, A. 2000. Disinfection with flash lamps. *PDAJ. Pharm. Sci. Tech.* 54: 264–276.

Wekhof, A., F.J. Trompeter, and O. Franken. 2001. *Pulse UV disintegration (PUVD): A new sterilisation mechanism for packaging and broad medical-hospital applications*. In *First International Conference on Ultraviolet Technologies*. Washington, DC: Xenon Corp, pp. 1–15. 2005. Achieving faster cure time with pulsed ultraviolet. Available at http://old.polytec.com/fileadmin/user_uploads/Products/Lichtquellen/Xenon_Blitzlampen/Documents/PH_HL_XEN_Sanitization_and_Sterilization_01.pdf. Accessed on 16 February 2020.

Xu, W., and C. Wu. 2016. The impact of pulsed light on decontamination, quality, and bacterial attachment of fresh raspberries. *Food Microbiol.* 57: 135–153. doi:10.1016/j.fm.2016.02.009

Yi, J.Y., N.-H. Lee, and M.-S. Chung. 2016. Inactivation of bacteria and murine norovirus in untreated groundwater using a pilot-scale continuous-flow intense pulsed light (IPL) system. *LWT - Food Sci. Technol.* 66: 108–113. doi:10.1016/j.lwt.2015.10.027

Zhang, Z.-H., L.-H. Wang, X.-A. Zeng, Z. Han, and C.S. Brennan. 2019. Non-thermal technologies and its current and future application in the food industry: A review. *Int. J. Food Sci. Tech.* 54: 1–13. doi:10.1111/ijfs.13903

Zhu, Y., C. Li, H. Cui, and L. Lin. 2019. Antimicrobial mechanism of pulsed light for the control of Escherichia coli O157: H7 and its application in carrot juice. *Food Control.* 106: 106751. doi:10.1016/j.foodcont.2019.106751

6 Application of Membrane Technology in Food-Processing Industries

Ananya Bardhan, Senthilmurugan Subbiah, and Kaustubha Mohanty
Indian Institute of Technology, Guwahati, Assam, India

CONTENTS

6.1 Introduction 124
6.1.1 Overview of the Membrane Process in Food Processing 125
6.1.1.1 Microfiltration (MF) 125
6.1.1.2 Ultrafiltration (UF) 126
6.1.1.3 Nanofiltration (NF) 126
6.1.1.4 Reverse Osmosis (RO) 126
6.1.1.5 Electrodialysis (ED) 127
6.1.1.6 Pervaporation (PV) 127
6.1.1.7 Membrane Distillation (MD) 127
6.1.1.8 Osmotic Distillation (OD) 127
6.1.1.9 Forward Osmosis (FO) 127
6.1.2 Membrane Materials for Food and Beverage Processing 128
6.2 Membrane Processes Used in Food Processing 129
6.2.1 Dairy Industries 130
6.2.1.1 Raw Milk Concentration and Processing 130
6.2.1.2 Removal and Reduction of Microbial Load (Cold Pasteurization) 133
6.2.1.3 Concentration of Milk 134
6.2.1.4 Fractionation and Separation of Milk Components 134
6.2.1.5 Cheese Processing 134
6.2.1.6 Whey Processing 135
6.2.2 Beverage Industries 137
6.2.2.1 Non-Alcoholic Beverages 137
6.2.2.2 Alcoholic Beverages 142

6.2.3 Treatment of Waste Effluents Generated by Food Industries............. 144
6.2.3.1 Treatment of Waste Effluent Generated by Dairy Processing 145
6.2.3.2 Treatment of Wastewater Generated by Beverage Processing Industries.. 146
6.3 Challenges 147
6.3.1 Concentration Polarization.. 147
6.3.2 Membrane Fouling.. 148
6.3.2.1 Fouling Mechanisms 148
6.3.2.2 Control and Reduction of Fouling...................................... 149
6.4 Recent and Emerging Trends in Membrane Processes in the Food-Processing Industry.. 152
6.5 Conclusion 153
References.. 154

6.1 INTRODUCTION

The membrane separation process involves selective removal of solute, solvent, and suspended solids from solution using a selective semi-permeable membrane. Membrane processes are widely acknowledged in food and beverage processing industries, accounting for 20–30% of total worldwide production (Kotsanopoulos and Arvanitoyannis 2015). In food-processing industries, this process has been extensively used for clarification, concentration, and recovery of value-added components from the waste effluent (Cui and Muralidhara 2010; Galanakis et al. 2016; Youn et al. 2004; Andrade et al. 2014). Conventional clarification and separation processes are reported to involve application of diatomaceous earth, which results in numerous health and safety problems (Ionics Inc 2004). Concentration of liquid food is widely practiced to reduce packaging, storage, transportation, and handling costs. Thermal evaporation is the technique predominantly practiced for preparation of liquid food concentrate. However, higher operating temperature alters the natural color, textural, and nutritional characteristics of the food; higher operating conditions cause degradation and alteration of most aroma compounds present in liquid food (Petrotos and Lazarides 2001). Vacuum evaporation and freeze drying induce less damage to and alteration of essential aromatic compounds. However, these processes are energy intensive, making them ultimately unsuitable for large-scale production. Membrane separation processes are typically operated at ambient temperature, which reduces thermal damage and improves product quality. This chapter provides a brief overview of the membrane separation process in food processing, with a particular focus on dairy and beverage processing. It also considers its application, challenges, and strategies, and discusses recent trends and the different kinds of membrane techniques used in the food and beverage industries.

Different types of membrane modules are extensively used in food processing in view of the fouling potential of aqueous food and beverages. Flat-sheet, tubular, hollow-fiber, and spiral-wound are some of the widely used membrane modules in food processing (Cui and Muralidhara 2010). The merits and demerits of these modules are summarized in Table 6.1. These modules are designed to improve performance in

TABLE 6.1
Merits and Demerits of Different Membrane Modules

Module Configuration	Merits	Demerits
Tubular	(i) Capable of dealing with feed stream with large particle size, due to large internal diameter (ii) Can be efficiently cleaned using chemical and mechanical methods of cleaning	(i) Usually operated in turbulent flow conditions (ii) Large floor space for operation (iii) Low packing density
Hollow fiber	(i) Usually operated in laminar flow region (ii) Highest high packing density (iii) Can be operated in low pressure (iv) Good backwash capacity (v) Easy to clean (vi) Most economical in terms of energy consumption	(i) Susceptible to blockage due to thin fiber when operated in inside-out mode
Plate-and-frame	(i) Easy to dismantle and mechanical cleaning (ii) Lower investment and operating cost	(ii) Lack of sufficient membrane support (iii) Low packing density
Spiral-wound	(i) Medium packing density (ii) Lowest capital cost	(i) Relatively high pressure drop (ii) Requires clean feed with minimum solid content

terms of total energy consumption, flux, filtration area per unit module volume, and hydrodynamic conditions for fouling-resistant operations. In the plate-and-frame type of module, multiple flat-sheet membranes are arranged between porous support layers to create a flow channel for both permeate and feed streams. In tubular membrane modules, at high velocity the feed solution is pumped along the membrane surface with a low pressure drop. This high cross-flow velocity leads to stable permeate flux with low fouling issues. Both spiral-wound and hollow-fiber modules are extensively used in water and desalination applications due to their high packing density.

6.1.1 Overview of the Membrane Process in Food Processing

The following sections discuss commonly used membrane processes and their application in food processing.

6.1.1.1 Microfiltration (MF)

The membrane pore size is 0.1–2 μm and the operating pressure of the MF process is lower than 0.2 MPa. This process is capable of separating molecules ranging from 0.025 to 10 μm. In dairy industries, this is a well-recognized process for cheese whey

clarification and elimination of large suspended solids, fat content, and microbial load from milk (Merin 1986). The application of MF in beverage processing results in a clarified product free from spoilage-causing microorganisms and fibrous pulp (de Cardoso et al. 2012; Vaillant et al. 1999).

6.1.1.2 Ultrafiltration (UF)

Membranes with pore size ranging from 0.01 to 0.1 μm (500–100,000 Da) are grouped under a UF membrane operated at maximum pressure of 1 MPa. The UF process is widely used for clarification (Kawakatsu et al. 1995; Evans et al. 2008; Youn et al. 2004), concentration (Yan et al. 1979), and fractionation (Brans et al. 2004) in dairy and beverage processing. The fractionation of liquid food is achieved by integrating UF membranes of different pore sizes in series to produce high molecular weight components from the reject of large pore size UF membranes and low molecular weight components from the reject of low pore size UF membranes (Brans et al. 2004). Vyas and Tong (2010) reported that the concentration of skim milk using the UF process produces a concentrated milk product with high calcium and protein content. Similarly, Limsawat and Pruksasri (2010) demonstrated the segregation of lactose from milk using a UF (MWCO, 5 kDa) membrane. In beverage processing, UF is used for the clarification of alcoholic and non-alcoholic beverages by excluding protein, polysaccharides, colloids, tannins, spores, and microorganisms such as yeast and molds (Cassano et al. 2007a; Mohammad et al. 2012; Kawakatsu et al. 1995).

6.1.1.3 Nanofiltration (NF)

The NF process is used for concentration, fractionation, and purification of various valuable products from waste streams. For the NF membrane, pore size ranges between 0.5 and 10 nm (100–500 Da). Many researchers have investigated the combination of UF and NF processes for fractionation and purification of liquid food products; it is very important to choose the appropriate pore size for the UF and NF membranes. The NF membrane allows permeation of solvent while retaining organic solute. In food processing, the high organic rejection by the NF membrane has been used for concentration of beverages (Banvolgyi et al. 2006; Vincze and Vatai 2004; Warczok et al. 2004) and recovery of bioactive components from juice (Arriola et al. 2014). The NF process can also be utilized for demineralization of whey protein (Okawa et al. 2015) and separation of lactose from whey protein (Cuartas-Uribe et al. 2009).

6.1.1.4 Reverse Osmosis (RO)

The RO membrane is characterized by pore size <0.5 nm, and operates in a pressure range of 4–10 MPa. The RO membrane can concentrate particles with molar mass below 350 Da. Applications of the RO process include dewatering of fruit juice, tea extract, and alcoholic beverages. It can also be used for dealcoholization of alcoholic beverages and desalination. The concentrated liquid food product is expected to induce high concentration polarization and osmotic pressure. The RO process therefore has to be operated at very high pressure to achieve the desired

concentration level for liquid food application, but due to the limitations of the technology, RO may not be viable when the osmotic pressure of the feed exceeds 100 bar.

6.1.1.5 Electrodialysis (ED)

This process is exclusively used for demineralization and modification of properties of foods (Mikhaylin et al. 2015). Applications include demineralization of milk and whey (Andrés et al. 1995; Houldsworth 2007), and deacidification of fruit juice and whey (Vera et al. 2003; Chen et al. 2016; Greiter et al. 2002). The ED membrane module and the concentrate and dilute solution compartments are fabricated by assembling both anion and cation membranes. This charged membrane facilitates the transport of ions from one compartment to another, enabling the removal of minerals.

6.1.1.6 Pervaporation (PV)

PV can be defined as a membrane process essentially used for separating binary or multi-component mixtures of organic fluids based on boiling point difference and permeation rate through the PV membrane (Wijmans et al. 1991). The PV process is applied for aroma recovery (She and Hwang 2006; Blanco et al. 2014; Matson 1985; Lipnizki 2010), and treatment of waste effluent (Wijmans et al. 1991).

6.1.1.7 Membrane Distillation (MD)

The MD process can be defined as a thermally driven membrane separation process that employs a non-wetting and microporous hydrophobic membrane with liquid feed on one side of the membrane. Vapor is generated in the membrane pores and condensed on the other membrane surface, taking account of the thermal gradient across membrane. The driving force for vapor generation and transportation is the solvent partial pressure gradient across the membrane. MD is used for the dewatering of fruit juice (Sotofta et al. 2012; Alves and Coelhoso 2006b; Kujawski et al. 2013), and dealcoholization of beverages (Purwasasmita et al. 2015).

6.1.1.8 Osmotic Distillation (OD)

This process involves transport of volatile compounds from one liquid (feed) solution to another aqueous (stripping) solution. The vapor pressure gradient of volatile components across the microporous hydrophobic membrane is the driving force. In beverage processing, the OD process is used to prepare concentrate including grape (Bailey et al. 2000), kiwifruit (Cassano et al. 2006), and orange (Quist-Jensen et al. 2016).

6.1.1.9 Forward Osmosis (FO)

The FO process can be defined as a low pressure-operated membrane technology that takes advantage of the osmotic pressure gradient between two aqueous solutions separated by a hydrophilic perm-selective membrane. The FO process can be effectively used to concentrate liquid food without significant membrane fouling, unlike RO and NF (Rastogi 2017). In contrast to pressure-driven processes (such as MF, NF, UF, and RO), the minimal operating pressure in FO allows the

concentration of liquor with high solid content. The practicability of FO for concentrating fruit juice (Popper et al. 1966; Patino 2017; An et al. 2018; Rastogi 2017; Garcia-Castello and McCutcheon 2011), milk, and whey (Chen et al. 2019a) has been successfully studied and demonstrated by various researchers 6.1.1.10. Osmotic Evaporation (OE)

OE is a membrane process that allows concentration of heat-sensitive beverages while preserving their organoleptic and nutritional properties. In the OE process, the aqueous solution to be dewatered is separated from hypertonic solution using a membrane. The hydrophobic nature of the membrane prevents permeation of water through the membrane pores. Similar to the FO process, the driving force is concentration gradient. In food processing, the OE process has been widely investigated for the concentration of thermolabile aqueous solutions such as juice (fruits/vegetables) (Alves and Coelhoso 2006a; Souza et al. 2013; Hongvaleerat et al. 2008), tea (Marques et al. 2016), and coffee (Paiva et al. 2018).

6.1.2 Membrane Materials for Food and Beverage Processing

Coutinho et al. (2009) classified membrane materials into four generations (Table 6.2).

Membranes used in food and beverage processing can be broadly classified as organic (polymeric) and inorganic (ceramic) membranes based on their material. Table 6.3 summarizes the advantages and limitations of polymeric and ceramic membrane materials with examples.

Due to their superior thermal, mechanical, and chemical properties, ceramic membranes have been recognized as an effective alternative to polymeric membranes. These inorganic membranes can be manufactured with tubular, flat, and

TABLE 6.2
Classification of Membrane Material Based on Technical Development

Generation	Example	Description
First	Natural polymer, cellulose acetate	(i) Susceptible to microorganisms and disinfectants (ii) Sensitive to pH (3–8) (iii) Can withstand maximum 50 °C temperature
Second	Synthetic polymer, polysulfone, or polyolefin derivatives	(i) More resistant towards hydrolysis, pH and temperature (ii) Lower resistance to mechanical compacting
Third	Ceramic material based on alumina oxide or zirconium accumulated on a graphite surface	(i) Greater mechanical strength and chemically inert (ii) Can be operated at high pressures (iii) Can withstand wide pH range (0–14) and temperature range (>400 °C). (iv) Long operation life (v) Expensive compared to above two generations of membranes
Fourth	Hybrid process, integration of two (or more) membrane processes	(i) Reduced energy and processing (i.e., operational) costs (ii) Improves product quality

TABLE 6.3
Comparison of Polymer and Ceramic Membranes

	Polymer Membrane	Ceramic Membrane
Advantage	(i) Relatively less expensive (ii) Less complicated manufacturing process (iii) Accessible in a wide range of pore sizes	(i) High mechanical strength, with high thermal and chemical stability (ii) Resistant to harsh operating conditions such as pH, temperature, and pressure (iii) Extended lifetime
Limitation	(i) Low thermal, mechanical, and chemical stability (ii) Low chlorine tolerance	(i) Brittle in nature (ii) Mostly available in UF and MF ranges only (iii) Much more expensive than polymeric membranes
Examples	(i) Cellulose acetate (CA) (ii) Polyamide (PA) (iii) Polysulfone (PS) (iv) Polyethersulfone (PES) (v) Polyvinylidene fluoride (PVDF) (vi) Polypropylene (PP).	(i) γ-alumina/α-alumina (ii) Borosilicate glass (iii) Pyrolyzed carbon (iv) Zirconia/stainless steel, or zirconia/carbon

monolithic (or multichannel) geometry (Mancinelli and Hallé 2015). Commercial applications of ceramic membranes involve harsh operating conditions including filtration of hot liquids. Typical applications of UF and MF processes using ceramic membranes can be found in the pharmaceutical, food, paper, paint, and water industries. Table 6.4 summarizes the major application of ceramic membranes in food and beverage processing.

6.2 MEMBRANE PROCESSES USED IN FOOD PROCESSING

Membrane techniques play an essential role in food and beverage processing, and the use of membranes is expected to increase at a compound annual growth rate of 6.7% between 2015 and 2020 (Abdel-fatah 2018) because of the following multiple advantages (Table 6.5):

(i) Similar to desalination applications, the membrane process for food processing is expected to be more energy efficient than other conventional thermal processes adopted in food industry.
(ii) The modular design concept of the membrane process enables seamless integration with other separation processes and allows easy scale-up from lab scale to industrial production capacity.
(iii) The properties of processed liquid can be adjusted without loss to the environment by adopting appropriate membrane and process-operating conditions.
(iv) The membrane process uses fewer additives than conventional food-processing technologies, and these are sometimes not required at all.

TABLE 6.4
Applications of Ceramic Membrane in Food Processing

Application	Food/ Beverages	Pore Size	Membrane Module	Membrane Process	References
Clarification (removal of pectin, protein, tannin, and suspended particles) and concentration	Apple	0.2 μm	Tubular (ZrO_2)	UF	Bruijn et al. (2003)
	Orange and lemon	0.5–0.8 μm	Tubular		Capannelli et al. (1994)
	Mosambi	0.2 μm	Tubular	MF	Nandi et al. (2009)
	Sugarcane	20 nm	Tubular (Al_2O_3/ZrO_2)		Shi et al. (2019)
	Carrot	0.2 μm	Multi-tubular (ZrO_2–TiO_2)	MF	Ennouri et al. (2015)
	Blackcurrant	0.45 μm	Tubular	MF	Ennouri et al. (2015)
To reduce fat content and microbial (bacteria and spore) load	Milk	1.4 μm	Tubular	MF	Pafylias et al. (1996); García and Rodríguez (2015)
Fractionation of protein (α-lactalbumin, β-lactoglobulin, BSA, IgG and lactoferrin)	Whey		Tubular (ZrO_2–TiO_2)	UF	Almécija et al. (2007)
Fractionation of protein (casein and whey protein) to produce infant milk formula	Skim milk	0.1 μm		MF	McCarthy et al. (2017); Karasu et al. (2010)
Removal of crude protein, suspended solids and turbidity	Raw rice wine	200/500/200 nm	Multi-tubular	MF	M. Li et al. (2010)

6.2.1 Dairy Industries

Dairy industries account for the largest share of membrane applications in food processing (Figure 6.1). Membrane processing has been successfully incorporated in various stages of manufacturing, from raw milk processing and concentration to waste-effluent treatment. Milk is a complex colloidal system ranging from small ions to large fat globules. The average composition of bovine milk is shown in Table 6.6.

The pore size and molecular weight cut-off (MWCO) of different membrane techniques plays an essential role in determining their application in dairy processing (Table 6.7).

6.2.1.1 Raw Milk Concentration and Processing

The idea of concentrating raw milk using the UF process was developed in the early 1970s (Yan et al. 1979) but was first incorporated on a commercial level in 1996 in

TABLE 6.5
Membrane Processes and their Application in Food Industries (Daufin et al., 2001)

Membrane Process	Pore Size (μm)	Application	Product
Microfiltration	(0.1–10)	Clarification, cold sterilization, and fractionation	Milk, beer, wine, juice, coconut water
Ultrafiltration	(0.01–0.1)	Clarification, cold sterilization, and fractionation	Milk, egg, juice, coconut water, beer, wine, tea, and coffee
Nanofiltration	<0.001	Product recovery, fractionation, demineralization, and concentration	Milk, whey, waste effluent, juice
Reverse osmosis		Concentration/dehydration	Whey protein concentrate, liquid food, and beverage concentrate
Pervaporation		Aroma recovery and dealcoholization	Tea, coffee, wine, and juice
Electrodialysis		Desalting or demineralization, and product recovery	Juice, milk, and whey
Membrane distillation		Concentration/dehydration	Tea, coffee, juice
Forward osmosis		Desalting and concentration/dehydration	Juice, milk, and whey

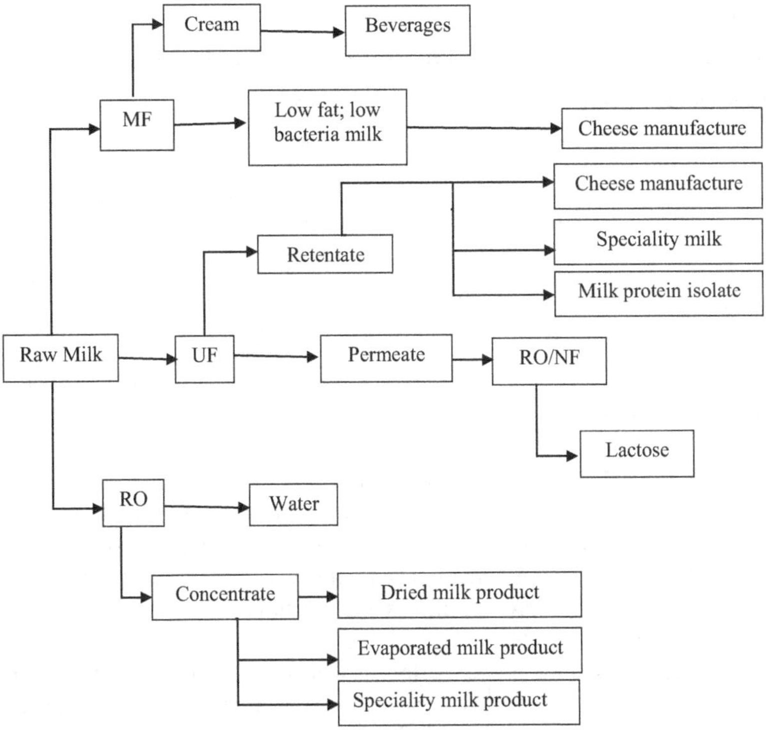

FIGURE 6.1 Diagrammatic representation of membrane processing for raw milk

TABLE 6.6
Composition of Cow's Milk

Component	Concentration (g L^{-1})	Size Range
Water	87.1	–
Fat globules	4.0	0.1–15 μm, average 3.4 μm
Dry matter	12.9	–
Serum proteins	0.7	3–6 nm
α-Lactalbumin	0.12	14 kDa
β-Lactoglobulin	0.32	18 kDa
Bovine serum albumin	0.04	66 kDa
Lactoferrin	0.01	86 kDa
Proteose-pepton	0.08	4–40 kDa
Immunoglobulins	0.08	150–900 kDa
Transferrin	0.01	76 kDa
Others	0.04	–
Casein	2.6	20–300 nm, average 110 nm
Lactose	4.6	0.35 kDa
Organic acids	0.17	–
Mineral	0.7	–
Other	0.15	–

TABLE 6.7
Membrane Application in Dairy Processing (Chen et al., 2019b)

Membrane Process	Pore Size	MWCO	Application
MF	0.2–2 μm	>200 kDa	• Production of skim milk and cheese • Clarification of dextrose • Removal of bacteria and bacterial spores
UF	1–500 μm	1–200 kDa	• Standardization of milk • Reduction of lactose and calcium level • Protein, milk, and whey concentration
NF	0.5–2 nm	300–1,000 Da	• Demineralization of whey • Production of lactose-free milk • Dewatering of milk
RO	Non-porous	100 Da	• Recovery of total solids and water

New Mexico, USA. The UF process allows the separation of milk components based on their molecular size. The membrane allows permeation of small units such as lactose, water, and soluble minerals through the membrane while restraining significant molecular-size milk components. The retentate stream consists mainly of whey and casein protein. The UF process allows manufacturers to produce lower-carbohydrate/high-protein dairy products.

Most importantly, the UF process offers a clean and neutral flavor. Along with the UF process, the RO and NF processes have also been explored for the concentration of raw milk. In the dairy industry, the RO process is usually used for dewatering and recovery of milk solids and whey.

6.2.1.2 Removal and Reduction of Microbial Load (Cold Pasteurization)

In dairy processing, heat treatment is a commonly employed technique for the destruction and inactivation of pathogen- and spoilage-causing microorganisms. Conventional techniques for reduction of microbial loading are pasteurization, ultra-high-temperature (UHT) treatment, and sterilization. Heat processes result in thermally induced chemical and physicochemical alterations, protein denaturation, damage to creaming properties, and altered sensory characteristics such as oxidized and cooked flavor. Pathogenic (*Listeria*, *Brucella*, *Mycobacterium*, or *Salmonella* spp.) and non-pathogenic microbial flora present in milk necessarily need to be removed before further processing (especially in cheese processing). The MF process is an on-thermal alternative for elimination of bacteria and bacterial spores from raw milk before further processing. The application of a membrane to remove bacteria and spores was first developed by Tetra Pak and marketed by Alfa Laval in 1986 and the same as "Tetra Al cross_Bactocatch"

Figure 6.2 illustrates the Bactocatch process for preparation of bacteria-free milk. Using a cream separator, the cream is first removed from raw milk, followed by microfiltration of the liquid stream containing skim milk using a ceramic (pore size, 1.4 μm) membrane at constant transmembrane pressure (TMP). The retentate stream consists of almost all the bacteria and spores, while the concentration of bacterial species in the permeate stream is less than 0.5% of its initial value in the milk. The retentate stream is then combined with a standardized quantity of cream and is further exposed to a traditional high-temperature thermal treatment at 130°C for 4 s. The resultant stream is reintroduced to the UF permeate and then pasteurized. Less than 10% is heat treated, resulting in milk with improved sensory quality.

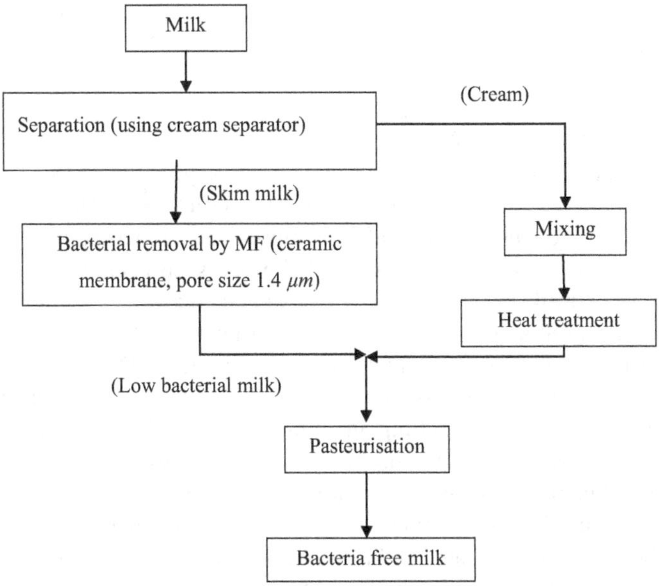

FIGURE 6.2 Bactocatch process patented by Alfa Laval (Sweden)

6.2.1.3 Concentration of Milk

In this process, milk is conventionally concentrated using evaporative techniques and membrane processes such as RO. Concentrated milk is widely used for ice-cream processing since all the solids are retained in concentrate while 70% of water is rejected. Membrane processes such as MF and UF can potentially concentrate milk with 50–58% protein.

6.2.1.4 Fractionation and Separation of Milk Components

Milk is a complex colloidal system consisting of species ranging from small ions to large fat globules. Membranes can separate milk components such as casein, lactose, saccharides, and fat globules. Apart from its nutritional properties, milk also contains natural bioactive compounds (such as immunoglobulins, antibacterial peptides, antimicrobial proteins, oligosaccharides, and lipids). The proper separation of such components provides scope for the development of new components with desired functional properties (such as edible coatings and bioactive peptides). The lactose present in milk causes gastrointestinal problems. The removal of lactose from milk can also improve solubility, storage stability, and functionality. A membrane separation process such as UF allows the passage of lactose while retaining all fats (1–10 µm diameter) and milk proteins (casein protein and whey protein, 3–6 nm). Limsawat and Pruksasri (2010) claimed partial segregation of lactose from milk using a cross-flow hollow-fiber UF (MWCO 5,000 Da) system with approximately 13% and 100% rejection of lactose and protein respectively.

The milk retentate from the UF process can be regarded as concentrated milk suitable for yogurt and cheese processing. Dairy protein consists of certain valuable nutritional products that can be further recovered using membrane processes. These recovered milk proteins can be used as food additives, nutraceuticals, and pharmaceuticals.

6.2.1.5 Cheese Processing

The traditional method of cheese making involves coagulation of caseins of pasteurized milk using rennet enzymes and lactic acid bacteria (LAB). The coagulated curd is drained to remove whey, then molded, compressed, submerged in brine solution, and ripened under suitable conditions. The UF process has been reported to improve concentration of milk by a factor of 1.2–2 and increase its casein:protein ratio, resulting in improved and better-quality cheese. Maubois (1997) patented a process of cheese production from UF retentate and named it the MMV method (after inventors Maubois, Mocquot, and Vassal). Using the MMV method, milk can be concentrated to a factor of 5–7, resulting in high-quality curd. UF-concentrated milk reduces the processing cost of semi-hard and hard cheese and offers advantages over traditional methods including increased total solids, reduced rennet and starter culture (Mistry and Maubois 2017), increased nutritional quality, and reduced volume of whey. MF-concentrated milk can be exclusively used for preparation of cheese due to its optimized microbial quality, casein content, and amount of major milk product.

6.2.1.6 Whey Processing

Whey is an essential by-product of cheese making consisting of 5–6%, 0.8–1%, and 0.06% lactose, protein, and fat respectively (de Souza et al. 2010). With its high organic and nutrient load, whey should not be directly discharged into the environment without proper treatment.

Fractionating essential whey constituent can help minimize organic load before discharge into the environment, helping to reduce environmental impact. It also provides the dairy industry with an economic incentive in the form of the possible sale of these processed/recovered products (Ganju and Gogate 2017). Whey consists of a mixture of globular proteins such as of β-lactoglobulin, α-lactalbumin, vlactoferrin, bovine serum albumin, glycomacropeptides, immunoglobulins (Ig), and lactoperoxidase. These individual whey components have promising potential as components of health-promoting functional foods. The β-lactoglobulin and α-lactalbumin can be employed as gelling agents and for formulation of peptides with physiological properties (Lipnizki 2010).

Figure 6.3 provides an overview of membrane applications in whey processing.

6.2.1.6.1 Whey Protein Concentrate (WPC)

Whey processing was the first successful commercial application of membranes in dairy processing. For pre-concentration, the use of the RO process instead of a vacuum evaporator saved a large amount of energy. The membrane separation process enables recovery of valuable whey constituents without a substantial energy cost. Membrane processes are relatively simple, better for the environment, and

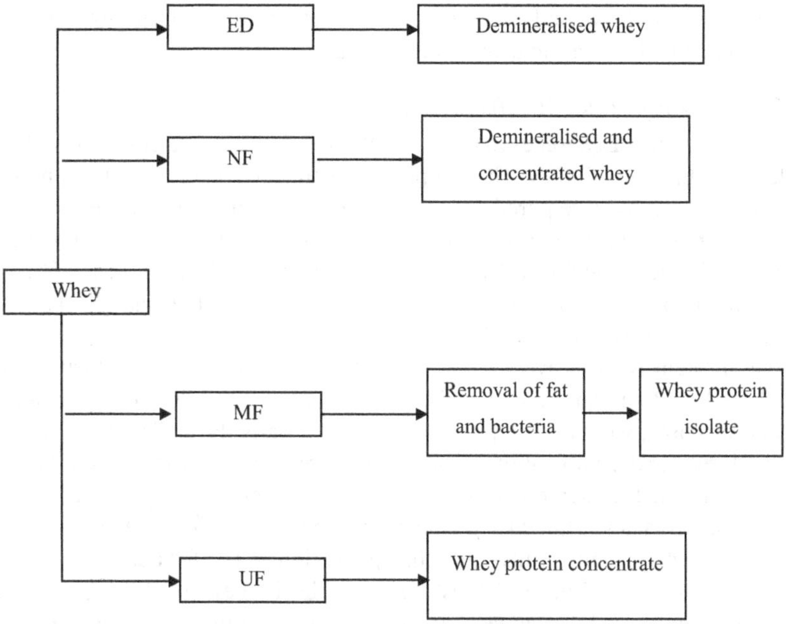

FIGURE 6.3 Application of membranes in whey processing

most importantly, the separation can take place without phase change. Membrane separation processes such as MF, UF, RO, and NF have been extensively used to recover, concentrate, and demineralize whey, and to fractionate essential whey protein (Lipnizki 2010; Brans et al. 2004; Schroën et al. 2010). UF and RO processes have been widely investigated for concentration and fractionation of essential whey protein. NF can be an effective alternative for separation and concentration of minerals. Aydiner et al. (2012) investigated the effectiveness of a lab-scale FO–RO integrated membrane process for concentration and recovery of whey. Whey proteins are rich in essential amino acids and are therefore an essential additive in food processing.

Compared with other conventional protein sources (such as casein, eggs, fish, meat, and soy), whey protein has higher biological value. Individual whey proteins have their own specific nutritional, functional, and biological properties. For example, α-lactalbumin is rich in tryptophan and is widely used in infant formula. β-lactoglobulin can be potentially used for gelling, foaming, and emulsification. Similarly, bovine serum albumin (BSA) and bovine immunoglobin are used in infant formula and have therapeutic applications as well. The proper fractionation of whey protein enables better utilization of components than whey protein concentrates as a whole. The UF and MF membrane processes are the most widely used for protein fractionation, along with other techniques such as precipitation and chromatography. Since the molecular sizes of α-lactalbumin and β-lactoglobulin are the same, fractionation by single-step membrane filtration is not possible. Therefore, the protein fraction needs to be manipulated before processing to enhance yield in terms of purity. At a temperature of 55°C and pH 3.8, lactalbumin loses its stability and calcium within 30 min. Under the same conditions, as calcium dissolves in solution the β-lactoglobulin unfolds and precipitates. The precipitated β-lactoglobulin can be recovered from UF permeate using diafiltration.

6.2.1.6.2 Whey Protein Demineralization

In dairy processing, ED, ion exchange, and NF membrane processes are widely used for demineralization of whey. The NF process is the most economical when moderate demineralization is required. The NF membrane can potentially remove lactic acid from whey, resulting in downstream separation of lactose and whey protein. ED can also be implemented for a 60% reduction of minerals. The efficiency of the ED process can be improved by pre-concentrating (20% dry matter) using either RO or evaporation (P. Kumar et al. 2013).

Whey protein is extensively used in diabetic foods, infant foods, health drinks, animal feed, and similar value-added protein supplements. The mineral content of whey is very high, which gives it a very undesirable salty flavor. The NF process has been explored for deacidification, desalting, and demineralization of whey. Owing to the selectivity of the membrane, nearly all monovalent ions, organic acids, and a trace amount of lactose pass through the membrane. ED and ion exchange are generally used for whey demineralization. The application of ED for demineralization of whey was first reported in the 1970s. Demineralized whey powder is produced by spray-drying followed by selective extraction of the minerals from liquid whey. Since ED is both energy and capital intensive, its application for demineralization is only

limited when we need a lesser amount of minerals to be removed. Compared to ED, the NF process is found to be much more economical and environmentally feasible. The NF (MWCO, 200–1,000 Da) membrane separation process allows selective passage of monovalent salt, water, and low molecular weight organic molecules (K. Pan et al. 2011). The demineralization of whey using the NF process is advantageous for the manufacturing of infant formula as it allows retention of calcium phosphate while rejecting monovalent ions (Na^+, Cl^- and K^+). The permeate water consists of sodium, calcium, and non-protein lactose. Using the RO process, this permeate water can be further cleaned and used to produce clean water while rejecting monovalent salts, which can be further treated and used for the production of health beverages and sports drinks. The NF process is more economical than ED, since it can simultaneously concentrate and demineralize the whey.

6.2.2 Beverage Industries

Beverages can be defined as any potable liquid consumed by drinking to fulfill a craving, or provide nourishment, energy, and hydration. In food processing, beverages can be broadly classified as alcoholic or non-alcoholic beverages.

6.2.2.1 Non-Alcoholic Beverages

Any beverages that are not fermented or distilled can be defined as non-alcoholic. Water, fruit juice, tea/coffee, and sweetened carbonated beverages are a few of the most widely consumed non-alcoholic beverages.

6.2.2.1.1 Fruit and Vegetable Juice

Fruit and vegetables are rich sources of vitamins, protein and antioxidants which provide essential nutrients and energy necessary for healthy living. Freshly squeezed fruit or vegetable juices have high water and low solid content, making them highly susceptible to microbial degradation. Fruits and vegetables cannot be stored for a long duration due to their high moisture content; instead, they are processed and preserved to make them available throughout the year. The major challenge associated with any beverage processing is retention of original aroma and flavor. The membrane separation process is preferred to conventional processes in the beverage industry. While thermal treatment enables the deactivation of pathogenic and spoilage-causing microorganisms, the high temperature also results in degradation of flavor, aroma, and odor, and eventually in textural and nutritional loss. Low-temperature membrane separation allows retention of the original nutritional and organoleptic properties. The membranes can effectively retain large molecules such as lipids, proteins, colloids, and microorganisms, while allowing molecules such as water, salt, minerals, vitamins, and sugar to permeate through the membrane.

Clarification, depectinization, and stabilization are the unit operations for which membrane processes such as UF, MF, NF, and RO are most extensively used in the juice-processing industry. The MF membrane process allows removal of spoilage microorganisms, lignin, starch, and other high molecular weight components. MF enables separation of juice with slight fibrous matter free from spoilage

microorganisms. Due to high water activity, freshly produced fruit juices are highly susceptible to microbial degradation. Membrane techniques are broadly used for processing fruit (orange, grape and, apple, etc.) and vegetable (lemongrass, tomato, carrot, etc.) juices. The UF process allows clarification of fruit juice by aiding separation of fruit pulp and spoilage-causing microorganisms while retaining polyphenols, dissolved salts, essential vitamins, and minerals. Since the 1970s, MF/UF processing has proved an effective alternative for fining and filtering that balances economics and quality. The UF membrane process has been used for clarification of blood orange (Cassano et al. 2007b) and kiwifruit (Cassano et al. 2003a) juice. The UF process can effectively preserve total antioxidant activity and bioactive compounds such as anthocyanin, hydroxycinnamic acids, ascorbic acid, and flavanones. Applicability of the RO process for fruit juice concentrate such as blackcurrant (Pap et al. 2009), grape (Gurak et al. 2010), and apple (Gunathilake et al. 2014) has been evaluated on the basis of total antioxidant capacity and bioactive compounds. Gunathilake et al. (2014) reported that the RO process enhances bioactive concentration of blueberry, cranberry, and apple juices. However, the high pressure involved in the RO process can potentially damage the quality of fruit juice. The NF process facilitates removal of water while retaining essential components such as antioxidants, vitamins, minerals, polyphenols, and flavonoids, and was found to be an efficient and economical alternative to conventional evaporators. Arriola et al. (2014) studied the applicability of the NF process to the concentration of bioactive components present in watermelon juice. Table 6.8 summarizes the separation range (pore size, operating pressure, and molecular weight) of the pressure-driven membrane separation process.

The species retained by the conventional pressure-driven membrane process in fruit juice processing are summarized in Table 6.8. Manufacturers can therefore select from the tabulated membrane processes depending on the species to be retained.

TABLE 6.8
Separation Range of Pressure-Driven Membrane Process for Fruit Juice

Membrane Process	Pore Size	Operating Pressure (bar)	Molecular Weight of Solute	Retained Species
MF	>0.1 μm	0.1–3	>1,000,000	Bacteria, suspended solids, pulp, fiber, starch
UF	0.1 μm–2 nm	1–10	4,000–10,000	Proteins, viruses, molds, enzymes, tannins, polysaccharides
NF	<2 nm	10–50	100–500	Sugar, antibiotics, amino acids, organic acids, bioactive, phenolic, and aromatic compounds
RO	–	10–100	<800	Monovalent salts, metal ions, flavor compounds

Source: Conidi et al. (2020)

Pervaporation (PV) involves separation of feed mixture through a perm-selective membrane using partial vaporization. In fruit juice processing, PV is widely used for aroma recovery and restoration. Cassano et al. (2006) used PV to recover particular aromatic compounds from UF-filtered kiwifruit juice; later, the recovered aromatic components (such as methyl butanoate, ethyl butanoate, 1-hexen-1-ol, (E)-2-hexen-1-ol, 1-hexane) were added back to the concentrated juice. Similarly, Aroujalian and Raisi (2007) recovered volatile aromatic compounds (such as ethyl acetate, ethyl butyrate, hexanal, α-terpineol, limonene, and linalool) from orange juice using PV. Electrodialysis (ED) is used in fruit juice processing for deacidification of fruit juices that are too sour to be tolerable. The acidity of fruit juice can be weakened by precipitating acidic compounds using salts (such as calcium salts). However, the addition of chemicals to fruit juice can be avoided by the application of ED.

After clarification, deacidification, and aromatic recovery, the next essential process in beverage processing is concentration (i.e., dewatering). The advantages of concentrating fruit juices include reduction in weight and volume, as well as storage, packaging, and transportation costs. Conventionally used thermal processes involve large energy consumption, reduced nutritional and bioactive compounds, along with altered color, flavor and aroma. RO, NF, MD, and FO are among the widely investigated membrane processes for production of concentrated juices at mild operating conditions, reducing energy consumption and preserving aroma, bioactive compounds, and nutritional value.

Membrane distillation (MD) is a thermally driven membrane separation technique. The process is driven by the vapor pressure gradient between the porous hydrophobic membrane surface. In juice processing the commonly used configurations of the MD process are vacuum membrane distillation (VMD), direct contact membrane distillation (DCMD), and osmotic membrane distillation (OMD). In the DCMD configuration, the aqueous solution on both sides of the membrane is in direct contact with the hydrophobic membrane surface. The feasibility of DCMD for the dewatering of orange (Calabro et al. 1994), apple (Lukanin et al. 2003), and grape juice (Bailey et al. 2000) has been examined and reported by various researchers at lab scale. In the OMD technique, using a hydrophobic porous membrane the aqueous solution is allowed to concentrate at constant temperature under atmospheric pressure. The vapor pressure difference between the two aqueous phases is the major driving force. Kujawski et al. (2013) studied the feasibility of osmotic membrane distillation for preparation of red grape juice concentrate. The OMD process successfully preserved physicochemical properties of the initial juice in terms of both total polyphenol content and total antioxidant content.

Direct osmosis (DO), or FO uses a dense hydrophilic semi-permeable membrane capable of separating two liquid solutions (i.e., feed and draw solutions). In this process, the major driving force across the membrane is the concentration gradient between the solutions. In fruit juice processing, the FO process can be executed at ambient temperature and pressure, resulting in higher retention of thermolabile components. The FO process has been successfully investigated for a wide range of juices such as sugarcane (Shalini and Nayak 2016), pineapple, orange (Garcia-Castello and McCutcheon 2011; Versari et al. 2011), grape, tomato (Petrotos et al. 1998), and raspberry (Wrolstad et al. 1993). Similar to FO, the osmotic evaporation (OE)

process allows dewatering by a driving force generated under atmospheric temperature and pressure conditions. The OE process has been extensively investigated for concentration of redgrape (Kujawski et al. 2013), apple (Lukanin et al. 2003), orange (Alves and Coelhoso 2006a), and pineapple (Hongvaleerat et al. 2008) juice. The operating conditions are the same for FO/DO and OE, the only difference being the type of membrane used. In the OE process, hydrophobic membranes are used, whereas in DO/FO the membrane is hydrophilic.

6.2.2.1.2 Sugarcane Processing

Sugarcane processing is one of the most energy-exhaustive processes in food processing. It involves crushing/milling, clarification, concentration, crystallization, and refining. After crushing, the extracted sugarcane juice contains certain impurities such as reducing sugars, organic/non-organic acid, gums, amino acids, minerals, starches, coloring matter, proteins, waxes, and other suspended matter that causes undesirable color and turbidity in the juice (Gil et al. 1994; Abdel-Rahman 2015; Verma et al. 1996; Jegatheesan et al. 2012a). The membrane process shows significant potential for producing superior-quality filtered sugarcane juice with improved productivity and organoleptic properties by maintaining the phytonutrient composition intact (Jain and De 2019). Verma et al. (1996) investigated the feasibility of UF for clarification of sugarcane juice using an organic membrane (MWCO~10,000–30,000) and a hollow-fiber UF membrane (MWCO ~20,000) respectively. MF and UF (Jegatheesan et al. 2012b; Verma et al. 1996; Hamachi et al. 2003; Bhattacharya et al. 2001; Shi et al. 2019) are the most widely investigated techniques for clarification of sugarcane juice. In a pilot-scale demonstration of UF of sugarcane juice (at 91–97°C), Ghosh and Balakrishnan (2003) used a polymeric spiral-wound membrane which exhibited low (7 $Lm^{-2} h^{-1}$) permeate flux. Li et al.'s (2017) pilot-scale demonstration of a UF set-up for clarification of sugarcane juice to produce raw sugar used a tubular ceramic membrane (pore size, 0.05 μm). Compared to a polymeric membrane, the ceramic membrane gave superior flux (119.13–142.43 $Lm^{-2} h^{-1}$) with 99.96% and 10.42% reduction in turbidity and color, respectively. Shi et al. (2019) identified poor permeate flux and membrane fouling as two major issues hindering the commercial application of the membrane filtration technique in sugar processing. Heating of juice is frequently used in juice clarification. However, these authors claimed that high operating temperature increases coagulated protein and polysaccharides, eventually causing deposition of crystalline and microcrystalline compounds on the membrane surface and pore channels. They proposed a cake filtration and complete blocking fouling model to accurately predict the performance of filtration process using a tubular ceramic (Al_2O_3 and ZrO_2, pore size 20 nm) membrane module.

After clarification of the extracted sugarcane juice, the next challenge in sugar processing is concentration. Traditional thermal processing results in high energy consumption, altered sensory and nutritive properties, and dark color. RO (Madaeni et al. 2004), MD (Nene et al. 2002), and FO (Shalini and Nayak 2016) are among the widely investigated membrane separation processes for preparation of clarified sugarcane juice concentrate (Figure 6.4). FO has emerged as a promising technology for preparation of concentrated sugarcane juice. Using a commercially available FO

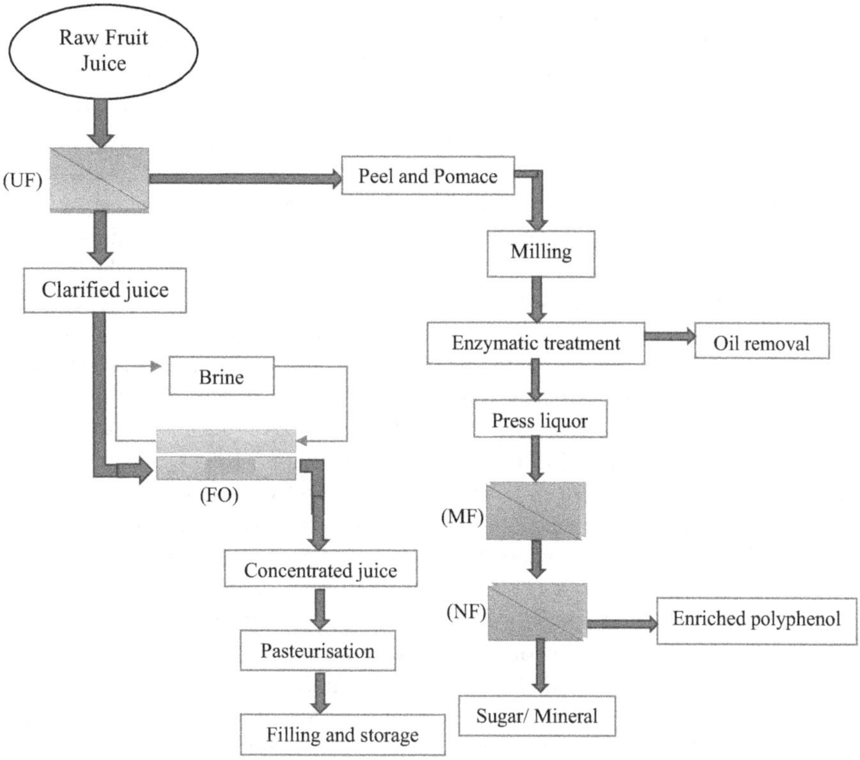

FIGURE 6.4 Conventional membrane separation process for juice processing (Bhattacharjee et al., 2017)

membrane (Osmotek Inc., USA), Shalini and Nayak (2016) concentrated sugarcane juice from 250 to 99.6 mL in 18 hours using 6 M NaCl as osmotic agent. Using a self-fabricated thin-film polyamide (active area, 0.0057 m^2) FO membrane, Mondal et al. (2015) effectively dewatered freshly extracted clarified sugarcane juice with waste salt bittern as an osmotic agent. According to these authors, sucrose concentration was improved from 10.5 to 40.6% (w/v) over 4 h, with average flux of 13 Lm^{-2} h^{-1} and insignificant back diffusion of inorganic constituent of draw solution.

6.2.2.1.3 Tea and Coffee

Consumption of tea and coffee is mostly related to their refreshing flavor and nutraceutical properties. The volatile components present in tea leaves and coffee beans provide tea and coffee with their characteristic flavor and aroma. During processing, any quantitative and qualitative change in these compounds exerts a negative impact on the desirability of the product.

Tea: Catechins are a group of polyphenolic compounds, the consumption of which may prevent diseases such as diabetes, obesity, and cancer. Catechins naturally present in tea extract include epicatechin (EC), epicatechin gallate (ECG), epigallocatechin (EGC), gallocatechin gallate (GCG), epigallocatechin gallate (EGCG), and

catechin (C). Due to their antioxidant properties, catechins are used in the food-processing and pharmaceutical industries. In tea processing, membranes are used for clarification, fractionation, purification, and concentration of tea extract. The clarification of tea extract using the UF (MWCO 30–100 kDa) and MF (pore size 0.5–10 μm) process results in 30–50% removal of pectin. The partial removal of pectin results in reduced cloudiness and sedimentation in the final product (Kawakatsu et al. 1995). Todisco et al. (2002) investigated black tea extract clarification using a 40 kDa ceramic tubular membrane to eliminate a protein that interacts with soluble tannins and precipitates during storage. The major problem in tea processing is the formation of a colored precipitation of hot/cold tea extract during cooling, which is called tea cream. Tea cream gives a hazy and turbid appearance to bottled ready-to-drink tea. The polyphenols react mostly with proteins to form tea cream, and similar polyphenol–protein interactions have been reported in beverages such as beer, fruit juice, and wine. Evans and Bird (2006) examined clarified black tea extract using two flat-sheet polymeric (fluoropolymer, FP and regenerated cellulose, RC) UF membrane processes. The FP (MWCO, 10/30/100 Da) and RC (MWCO, 10/30/100 Da) polymeric membranes effectively rejected 21% and 27% of solids respectively, with significant cream and haze reduction. Ramarethinam et al. (2006) studied clarification and concentration of polyphenols of tea using pre-filter, MF, UF, and RO membrane processes. They concluded the integration of the UF and RO process is suitable for clarification and concentration of green tea extract.

Coffee: The aroma compounds in coffee are complex, consisting of more than 800 volatile compounds responsible for the characteristic flavor and aroma of coffee. Industrially, coffee is produced by roasting and grinding the coffee beans followed by extraction and concentration of coffee extract. The NF process allows complete rejection of organic compounds. Vincze and Vatai (2004) investigated the applicability of the NF instead of the RO process for the preparation of concentrated coffee extract. Using a flat-sheet NF45 membrane, the authors concentrated the coffee extract from 14 to 45 gL^{-1}. Pan et al. (2013) screened six commercially available NF membranes using pure water and coffee extract. The authors concluded the NF2 (0.57 nm, pore size) membrane could concentrate coffee extract up to a particular concentration and could partially replace the conventionally used evaporation process. Weschenfelder et al. (2015) investigated the application of the pervaporation process for the restoration of aromatic compounds from the soluble coffee extract obtained by industrial extraction.

6.2.2.2 Alcoholic Beverages

Alcoholic beverages are any beverages that contain ethanol. Indian standards classify any beverage with ethanol content 0.5–42.8% (*v/v*) as alcoholic, while under international standards the content may vary from 0.5 to 95% (*v/v*). Beer and wine are two important alcoholic beverages produced using fermentation. Beer (alcohol content around 5%) is produced by fermenting malt and hops, and similarly wine (alcohol content 15%) is produced by fermenting grape juice using yeast (*Saccharomyces cerevisiae*).

In alcoholic beverage processing, the application of membrane technology involves clarification/filtration and dealcoholization.

6.2.2.2.1 *Clarification*

Clarification is an essential step in beverage processing. Effective clarification involves rejection of microorganisms, proteins, lipids, and colloids while allowing permeation of small solutes such as salts, vitamins, and sugars with water (Jain and De 2019).

Beer: Beer can be defined as a beverage consisting of carbon dioxide and alcohol. It has significant vitamin B content (folate, niacin, riboflavin, thiamine, vitamin B6 and B12) that can potentially weaken the risk of beriberi and neural disease. Beer contains significant phenolic compounds such as benzoic acid, coumarins, catechins, cinnamic acid derivatives, di-, tri-, and oligo-meric proanthocyanidins, polyphenols, and flavonoids (Sohrabvandi et al. 2010). In beer processing, perfect balance needs to be achieved in filtration by allowing the passage of the dissolved macromolecules that give beer its characteristic flavor and functional properties, while removing particles such as yeast cells, protein-polyphenol complexes, proteins, protein tannins, hop resins, protein polyphenol, and β-glucan. These colloidal particles need to be removed to avoid turbidity of the final bottled product. Efficient elimination of spoilage-causing microorganisms such as bacteria and yeast prolongs shelf life of the final product. Sterile filtration using cross-filtration microfiltration (CFMF) (Fillaudeau et al. 2007) is an interesting alternative to conventional thermal pasteurization. It eliminates spoilage-causing microorganisms while ensuring stability of the final product in terms of aroma, color, flavor, and biological and foam stability.

Wine: Wine contains antioxidants such as hydroxycinnamic acids, hydroxytyrosol, and tyrosol, and nutritive bioactive compounds such as catechin, caffeic acid, resveratrol, hydroxytyrosol, and melatonin (Lugasi and Hóvári 2003; Cao and Zhu 2006; Norata et al. 2007). Consumption of moderate quantities of wine is associated with numerous health benefits (Table 6.9). More significant health benefits can be achieved by lowering or completely removing the alcohol content to a safe level while maintaining the original taste, aroma, and flavor of the compound.

6.2.2.2.2 *Dealcoholization*

Several techniques are available in in the post- or pre-production phases of be vouterage processing. The dealcoholization step involves the removal of only the alcohol content with altering the aroma, flavor, and nutritional content. Traditional techniques for alcohol separation in post-production are distillation and evaporation. Dealcoholization using heat treatment demands a huge amount of energy and alters the flavor of the product due to thermal degradation. Membrane technologies such as

TABLE 6.9
Composition, Nutritive and Medicinal Properties of Wine

Compound Name	Nutritive and Medicinal Properties
Resveratrol (stilbene phenolic compound)	Cardio-protective, anti-aging, and anti-diabetic agent
Hydroxytyrosol (phenyl ethyl alcohol)	Antimicrobial activity
Melatonin (Indolamine)	Neuro-hormone (supports immune system)

RO, NF, OD, FO, dialysis, and PV are widely investigated as commercialized technologies for dealcoholization (Liguori et al. 2018).

Reverse osmosis (RO): RO is a potential substitute for heat treatment for dealcoholization. The smaller pore size of the RO membrane allows selective permeation of lower molecular weight ethanol and water while retaining large molecules responsible for the preservation of taste and nutritive properties (Güzel et al. 2020). The RO process is carried out at lower operating temperatures than traditional distillation, and it does not include any phase change for alcohol removal. A dealcoholized product (in terms of flavor and body) is created with minimal damage to temperature-sensitive compounds, However, the use of RO for dealcoholization has certain disadvantages such as high pressure, high energy consumption, and high production cost. According to the literature, RO is not an economically viable technique if the user is interested in reducing the alcohol content below 0.45% (Pilipovik and Riverol 2005).

Nanofiltration (NF): The NF process can be defined as a membrane process in which molecules of 0.0005 μm or bigger are not allowed to permeate. Catarino and Mendes (2011) identified NF as a promising technique for preparation of dealcoholized wine. Mangindaan et al. (2018) reported NF as a promising alternative to RO in brewing non-alcoholic beverages.

Osmotic distillation (OD): OD is among emerging membrane technologies with the greatest potential for reducing overall alcohol content in alcoholic beverages (Diban et al. 2008; Liguori et al. 2013). The OD process operates at moderate temperature and under atmospheric pressure to ensure lower energy consumption with negligible damage to aromatic components.

Dialysis: Ethanol removal from alcoholic beverages using dialysis is a very old membrane-based dealcoholization method. The main driving force for this system is the concentration gradient between the two sides of the membrane. Here, ethanol (alcohol) is removed via diffusion through a semi-permeable membrane in aqueous form. The operating temperature required for dialysis is very low (1–6°C).

Pervaporation (PV): PV has been widely investigated for aroma recovery in beverages. It has also been successfully implemented for alcohol (ethanol) removal from beverages such as beer (Blanco et al. 2014) and wine (Catarino and Mendes 2011). The PV process is a suitable substitute for typical separation processes such as steam distillation, liquid solvent extraction, and vacuum distillation. Compared to traditional distillation processes, PV provides certain advantages in its lower energy consumption and lower operating temperature, and there is no need for chemical additives (Raisi et al. 2008).

6.2.3 Treatment of Waste Effluents Generated by Food Industries

The food-processing industry is responsible for the consumption of a large volume of water. The type of effluent released by any processing industry depends upon the technology used, design of the processing plant, operating conditions, and nature and type of raw material. The nature of the effluent released by different processing industries requires different treatment processes. The stringent protocols established in the food-processing industry require a large quantity of water to be used for maintaining clean, hygienic, and sanitized processing conditions. The effluent released by

these industries involves liquid streams generated during processing, cleaning, and operational steps. The waste effluent stream released by food-processing plants contains organic constituents such as fatty acids and nutrients that are not biodegradable and may pass through conventional biological treatment processes.

Waste generated in food processing contains a large portion of food residues with valuable bioactive compounds. The organic load of the waste is quite high, and thus discharge of waste directly to the environment is not usually advised. The treatment of waste and by-products along with recovery of valuable products is beneficial in both environmental and commercial terms. Pressure-driven membrane separation processes such as MF and UF have been successfully introduced to reduce organic load and chemical oxygen demand in wastewater treatment. Conventionally used methods for recovery of valuable bioactive compounds such as polysaccharides, phenolic compounds, anthocyanin, carotenoids, fibers, traces of minerals, and vitamins include pulsed electric field, high voltage electric discharge, pulsed ohmic heating, high-pressure processing, supercritical fluid extraction, and solvent extraction. Compared with conventional wastewater treatment technology, membrane technology allows complete removal of pathogenic organisms while also providing disinfection.

Membrane processes are regarded as one of the most promising technologies for the treatment of waste effluent generated by the food-processing industry. Membrane processes for wastewater involve microbial bioreactor (MBR) and anaerobic or aerobic treatment followed by clarification. An MBR consists of a biological reactor associated with membrane separation, usually UF or MF membrane processes. The MF and UF membrane processes allow removal of suspended solids from effluent by physical straining. After colloidal particle and particulate matter are removed, the RO process is used to eliminate total dissolved solids (TDS), microorganisms, salts, and other micro-pollutants not removed by the pre-treatment system. The MBR-treated wastewater is free from total suspended solids, and the water can be directly used for unrestricted irrigation. For higher-quality water, high-pressure membrane processes such as RO and NF are typically used to remove dissolved solids and microorganisms. The NF and RO processes enable food-processing discharge to produce recycled water suitable for irrigation or for drinking. The use of NF has been reported in the treatment of wastewater generated by tannery, municipal, dairy, textile, and refinery processing industries. Recovered organic compounds are further purified, concentrated, and used in food, pharmaceuticals, and biotechnology. The volume and characteristics of the waste effluent generated by any processing industry depend on factors such as raw materials, processing technology, cleaning frequency, etc. Membrane processing is an attractive substitute for the treatment of waste generated by food processing, since it allows recovery of important bioactive compounds along with the generation of high-quality water. The following sections provide a brief overview of the extraction of bioactive compounds from waste generated by beverage- and dairy-processing industries using membrane technology.

6.2.3.1 Treatment of Waste Effluent Generated by Dairy Processing

The high concentration of organic matter in wastewater generated by dairy processing can be characterized as among the most serious polluters with COD over 6 gL^{-1}

(Smithers 2008). Dairy wastewater contains a high amount of organic and nutrient components, and if not treated appropriately, the direct release of effluent to a water source may cause a critical environmental problem. The treatment of dairy wastewater includes primary treatment (for removal of solid, oil, and fats), secondary treatment (for elimination of organic matter and nutrients) (Erkan et al. 2018), and tertiary treatment (for polishing).

The membrane process is the best substitute for the treatment and reuse of wastewater generated by dairy processing. Luo et al. (2011) investigated an integrated two-stage UF–NF process to recycle water and nutrients from synthetically made model dairy waste effluent. The NF270 membrane (MWCO 150) used for tertiary treatment of UF permeate exhibits high permeability, high lactose rejection, and low retention of salts. Andrade et al. (2014) investigated an MBR and NF as secondary and tertiary treatments respectively, and suggested MBR as a feasible system for treatment of dairy effluent with effective eradication of color, organic compounds, and nutrients. The quality of permeate generated by NF met all standards for applications that require lower-quality water such as steam generation, cooling or washing of external areas and transport vehicles. The overall chemical oxygen demand (COD) and biological oxygen demand (BOD) removal efficiency for this MBR and NF system was 99.9% and 93.1% respectively. Bortoluzzi et al. (2017) compared the effectiveness of integrated MF–NF and MF–RO for treatment of dairy waste streams, based on the efficiency of color, turbidity, total nitrogen, and total oxygen content. Brião et al. (2019) examined the probable uses of permeate and retentate to enhance the feasibility of the dairy industry in both economic and technical terms. They reported that RO is effective for restoring milk solids from real dairy rinse water, although the reclaimed water (COD, 40 mgL^{-1}) failed to achieve drinking-water quality requirements even after the second RO passage. Due to its high organic and volumetric load, whey is a principal pollutant of the milk-processing industry. Discarding the effluent directly to rivers or lakes could potentially alter their physical and chemical composition (see Section 6.2.1.6 for membrane processes for whey processing). The fractionation and extraction of whey protein effectively reduce organic load before the effluent stream is released to the environment.

6.2.3.2 Treatment of Wastewater Generated by Beverage Processing Industries

Fruit processing waste contains a wide range of bioactive compounds such as phenolic compounds, alkaloids, essential/volatile oil, gums, and oleoresin. Apart from essential bioactive compounds, the waste can also be treated to generate by-products such as pectin, natural color, and dyes. Waste generated by beverage processing includes peels, seeds, pomace, skin, starch, sugar, and fibers, along with a high volume of aqueous waste generated during cleaning and processing. Similar to dairy processing, waste effluent released by fruit/vegetable juice processing can be characterized by high BOD and COD. The membrane process for treatment of wastewater generated by food processing offers the advantage of higher effluent water quality, and simpler operating conditions than typical treatment processes.

The advantages of the membrane separation process over conventional techniques are:

(i) The membrane separation process is a non-thermal technology. The mild operating conditions minimize damage caused by high operating conditions, such as denaturation of protein, and change in color and flavor of a concentrated product.
(ii) The selectivity of the membrane enables effective separation of unwanted components such as microorganisms, sediments, and pectin.
(iii) The membrane process is easy to design, implement, and scale up.
(iv) The membrane separation process requires less energy than conventional technologies.

However, membrane application also presents certain challenges, which can be mitigated by strategies outlined in the next section.

6.3 CHALLENGES

6.3.1 Concentration Polarization

Concentration polarization (CP) can be described as a reversible aggregation of rejected solute and particulate matter in the mass transfer boundary layer adjoining the membrane surfaces (Abdelrasoul et al. 2016). In MF, UF, NF, RO, PV, FO, and MD processes, feed components are carried to the membrane surfaces by convection, and as the rate of permeation increases through the membrane, the rate of feed component transport to the membrane also increases (Singh 2005). Due to the selectivity of the membrane, the solvent passes through the membrane but solute molecules with molecular size are rejected and retained on the membrane surface. The accumulation of solute particles on the membrane surface emerges as a concentration gradient just above the membrane surface. CP is a natural and unavoidable consequence of membrane semi-permeability and selectivity that occurs in both pressure-driven and osmotically driven membrane processes. In pressure-driven membrane separation processes, the solute is partially (or completely) retained by the membrane when pressure is applied to the feed side of a membrane. This results in accumulation of solute particles and gradually increasing solute concentration on the surface due to convective flow, while the solvent passes freely through the membrane. At a certain point, the convective solute flow to the surface of the membrane will be counterbalanced by the solute flux through the membrane and the diffusive flow from the membrane surface to the bulk. In the osmotically driven membrane process, CP is caused by the concentration gradient between the feed and the draw solution through an asymmetric hydrophilic membrane. The CP provides resistance to the flow of solvent through the membrane and results in osmotic back pressure, which eventually weakens the transmembrane pressure. It is a reversible process that does not affect any intrinsic membrane properties.

6.3.2 Membrane Fouling

The inevitable process of membrane fouling is often regarded as the main limitation to successful membrane application in food processing. It occurs when large molecular-size feed stream components accumulate on the membrane surface or inside the pores of a membrane (Kumar and Ismail 2015). In food processing, deposition of mineral, protein, microorganism, and fine food residues on the membrane surface or inside the pore surface is primarily responsible for membrane fouling (Wilson 2018). These deposited aggregates provide additional substrate for growth and metabolism of microorganisms on the membrane surface, which eventually leads to a flux decline below the theoretical capacity of the membrane and may cause irreversible damage to membrane permeability. In any membrane process, the concentration and nature of solute/solvents, membrane type, surface characteristics, pore size distribution, membrane material, and hydrodynamics of the membrane modules are among the parameters determining the fouling rate. Reversible fouling can be easily eliminated by adopting appropriate cleaning methods, whereas irreversible fouling cannot be eradicated even after cleaning.

6.3.2.1 Fouling Mechanisms

The types of fouling that occur in food processing are:

Particulate fouling induced by deposition of particulates, biomass, and colloidal particles on the membrane surface (De Barros et al. 2003). Particulates adhering to the membrane surface can easily be detached by air scrubbing and back flushing. Particulate fouling combined with organic and inorganic foulants requires chemical cleaning to restore membrane properties.

Organic fouling is the accumulation of organic material (such as proteins, carbohydrates, and polysaccharides) on membrane surfaces. Due to their fluctuating and dynamic nature, the fouling characteristics of proteins are the most complicated. Proteins in solution interact with each other and aggregate, concentrating close to the membrane surface where they precipitate. In juice processing, pectin and cellulose fouling is primarily responsible for flux decline. Polydimethylsiloxane and polypropylene glycol (PPG)-based agents are often used in fermentation broth to control foaming generated by aeration in the bioreactor (Liew et al. 1997). Lipids are a group of generally hydrophobic molecules that include fats, oils, phospholipids, waxes, and similar compounds. In fermentation broth and oily waste effluent, fatty acids are found in solubilized form. The capillary condensation of fatty acids causes blockage/closure of membrane pores, the most severe type of fouling to occur in food processing.

Biofouling is the development of an unwanted microbial layer on the membrane surface, resulting in the production of soluble microbial products and extracellular polymeric substances of large molecular weight. These compounds are mostly composed of proteins, lipids, and polysaccharides. Microbial products change membrane surface properties and provide nutrition for further growth of microorganisms. Due to stringent disinfection processes, biofouling is rarely reported in the food industry, and the likelihood can be limited by frequent cleaning and removal of biodegradable components from the feed (Cui and Muralidhara 2010).

Membrane fouling in the UF process can be explained as follows (Mohammad et al. 2012):

Adsorption. Adsorption occurs due to the synergy between foulant and membrane surface. The interaction between membrane and foulant is because of weak van der Waals forces, whereas strong chemical bonding (electrostatic attraction) results from the functional groups involved. Chances of adsorption are the most severe in food processing, due to the presence of proteins, polysaccharides, and humic acids (Mänttäri et al. 2000). Material adsorbed on the membrane surface material may alter the surface characteristics (Evans and Bird 2006).

Pore blocking. Pore blocking occurs due to full or partial closure of membrane pores by solute particles, and this is the main cause of fouling in porous membranes. Pore blockage is usually observed during the initial filtration stage when the membrane surface is exposed and the approaching solute particles can directly interact with membrane pores. The fouling mechanism is different for porous and non-porous membranes. Pore blocking may be complete, partial, or internal, followed by cake layer formation (Cui and Muralidhara 2010).

Gel/cake formation. Cake formation is the layer-by-layer build-up of a solute particle on the outer membrane surface. The additional resistance provided by the cake layer is called cake resistance. Flux decline and fouling reversibility are dictated by the morphology of the fouling layer and the synergy between cake layer and membrane surface, respectively. The flux at which gelation occurs is known as the limiting flux, i.e., the maximum stationary permeation flux that the system can reach.

The main factors affecting membrane fouling are:

(i) Membrane properties, such as material, membrane morphology, and surface properties, including smoothness, charge, and hydrophilicity (Evans and Bird 2006).
(ii) Feed solution properties, such as concentration, pH, composition, and ionic conditions (Ramachandra Rao 2002; Marshall et al. 2003).
(iii) Operating conditions, including transmembrane pressure (Balakrishnan et al. 2000), filtration mode, temperature (Cassano et al. 2007b; Makardij et al. 1999), cross-flow velocity of feed solution (Bruijn et al. 2003; Choi et al. 2005), application of turbulence promoter (Krstic et al. 2004), back-pulsing (Sondhi and Bhave 2001), gas sparging (Fadaei et al. 2007), and ultrasound (Muthukumaran et al. 2006).

6.3.2.2 Control and Reduction of Fouling

While fouling cannot be avoided, it can be reduced to an acceptable level with an improved understanding of feed characterization and process operations. The following approaches have been developed to control the rate of fouling (Cui and Muralidhara 2010).

6.3.2.2.1 Hydrodynamic Management

Hydrodynamic management promotes local mixing adjacent to the membrane surface to buildup the back diffusion of rejected molecules. Corrugated membrane

surface, spacers, turbulence promoter, vibrating membrane, reverse flow, and the use of sponge balls/gas bubbles generates a secondary flow, which improves local mixing and hence mass transfer. Zhu et al. (2015) investigated high pulp content juice using a vibrating and rotating membrane system. The shear rate created by vibration and rotation resulted in reduced interface between foulant and membrane.

6.3.2.2.2 *Backflushing and Pulsing*

Backflushing reverses the permeate flow through the membrane, re-establishing the flux by dislodging foulants from the membrane surface. At industrial scale, backflushing is carried out periodically for feed with high solid content and high fouling tendency.

6.3.2.2.3 *Membrane Surface Modification*

Fouling can be decreased by modifying membrane surface properties such as hydrophilicity, charge, and pore size distribution.

6.3.2.2.4 *Feed Pre-Treatment*

Pre-treatment of feed solution allows removal of suspended solutes and helps reduce membrane fouling tendency. Particle charge can affect particle size distribution by promoting or demoting aggregation. Feed treatment before membrane processing, such as adjustment to salt concentration and pH, modifies the charge effect by decreasing the quantity of suspended particulate material, causing advancement in flux and contributing to superior concentration factors.

Enzyme pre-treatment is usually performed before membrane clarification of citrus fruit juice. Pectinase (enzyme) pre-treatment reduces the viscosity of juice, which eventually improves permeate flux and recovery rate. Vaillant et al. (1999) combined pectinase and cellulase enzymes to improve permeate flux in MF of passion fruit juice using a ceramic membrane. Domingues et al. (2014) analyzed the effect of pre-treatments such as centrifugation, chitosan coagulation, and enzymatic liquefaction for microfiltration of passion fruit using a hollow-fiber membrane module. Chitosan addition was found to be a promising alternative pre-treatment.

Other pre-treatments include protease enzyme for removal of protein (Pinelo et al. 2010); centrifugation before and after depectinization (Yousefnezhad et al. 2016); and fining agents such as bentonite and gelatin (Youn et al. 2004). Rai et al. (2007) examined pre-treatment of citrus fruit juice clarification using UF membrane with high permeate flux and low membrane fouling. With enzymatic treatment followed by adsorption using bentonite, maximum permeate flux was observed.

6.3.2.2.5 *Effective Membrane Cleaning*

Membrane cleaning involves interrupting adhesive and cohesive bonds between foulant and foulant–membrane surface. Cleaning agents should only interact with foulants, not with the membrane.

Depending upon foulant species and membrane properties, cleaning procedures may be either physical or chemical.

Physical cleaning. Rinsing the membrane module with clean water removes loosely bound foulants from the membrane surface. Backflushing through pores during filtration or cleaning removes particle cake and internal foulants in hollow-fiber and flat-sheet membrane modules. Physical cleaning can be further improved by aeration and pulsed flow to generate higher turbulence. Physical removal with sponge balls and air scrubbers can also be used in tubular membrane modules, but it is only useful for external fouling.

Chemical cleaning. Chemical cleaning to break down chemical bonds between foulant aggregates and foulant–membrane involves the following steps:

(i) dispersal and formation of cleaning agent
(ii) transport of cleaning agent to the fouled layers
(iii) transport through a fouled layer
(iv) cleaning reaction
(v) transportation of reacting products back to the interface
(vi) transportation of product back to the bulk.

Appropriate cleaning agents should be selected for application to specific membranes based on the type of fouling and the species involved. The commonly used cleaning agents used can be classified as:

Acid. Acids, such as HCl, HNO_3, and H_2SO_4, are used to dissolve precipitates of inorganic salts and metal oxide/hydroxide. HNO_3 is a potent oxidizing agent and is used for removal of calcium salt precipitates, biological and other organic materials. Citric acid is usually preferred over nitric acid because of its mildness, and most importantly, it rinses off easily and does not corrode surfaces. An acid wash is generally used to neutralize residual alkalinity and subsequent removal of deposited minerals formed during alkaline cleaning. Acid cleaner mixed with wetting agents (to generate pH of 2.5 or less) can be used to clean heat-denatured protein residues from the membrane surface.

Alkalis. Alkalis, such as sodium hydroxide, NaOH, potassium hydroxide, KOH, and their mixture, can potentially favor the rapid hydrolysis of protein and polysaccharides into small amides and sugars, as well as providing efficient saponification of fats and oils, particularly above 50°C. NaOH is predominantly used for removal of whey protein-fouled inorganic membranes (Bartlett et al. 1995).

Oxidants. Membranes need to be regularly disinfected in food and pharmaceutical industries. NaOCl is essentially used for hydrophilic foulants but never for inorganic and hydrophobic foulants.

Surfactants. Surfactants are amphiphilic compounds with hydrophobic and hydrophilic segments. Used in low amounts, they can lower surface tension and increase the solubility of the foulant layer by displacing it from the membrane surface. Sodium dodecyl sulfate (SDS) combined with NaOH is reported as an efficient cleaning agent for membranes fouled with milk components (Kazemimoghadam and Mohammadi 2007).

Anionic and cationic surfactants are mostly used in membrane processing (Cui and Muralidhara 2010).

Enzymes. Enzymes are a selective catalyst designed to attack specific targets. Enzymatic cleaning offer advantages over other types of chemical cleaning agents, such as low operating temperature and mild pH. Most importantly, the enzymes are safe for most membranes. Most enzymes are biodegradable and environmentally friendly. Enzymatic cleaning is more useful for biofilm removal than alkali and acid cleaners. Enzymatic cleaning agents such as α-chymotrypsin and protease can potentially clean membranes fouled with whey protein with nearly 99% flux recovery and 100% cleaning efficiency.

After selection of a suitable cleaning agent, cleaning protocols also need optimization in terms of frequency, temperature, and dosage. Type of deposit, water quality and membrane material are also significant, as are parameters such as temperature, concentration, flow rate, pressure, and time. Temperature affects the cleaning process, increasing temperature improves diffusion, solubility (of both foulant and cleaning agent), and reaction rate, and melts fat. An optimum temperature of 50–55°C is required for maximum flux, beyond which overall flux recovery declines (Makardij et al. 1999). Similarly, depending upon the feed, foulant, and membrane properties, the optimal process parameters can be estimated to attain maximum permeate flux, although frequent cleaning retards fouling and prolongs membrane life. However, excessive cleaning results in damage to the membrane structure, enlarged pore size, and fractured membrane surface. During chemical cleaning, chemicals alter membrane properties due to alteration of functional groups in polymeric chains.

6.4 RECENT AND EMERGING TRENDS IN MEMBRANE PROCESSES IN THE FOOD-PROCESSING INDUSTRY

During fruit juice processing using membranes, depectinization helps minimization of gel/cake layer formation, which eventually reduces membrane fouling. UF and MF membrane processes are effectively used to reduce turbidity by subsequent removal of suspended solids. For preparation of concentrated liquid foods and beverages, membrane processes such as MD, OD, and RO are widely used. The integration of membrane processes reduces waste generation and improves economic viability. In recent years, numerous researchers have analyzed the feasibility of integrating membrane-based processes. Cassano et al. (2007b) developed an integrated membrane process for clarification and concentration of blood orange juice. Clarification using UF showed a minimal change in the composition of total antioxidant activity and bioactive compounds. The concentration of the clarified juice was performed in two stages using RO and osmotic distillation (OD). During the concentration process, a minor decrease in total antioxidant activity (TAA) was observed due to partial degradation of anthocyanin and ascorbic acid. A combined membrane process for clarification and concentration of carrot and citrus fruit juices was investigated by Cassano et al. (2003b). The suggested process consists of clarification of

TABLE 6.10
Applications of Integrated Membrane Processes in Food Processing

Author	Integrated Membrane Process	Application
Cassano et al. (2003a)	UF and OD	Clarification and concentration of kiwi fruit juice
Cassano et al. (2006)	UF, PV, and OD	Recovery of aromatic compounds prior to clarification and concentration of kiwi fruit juice
Galaverna et al. (2007)	UF, RO, and OD	Clarification and concentration of blood orange juice
A. L. R. Souza et al. (2013)	MF, RO, and OD	Clarification and concentration of camu-camu fruit juice
Quist-Jensen et al. (2016)	UF and DCMD	Clarification and concentration of orange juice
Bortoluzzi et al. (2017)	MF and RO	Filtration of dairy wastewater effluent to retain total solids and organic components
An et al. (2018)	FO and MD	Concentration of clarified apple juice and regeneration of draw solute

fruit juice using UF, followed by pre-concentration of the UF-clarified juice using RO. The retentate stream is further concentrated using the OD process. Souza et al. (2013) explored an integrated membrane process for clarification followed by concentration of camu-camu fruit juice. The final concentrated juice had a Brix value of 60° and retained the highest percentage of vitamin C. Cisse et al. (2005) used the UF process for clarification and the OD process for concentration of clarified juice. Sotofta et al. (2012) designed an integrated membrane separation process for the preparation of blackcurrant juice concentrate. The integrated membrane system consists of vacuum membrane distillation (for recovery of aromatic compounds), RO, NF, and direct contact membrane distillation (DOMD) for water recovery. An et al. (2018) investigated the feasibility of an integrated FO–MD process combining the advantages of both for concentration of apple juice, draw solution and recovery of freshwater.

The advantages of integrated membrane processes can be summarized as:

(i) Reduction in clarification times, simplification of the clarification process, improved clarified juice volumes.
(ii) Possibility of operating at ambient temperature, helping to preserve the original freshness, aroma, and nutritional value of fresh juice.
(iii) Improvement of final quality and productive processes.

Table 6.10 summarizes integrated membrane process applications in food processing.

6.5 CONCLUSION

The potential of membrane processes in food processing is broadly recognized both by academics and in industry, and the increasing call for healthy food from diet- and fitness-conscious consumers is encouraging researchers to find alternatives to

conventional heat treatments. The recovery of valuable bioactive products from waste effluent generated by dairy, beverage, and other processing has been particularly widely investigated. The development of novel cleaning procedures, anti-fouling membrane materials, and new module designs is ensuring advances in membrane application in food and beverage processing.

Membrane science and technology offers a great many options for the design and optimization of innovative production, compared with other technologies. Membrane processes can support sustainable industrial growth by preserving energy, decreasing capital costs, and reducing environmental impact. Despite the challenges associated with membrane processing – principally membrane fouling and concentration polarization – membrane technology represents a promising alternative for secure dairy and beverage processing.

REFERENCES

Abdel-fatah, Mona A. 2018. Nanofiltration Systems and Applications in Wastewater Treatment : Review Article. *Ain Shams Engineering Journal*, 9(4): 3077–3092. doi:10.1016/j.asej.2018.08.001

Abdel-rahman, Ali Kamel. 2015. *Productionof White Sugar Using Membrane Technology.* In *Conference: The Proceeding of the International Mechanical Engineering Conference (IMEC 2004)*, December 2004, pp. 560–575. https://www.researchgate.net/publication/283320211_PRODUCTION_OF_WHITE_SUGAR_USING_MEMBRANE_TECHNOLOGY.

Abdelrasoul, Amira, Huu Doan, Ali Lohi, and Chil-Hung Cheng. 2016. Mass Transfer Mechanisms and Transport Resistances in Membrane Separation Process. *Mass Transfer - Advancement in Process Modelling*, 1: 15–40. doi:10.5772/60866

Almécija, M. Carmen, Rubén Ibáñez, Antonio Guadix, and Emilia M. Guadix. 2007. Effect of pH on the Fractionation of Whey Proteins with a Ceramic Ultrafiltration Membrane. *Journal of Membrane Science*, 288(1–2): 28–35. doi:10.1016/j.memsci.2006.10.021

Alves, V.D., and I.M. Coelhoso. 2006a. Orange Juice Concentration by Osmotic Evaporation and Membrane Distillation: A Comparative Study. *Journal of Food Engineering*, 74(1): 125–133. doi:10.1016/j.jfoodeng.2005.02.019

Alves, V.D., and I.M. Coelhoso. 2006b. Orange Juice Concentration by Osmotic Evaporation and Membrane Distillation: A Comparative Study, 125–33. doi:10.1016/j.jfoodeng.2005.02.019

An, Xiaochan, Yunxia Hu, Ning Wang, Zongyao Zhou, and Zhongyun *Liu*, 2018. Continuous Juice Concentration by Integrating Forward Osmosis with Membrane Distillation Using Potassium Sorbate Preservative as a Draw Solute. *Journal of Membrane Science*, 573: 192–199. doi:10.1016/j.memsci.2018.12.010

Andrade, L.H., F.D.S. Mendes, J.C. Espindola, and M.C.S. Amaral. 2014. Nanofiltration as Tertiary Treatment for the Reuse of Dairy Wastewater Treated by Membrane Bioreactor. *Separation and Purification Technology*, 126: 21–29. doi:10.1016/j.seppur.2014.01.056

Andrés, L.J., F.A. Riera, and R. Alvarez. 1995. Skimmed Milk Demineralization by Electrodialysis: Conventional versus Selective Membranes. *Journal of Food Engineering*, 26(1): 57–66. doi:10.1016/0260-8774(94)00042-8

Aroujalian, Abdolreza, and Ahmadreza Raisi. 2007. Recovery of Volatile Aroma Components from Orange Juice by Pervaporation. *Journal of Membrane Science*, 303(1–2): 154–161. doi:10.1016/j.memsci.2007.07.004

Arriola, Nathalia Aceval, Gielen Delfino dos Santos, Elane Schwinden Prudêncio, Luciano Vitali, José Carlos Cunha Petrus, and Renata D.M. Castanho Amboni. 2014. Potential of Nanofiltration for the Concentration of Bioactive Compounds from Watermelon Juice. *International Journal of Food Science & Technology*, 49(9): 2052–2060. doi:10.1111/ijfs.12513

Aydiner, Coskun, Semra Topcu, Caner Tortop, Ferihan Kuvvet, Didem Ekinci, Nadir Dizge, and Bulent Keskinler. 2012. A Novel Implementation of Water Recovery from Whey: 'Forward–Reverse Osmosis' Integrated Membrane System. *Desalination and Water Treatmen*, 51(January): 1–14. doi:10.1080/19443994.2012.693713

Bailey, A.F.G., A.M. Barbe, P.A. Hogan, R.A. Johnson, and J. Sheng. 2000. The Effect of Ultrafiltration on the Subsequent Concentration of Grape Juice by Osmotic Distillation. *Journal of Membrane Science*, 164(1–2): 195–204. doi:10.1016/S0376-7388(99)00209-4

Balakrishnan, M., M. Dua, and J.J. Bhagat. 2000. Effect of Operating Parameters on Sugarcane Juice Ultrafiltration: Results of a Field Experience. *Separation and Purification Technology*, 19(3): 209–220. doi:10.1016/S1383-5866(00)00054-X

Banvolgyi, Szilvia, Istvan Kiss, Erika Bekassy-molnar, and Gyula Vatai. 2006. Concentration of Red Wine by Nanofiltration. 198 (September 2005): 8–15. doi:10.1016/j.desal.2006.09.003

Barros, S.T.D. De, C.M.G. Andrade, E.S. Mendes, and L. Peres. 2003. Study of Fouling Mechanism in Pineapple Juice Clarification by Ultrafiltration. *Journal of Membrane Science*, 215(1–2): 213–224. doi:10.1016/S0376-7388(02)00615-4

Bartlett, M, M.R. Bird, and J.A. Howell. 1995. An Experimental Study for the Development of a Qualitative Membrane Cleaning Model 105: 147–157.

Bhattacharjee, C., Saxena, V.K., and Dutta, S. 2017. Fruit juice processing using membrane technology: A review. *Innovative Food Science & Emerging Technologies*, 43: 136–153.

Bhattacharya, P.K., Shilpi Agarwal, S. De, and U.V.S. Rama Gopal. 2001. Ultrafiltration of Sugar Cane Juice for Recovery of Sugar: Analysis of Flux and Retention. *Separation and Purification Technology*, 21(3): 247–259. doi:10.1016/S1383-5866(00)00209-4

Blanco, Carlos A., Laura Palacio, Pedro Prádanos, and Antonio Hernández. 2014. Pervaporation Methodology for Improving Alcohol-Free Beer Quality through Aroma Recovery 133: 1–8. doi:10.1016/j.jfoodeng.2014.02.014

Bortoluzzi, Airton C., Julio A. Faitão, Marco Di Luccio, Rogério M. Dallago, Juliana Steffens, Giovani L. Zabot, and Marcus V. Tres. 2017. Dairy Wastewater Treatment Using Integrated Membrane Systems. *Journal of Environmental Chemical Engineering*, 5(5): 4819–4827. doi:10.1016/j.jece.2017.09.018

Brans, G., C.G.P.H. Schroën, R.G.M. Van Der Sman, and R.M. Boom. 2004. Membrane Fractionation of Milk: State of the Art and Challenges. *Journal of Membrane Science*, 243(1–2): 263–272. doi:10.1016/j.memsci.2004.06.029

Bruijn, J, A. Venegas, J.A. Martínez, and R. Borquez. 2003. Ultrafiltration Performance of Carbosep Membrane for the Clarification of Apple Juice. *LWT - Food Science and Technology*, 36(June): 397–406. doi:10.1016/S0023-6438(03)00015-X

Calabro, Vincenza, Bi Lin Jiao, and Enrico Drioli. 1994. Theoretical and Experimental Study on Membrane Distillation in the Concentration of Orange Juice. *Industrial & Engineering Chemistry Research*, 33(7): 1803–1808. doi:10.1021/ie00031a020

Capannelli, G., A. Bottino, S. Munari, D.G. Lister, G. Maschio, and I. Becchi. 1994. The Use of Membrane Processes in the Clarification of Orange and Lemon Juices. *Journal of Food Engineering*, 21(4): 473–483. doi:10.1016/0260-8774(94)90067-1

Cardoso de Oliveira, Ricardo, Roselene Caleffi Docê, and Sueli Teresa Davantel de Barros. 2012. Clarification of Passion Fruit Juice by Microfiltration: Analyses of Operating Parameters, Study of Membrane Fouling and Juice Quality. *Journal of Food Engineering*. doi:10.1016/j.jfoodeng.2012.01.021

Cassano, A., L. Donato, and E. Drioli. 2007b. Ultrafiltration of Kiwifruit Juice: Operating Parameters, Juice Quality and Membrane Fouling. *Journal of Food Engineering*, 79(2): 613–621. doi:10.1016/j.jfoodeng.2006.02.020

Cassano, A., E. Drioli, G. Galaverna, R. Marchelli, G. Di Silvestro, and P. Cagnasso. 2003a. Clarification and Concentration of Citrus and Carrot Juices by Integrated Membrane Processes. *Journal of Food Engineering*, 57(2): 153–163. doi:10.1016/S0260-8774(02)00293-5

Cassano, A., A. Figoli, A. Tagarelli, G. Sindona, and E. Drioli. 2006. Integrated Membrane Process for the Production of Highly Nutritional Kiwifruit Juice. *Desalination* 189(1–3): 21–30. doi:10.1016/j.desal.2005.06.009

Cassano, A., B. Jiao, and E. Drioli. 2003b. Production of Concentrated Kiwifruit Juice by Integrated Membrane Process. *Food Research International*, 37(2): 139–148. doi:10.1016/j.foodres.2003.08.009

Cassano, A, M. Marchio, and E. Drioli. 2007a. Clarification of Blood Orange Juice by Ultrafiltration : Analyses of Operating Parameters, Membrane Fouling and Juice Quality. *Desalination*, 212: 15–27. doi:10.1016/j.desal.2006.08.013

Catarino, Margarida, and Adélio Mendes. 2011. Dealcoholizing Wine by Membrane Separation Processes. *Innovative Food Science and Emerging Technologies*, 12(3): 330–337. doi:10.1016/j.ifset.2011.03.006

Chen, George Q., Anna Artemi, Judy Lee, Sally L. Gras, and Sandra E. Kentish. 2019a. A Pilot Scale Study on the Concentration of Milk and Whey by Forward Osmosis. *Separation and Purification Technology*. doi:10.1016/j.seppur.2019.01.050

Chen, George Q., Franziska I.I. Eschbach, Mike Weeks, Sally L. Gras, and Sandra E. Kentish. 2016. Removal of Lactic Acid from Acid Whey Using Electrodialysis. *Separation and Purification Technology*, 158: 230–237. doi:10.1016/j.seppur.2015.12.016

Chen, George Q, Thomas S.H. Leong, Sandra E. Kentish, Muthupandian Ashokkumar, and Gregory J.O. Martin. 2019b. Membrane Separations in the Dairy Industry. In: Charis M. Galanakis (ed.), *Separation of Functional Molecules in Food by Membrane Technology*. Academic Press, pp. 267–304. doi:10.1016/B978-0-12-815056-6.00008-5

Choi, Hyeok, Kai Zhang, Dionysios D. Dionysiou, Daniel B. Oerther, and George A. Sorial. 2005. Influence of Cross-Flow Velocity on Membrane Performance during Filtration of Biological Suspension. *Journal of Membrane Science*, 248(1–2): 189–199. doi:10.1016/j.memsci.2004.08.027

Cisse, Mady, Fabrice Vaillant, Ana Perez, Manuel Dornier, and Max Reynes. 2005. The Quality of Orange Juice Processed by Coupling Crossflow Microfiltration and Osmotic Evaporation. *International Journal of Food Science & Technology*, 40(1): 105–116. doi:10.1111/j.1365-2621.2004.00914.x

Conidi, Carmela, Roberto Castro-Muñoz, and Alfredo Cassano. 2020. Membrane-Based Operations in the Fruit Juice Processing Industry: A Review. *Beverages*, 6(1): 18. doi:10.3390/beverages6010018

Cuartas-Uribe, B., M.I. Alcaina-Miranda, E. Soriano-Costa, J.A. Mendoza-Roca, M.I. Iborra-Clar, and J. Lora-García. 2009. A Study of the Separation of Lactose from Whey Ultrafiltration Permeate Using Nanofiltration. *Desalination*, 241(1–3): 244–255. doi:10.1016/j.desal.2007.11.086

Cui, Z.F., and H.S. Muralidhara. 2010. *Membrane Technology*.

Daufin, G., Escudier, J.P., Carrère, H., Bérot, S., Fillaudeau, L., and Decloux, M. 2001. Recent and emerging applications of membrane processes in the food and dairy industry. *Food and Bioproducts Processing*, 79(2): 89–102.

Bruijn, Johannes De, Alejandro Venegasb, and Rodrigo Borquezc. 2002. Influence of Crossflow Ultrafiltration on Membrane Fouling and Apple Juice Quality. *Desalination*, 148: 131–136. doi:10.1016/S0011-9164(02)00666-5

Morais Coutinho, Cesar de, Ming Chih Chiu, Rodrigo Correa Basso, Ana Paula Badan Ribeiro, Lireny Aparecida Guaraldo Gonçalves, and Luiz Antonio Viotto. 2009. State of Art of the Application of Membrane Technology to Vegetable Oils: A Review. *Food Research International*, 42(5–6): 536–550. doi:10.1016/j.foodres.2009.02.010

Souza, Rosane Rosa De, Rosângela Bergamasco, Sílvio Cláudio da Costa, Xianshe Feng, Sergio Henrique Bernardo Faria, and Marcelino Luiz Gimenes. 2010. Recovery and Purification of Lactose from Whey. *Chemical Engineering and Processing: Process Intensification*, 49(11): 1137–1143. doi:10.1016/j.cep.2010.08.015

Diban, Nazely, Violaine Athes, Magali Bes, and Isabelle Souchon. 2008. Ethanol and Aroma Compounds Transfer Study for Partial Dealcoholization of Wine Using Membrane Contactor 311: 136–46. doi:10.1016/j.memsci.2007.12.004

Domingues, Rui Carlos Castro, Amanda Araújo Ramos, Vicelma Luiz Cardoso, and Miria Hespanhol Miranda Reis. 2014. Microfiltration of Passion Fruit Juice Using Hollow Fibre Membranes and Evaluation of Fouling Mechanisms. *Journal of Food Engineering*, 121(1): 73–79. doi:10.1016/j.jfoodeng.2013.07.037

Ennouri, Monia, Ines Ben Hassan, Hanen Ben Hassen, Christine Lafforgue, Philippe Schmitz, and Abdelmoneim Ayadi. 2015. Clarification of Purple Carrot Juice: Analysis of the Fouling Mechanisms and Evaluation of the Juice Quality. *Journal of Food Science and Technology*, 52(5): 2806–2814. doi:10.1007/s13197-014-1323-9

Erkan, Hanife Sari, Gorkem Gunalp, and Guleda Onkal Engin. 2018. Application of Submerged Membrane Bioreactor Technology for the Treatment of High Strength Dairy Wastewater. *Brazilian Journal of Chemical Engineering*, 35(1): 91–100. doi:10.1590/0104-6632.20180351s20160599

Evans, P.J., and M.R. Bird. 2006. Solute-Membrane Fouling Interactions during the Ultrafiltration of Black Tea Liquor. *Food and Bioproducts Processing* 84(4 C): 292–301. doi:10.1205/fbp06030

Evans, Philip J, Michael R Bird, Arto Pihlajam, and Marianne Nystr. 2008. The Influence of Hydrophobicity, Roughness and Charge upon Ultrafiltration Membranes for Black Tea Liquor *Clarification*, 313: 250–262. doi:10.1016/j.memsci.2008.01.010

Fadaei, H., S.R. Tabaei, and R. Roostaazad. 2007. Comparative Assessment of the Efficiencies of Gas Sparging and Back-Flushing to Improve Yeast Microfiltration Using Tubular Ceramic Membranes. *Desalination*, 217(1–3): 93–99. doi:10.1016/j.desal.2007.02.008

Fillaudeau, Luc, Benjamin Boissier, Anne Moreau, Pascal Blanpain-Avet, Stanislav Ermolaev, and Nicolas Jitariouk. 2007. Investigation of Rotating and Vibrating Filtration for Clarification of Rough Beer. *Journal of Food Engineering*, 80(May): 206–217. doi:10.1016/j.jfoodeng.2006.05.022

Galanakis, C.M., R. Castro-Muñoz, A. Cassano, and C. Conidi. 2016. Recovery of High-Added-Value Compounds from Food Waste by Membrane Technology. *Membrane Technologies for Biorefining*, 1: 189–215. doi:10.1016/B978-0-08-100451-7.00008-6

Galaverna, Gianni, Gianluca Di Silvestro, Alfredo Cassano, Stefano Sforza, Arnaldo Dossena, Enrico Drioli, and Rosangela Marchelli. 2007. A New Integrated Membrane Process for the Production of Concentrated Blood Orange Juice : Effect on Bioactive Compounds and Antioxidant Activity. *Food Chemistry*, 106(3): 1021–1030. doi:10.1016/j.foodchem.2007.07.018

Ganju, Sparsh, and Parag R. Gogate. 2017. A Review on Approaches for Efficient Recovery of Whey Proteins from Dairy Industry Effluents. *Journal of Food Engineering*, 215: 84–96. doi:10.1016/j.jfoodeng.2017.07.021

García, L. Fernández and F.A.R. Rodríguez. 2015. Microfiltration of Milk with Third Generation Ceramic Membranes Microfiltration of Milk with Third Generation Ceramic Membranes. *Chemical Engineering Communications*. doi:10.1080/00986445.2014.950731

Garcia-Castello, Esperanza M., and Jeffrey R. McCutcheon. 2011. Dewatering Press Liquor Derived from Orange Production by Forward Osmosis. *Journal of Membrane Science*, 372(1–2): 97–101. doi:10.1016/j.memsci.2011.01.048

Ghosh, A M, and M. Balakrishnan. 2003. Pilot Demonstration of Sugarcane Juice Ultrafiltration in an Indian Sugar Factory. *Journal of Food Engineering*, 58(2): 143–150. doi:10.1016/S0260-8774(02)00340-0

Gil, Inventors Enrique G., Helene P. Wright, and Second Ave. 1994. Process for producing refined sugar from raw juice, issued 1994.

Greiter, Michael, Senad Novalin, Martin Wendland, Klaus-Dieter Kulbe, and Johann Fischer. 2002. Desalination of Whey by Electrodialysis and Ion Exchange Resins: Analysis of Both Processes with Regard to Sustainability by Calculating Their Cumulative Energy Demand. *Journal of Membrane Science - J Membrane Sci*, 210(December): 91–102. doi:10.1016/S0376-7388(02)00378-2

Gunathilake, K.D.P.P., Li Juan Yu, and H.P. Vasantha Rupasinghe. 2014. Reverse Osmosis as a Potential Technique to Improve Antioxidant Properties of Fruit Juices Used for Functional Beverages. *Food Chemistry*, 148: 335–341. doi:10.1016/j.foodchem.2013.10.061

Gurak, Poliana D., Lourdes M.C. Cabral, Maria Helena M Rocha-leão, Virgínia M. Matta, and Suely P. Freitas. 2010. Quality Evaluation of Grape Juice Concentrated by Reverse Osmosis. *Journal of Food Engineering*, 96(3): 421–426. doi:10.1016/j.jfoodeng.2009.08.024

Güzel, Nihal, Mustafa Güzel, and K. Savaş Bahçeci. 2020. Nonalcoholic Beer. *Trends in Non-Alcoholic Beverages*. doi:10.1016/b978-0-12-816938-4.00006-9

Hamachi, M., B.B. Gupta, and R. Ben Aim. 2003. Ultrafiltration: A Means for Decolorization of Cane Sugar Solution. *Separation and Purification Technology*, 30(3): 229–239. doi:10.1016/S1383-5866(02)00145-4

Hongvaleerat, Chularat, Lourdes M.C. Cabral, Manuel Dornier, Max Reynes, and Suwayd Ningsanond. 2008. Concentration of Pineapple Juice by Osmotic Evaporation. *Journal of Food Engineering*, 88(4): 548–552. doi:10.1016/j.jfoodeng.2008.03.017

Houldsworth, D.W. 2007. Demineralization of Whey by Means of Ion Exchange and Electrodialysis. *International Journal of Dairy Technology*, 33. doi:10.1111/j.1471-0307.1980.tb01470.x

Ionics Inc. 2004. Membrane Technology Benefits the Food Processing Industry. *Filtration and Separation*, October 2004. doi:10.1016/S0015-1882(05)00411-8

Jain, Amit, and Sirshendu De. 2019. Processing of Beverages by Membranes. In *Processing and Sustainability of Beverages*, pp. 517–560. Elsevier Inc. doi:10.1016/B978-0-12-815259-1.00015-X

Jegatheesan, V., L. Shu, G. Keir, and D.D. Phong. 2012a. Evaluating Membrane Technology for Clarification of Sugarcane Juice. *Reviews in Environmental Science and Biotechnology*, 11(2): 109–124. doi:10.1007/s11157-012-9271-1

Jegatheesan, Veeriah, Li Shu, Diep Dinh Phong, Dimuth Navaratna, and Adam Neilly. 2012b. Clarification and Concentration of Sugar Cane Juice through Ultra, Nano and Reverse Osmosis Membranes. *Membrane Water Treatment*, 3(2): 99–111. doi:10.12989/mwt.2012.3.2.099

Karasu, Kensuke, Nicole Glennon, Nicole D. Lawrence, Geoffrey W. Stevens, Andrea J. O'Connor, Andrew R. Barber, Shiro Yoshikawa, and Sandra E. Kentish. 2010. A Comparison between Ceramic and Polymeric Membrane Systems for Casein Concentrate Manufacture. *International Journal of Dairy Technology*, 63(2): 284–289. doi:10.1111/j.1471-0307.2010.00582.x

Kawakatsu, Takahiro, Toshiaki Kobayashi, Yoh Sano, and Mitsutoshi Nakajima. 1995. Clarification of Green Tea Extract by Microfiltration and Ultrafiltration. *Bioscience, Biotechnology, and Biochemistry*, 59(6): 1016–1020. doi:10.1271/bbb.59.1016

Kazemimoghadam, Mansoor, and Toraj Mohammadi. 2007. Chemical *Cleaning of Ultrafiltration Membranes in the Milk Industry*, 204(May 2006): 213–218. doi:10.1016/j.desal.2006.04.030

Kotsanopoulos, Konstantinos V., and Ioannis S. Arvanitoyannis. 2015. Membrane Processing Technology in the Food Industry: Food Processing, Wastewater Treatment, and Effects on Physical, Microbiological, Organoleptic, and Nutritional Properties of Foods. *Critical Reviews in Food Science and Nutrition*, 55(9): 1147–1175. doi:10.1080/10408398.2012.685992

Krstic, Darko M., Miodrag N. Tekic, Marijana D. Caric, and Spasenija D. Milanovic. 2004. Static Turbulence Promoter in Cross-Flow Microfiltration of Skim Milk. *Separation Science and Technology*, 163(7): 297–309. doi:10.1081/ss-120019092

Kujawski, W., A. Sobolewska, K. Jarzynka, C. Güell, M. Ferrando, and J. Warczok. 2013. Application of Osmotic Membrane Distillation Process in Red Grape Juice Concentration. *Journal of Food Engineering*, 116(4): 801–808. doi:10.1016/j.jfoodeng.2013.01.033

Kumar, Pavan, Neelesh Sharma, Rajeev Ranjan, Sunil Kumar, Z.F. Bhat, and Dong Kee Jeong. 2013. Perspective of Membrane Technology in Dairy Industry: A Review. *Asian-Australasian Journal of Animal Sciences*, 26. doi:10.5713/ajas.2013.13082

Kumar, Rajesha, and A.F. Ismail. 2015. Fouling Control on Microfiltration/Ultrafiltration Membranes: Effects of Morphology, Hydrophilicity, and Charge. *Journal of Applied Polymer Science*, 132(21). doi:10.1002/app.42042

Li, Meisheng, Yijiang Zhao, Shouyong Zhou, and Weihong Xing. 2010. Clarification of Raw Rice Wine by Ceramic Microfiltration Membranes and Membrane Fouling Analysis. *Desalination*, 256(1): 166–173. doi:10.1016/j.desal.2010.01.018

Li, Wen, Guo Qing Ling, Chang Rong Shi, Kai Li, Hai Qin Lu, Fang Xue Hang, Yu Zhang, Cai Feng Xie, Deng Jun Lu, and Hong Li. 2017. Pilot Demonstration of Ceramic Membrane Ultrafiltration of Sugarcane Juice for Raw Sugar Production. *Sugar Tech*, 19(1): 83–88. doi:10.1007/s12355-016-0434-1

Li, Yunbo, Zhuoxiao Cao, and Hong Zhu. 2006. Upregulation of Endogenous Antioxidants and Phase 2 Enzymes by the Red Wine Polyphenol, Resveratrol in Cultured Aortic Smooth Muscle Cells Leads to Cytoprotection against Oxidative and Electrophilic Stress. *Pharmacological Research*, 53(1): 6–15. doi:10.1016/j.phrs.2005.08.002

Liew, M.S., A.G. Fane, and P.L. Rogers. 1997. Fouling of Microfiltration Membranes by Broth–Free Antifoam Agents. *Biotechnology and Bioengineering*, 56. doi:10.1002/(SICI)1097-0290(19971005)56:1<89::AID-BIT10>3.0.CO;2-5

Liguori, Loredana, Paola Russo, Donatella Albanese, and Marisa Di. 2013. Evolution of Quality Parameters during Red Wine Dealcoholization by Osmotic Distillation. *Food Chemistry*, 140(1–2): 68–75. doi:10.1016/j.foodchem.2013.02.059

Liguori, Loredana, Paola Russo, Donatella Albanese, and Marisa Di Matteo. 2018. Production of Low-Alcohol Beverages: Current Status and Perspectives. In: Alexandru Mihai Grumezescu and Alina Maria Holban (eds), *Handbook of Food Bioengineering, Vol. 18: Food Processing for Increased Quality and Consumption*. Academic Press, pp. 347–382. doi:10.1016/B978-0-12-811447-6.00012-6

Limsawat, Pongsathon, and Suwattana Pruksasri. 2010. Separation of Lactose from Milk by Ultrafiltration. *Asian Journal of Food and Agro-Industry*, 3(02): 236–243.

Lipnizki, Frank. 2010. Cross-Flow Membrane Applications in the Food Industry. *Membrane Technology*, 3: 1–24. doi:10.1002/9783527631384.ch1

Lugasi, Andrea, and Judit Hóvári. 2003. Antioxidant Properties of Commercial Alcoholic and Nonalcoholic Beverages. *Food / Nahrung*, 47(2): 79–86. doi:10.1002/food.200390031

Lukanin, Oleksandr S, Sergiy M Gunko, Mikhaylo T Bryk, and Rinat R Nigmatullin. 2003. The Effect of Content of Apple Juice Biopolymers on the Concentration by Membrane Distillation. *Journal of Food Engineering*, 60(3): 275–280. doi:10.1016/S0260-8774(03)00048-7

Luo, Jianquan, Luhui Ding, Benkun Qi, Michel Y. Jaffrin, and Yinhua Wan. 2011. A Two-Stage Ultrafiltration and Nanofiltration Process for Recycling Dairy Wastewater. *Bioresource Technology*, 102(16): 7437–7442. doi:10.1016/j.biortech.2011.05.012

Madaeni, Sayed Siavash, Kiavan Tahmasebi, and Sayed Hatam Kerendi. 2004. Sugar Syrup Concentration Using Reverse Osmosis Membranes. *Engineering in Life Sciences*, 4(2): 187–190. doi:10.1002/elsc.200401801

Makardij, A., X.D. Chen, and M.M. Farid. 1999. Microfiltration and Ultrafitration of Milk: Some Aspects Of Fouling And Cleaning. 77(June).

Mancinelli, Daniele, and Cynthia Hallé. 2015. Nano-Filtration and Ultra-Filtration Ceramic Membranes for Food Processing: A Mini Review. *Journal of Membrane Science & Technology*, 05(02). doi:10.4172/2155-9589.1000140

Mangindaan, Dave, K. Khoiruddin, and I.G. Wenten. 2018. Beverage Dealcoholization Processes: Past, Present, and Future. *Trends in Food Science and Technology*, 71: 36–45. doi:10.1016/j.tifs.2017.10.018

Mänttäri, Mika, Liisa Puro, Jutta Nuortila-Jokinen, and Marianne Nyström. 2000. Fouling Effects of Polysaccharides and Humic Acid in Nanofiltration. *Journal of Membrane Science*, 165(1): 1–17. doi:10.1016/S0376-7388(99)00215-X

Marques, Marisa P., Vítor D. Alves, and Isabel M. Coelhoso. 2016. Concentration of Tea Extracts by Osmotic Evaporation: Optimisation of Process Parameters and Effect on Antioxidant Activity. *Membranes*, 7(1): 1. doi:10.3390/membranes7010001

Marshall, A.D., Peter Munro, and G. Trägårdh. 2003. Influence of Ionic Calcium Concentration on Fouling during the Cross-Flow Microfiltration of β-Lactoglobulin Solutions. *Journal of Membrane Science*, 217(June): 131–140. doi:10.1016/S0376-7388(03)00085-1

Matson, Stephen L. 1985. Process of treating process of treating alcoholic beverages by vapour arbitarated pervaporation. *United States Patent*, issued 1985. doi:10.1016/0375-6505(85)90011-2

Maubois, L.J. 1997. Current Uses and Future Perspectives of MF Technology in the Dairy Industry. In *Bulletin ofthe International Federation No. 320*, Brussels: International Dairy Federation.

McCarthy, Noel A., Heni B. Wijayanti, Shane V. Crowley, James A. O'Mahony, and Mark A. Fenelon. 2017. Pilot-Scale Ceramic Membrane Filtration of Skim Milk for the Production of a Protein Base Ingredient for Use in Infant Milk Formula. *International Dairy Journal*, 73: 57–62. doi:10.1016/j.idairyj.2017.04.010

Merin, U. 1986. Bacteriological Aspects of Microfiltration of Cheese Whey. *Journal of Dairy Science*, 69(2): 326–328. doi:10.3168/jds.s0022-0302(86)80409-x

Mikhaylin, Sergey, Laurent Bazinet, Functional Foods Inaf, and Université Laval. 2015. *Electrodialysis in Food Processing Electrodialysis in Food Processing. Reference Module in Food Science*. Elsevier. doi:10.1016/B978-0-08-100596-5.03116-4

Mistry, Vikram V., and J.-L. Maubois. 2017. Application of Membrane Separation Technology to Cheese Production. In *Cheese: Chemistry, Physics and Microbiology*, 4th ed. pp. 677–697. doi:101016/B978-0-12-417012-400027-2

Mohammad, Abdul Wahab, Ching Yin Ng, Ying Pei Lim, and Gen Hong Ng. 2012. Ultrafiltration in Food Processing Industry: Review on Application, Membrane Fouling, and Fouling Control. *Food and Bioprocess Technology*, 5(4): 1143–1156. doi:10.1007/s11947-012-0806-9

Mondal, Dibyendu, Sanna Kotrappanavar Nataraj, Alamaru Venkata Rami Reddy, Krishna K. Ghara, Pratyush Maiti, Sumesh C. Upadhyay, and Pushpito K. Ghosh. 2015. Four-Fold Concentration of Sucrose in Sugarcane Juice through Energy Efficient Forward Osmosis Using Sea Bittern as Draw Solution. *RSC Advances*, 5(23): 17872–17878. doi:10.1039/c5ra00617a

Muthukumaran, Shobha, Sandra E. Kentish, Geoff W. Stevens, and Muthupandian Ashokkumar. 2006. Application of Ultrasound in Membrane Separation Processes: A Review. *Reviews in Chemical Engineering*, 22(3): 155–194. doi:10.1515/REVCE.2006.22.3.155

Nandi, B K, B. Das, R. Uppaluri, and M.K. Purkait. 2009. Microfiltration of Mosambi Juice Using Low Cost Ceramic Membrane. *Journal of Food Engineering*, 95(4): 597–605. doi:10.1016/j.jfoodeng.2009.06.024

Nene, Sanjay, Suhkvinder Kaur, K. Sumod, Bhagyashree Joshi, and K.S.M.S. Raghavarao. 2002. Membrane Distillation for the Concentration of Raw Cane-Sugar Syrup and Membrane Clarified Sugarcane Juice. *Desalination*, 147(1–3): 157–160. doi:10.1016/S0011-9164(02)00604-5

Norata, Giuseppe Danilo, Patrizia Marchesi, Silvia Passamonti, Angela Pirillo, Francesco Violi, and Alberico Luigi Catapano. 2007. Anti-Inflammatory and Anti-Atherogenic Effects of Cathechin, Caffeic Acid and Trans-Resveratrol in Apolipoprotein E Deficient Mice. *Atherosclerosis*, 191(2): 265–271. doi:10.1016/j.atherosclerosis.2006.05.047

Okawa, Teiichiro, Masayuki Shimada, Yoshihiko Ushida, Nobuo Seki, Naoki Watai, Masatoshi Ohnishi, Yoshitaka Tamura, and Akira Ito. 2015. Demineralisation of Whey by a Combination of Nanofiltration and Anion-Exchange Treatment: A Preliminary Study. *International Journal of Dairy Technology*, 68(4): 478–485. doi:10.1111/1471-0307.12283

Pafylias, I., M. Cheryan, M.A. Mehaia, and N. Saglam. 1996. Microfiltration of Milk with Ceramic Membranes. *Food Research International*, 29(2): 141–146. doi:10.1016/0963-9969(96)00007-5

Paiva, Alexandre, Karen Ranocchia, Marisa Marques, Marco Gomes da Silva, Vítor Alves, Isabel Coelhoso, and Pedro Simões. 2018. Evaluation of the Quality of Coffee Extracts Concentrated by Osmotic Evaporation. *Journal of Food Engineering* 222: 178–184. doi:10.1016/j.jfoodeng.2017.11.020

Pan, Bingjie, Peng Yan, Lei Zhu, and Xianfeng Li. 2013. Concentration of Coffee Extract Using Nanofiltration Membranes. *Desalination*, 317: 127–131. doi:10.1016/j.desal.2013.03.004

Pan, Kai, Qi Song, Lei Wang, and Bing Cao. 2011. A Study of Demineralization of Whey by Nanofiltration Membrane. *Desalination*, 267(February): 217–221. doi:10.1016/j.desal.2010.09.029

Pap, N, S.Z. Kert, and E. Pongr. 2009. *Concentration of Blackcurrant Juice by Reverse Osmosis*, 241(September 2007): 2–6. doi:10.1016/j.desal.2008.01.069

Patino, Marcela. 2017. Evaluation of Forward Osmosis and Thermal Concentration for Quality Retention of Cherry Juice and Concentrate. Master's Thesis, Cornell University. Petrotos, Konstantinos B., and Harris N. Lazarides. 2001. Osmotic Concentration of Liquid Foods. *Journal of Food Engineering* 49(2–3): 201–206. doi:10.1016/S0260-8774(00)00222-3

Petrotos, Konstantinos B., Peter Quantick, and Heracles Petropakis. 1998. A Study of the Direct Osmotic Concentration of Tomato Juice in Tubular Membrane - Module Configuration. I. The Effect of Certain Basic Process Parameters on the Process Performance. *Journal of Membrane Science*, 150(1): 99–110. doi:10.1016/S0376-7388(98)00216-6

Pilipovik, M.V., and C. Riverol. 2005. Assessing Dealcoholization Systems Based on Reverse Osmosis. *Journal of Food Engineering*, 69(4): 437–441. doi:10.1016/j.jfoodeng.2004.08.035

Pinelo, Manuel, Birgitte Zeuner, and Anne S. Meyer. 2010. Juice Clarification by Protease and Pectinase Treatments Indicates New Roles of Pectin and Protein in Cherry Juice Turbidity. *Food and Bioproducts Processing*. doi:10.1016/j.fbp.2009.03.005

Popper, K., W.M. Camirand, F. Nury, and W.L. Stanley. 1966. Dialyzer Concentrates Beverages. *Food Eng.* 38(January): 102–104.
Purwasasmita, M., D. Kurnia, F.C. Mandias, and I.G. Wenten. 2015. Beer Dealcoholization Using Non-Porous Membrane Distillation. *Food and Bioproducts Processing*, 94: 180–186. doi:10.1016/j.fbp.2015.03.001
Quist-Jensen, C., F. Macedonio, C. Conidi, A. Cassano, S. Aljlil, O.A. Alharbi, and E. Drioli. 2016. Direct Contact Membrane Distillation for the Concentration of Clarified Orange Juice. *Journal of Food Engineering*, 187(May). doi:10.1016/j.jfoodeng.2016.04.021
Rai, P., G.C. Majumdar, S. Das Gupta, and S. De. 2007. Effect of Various Pretreatment Methods on Permeate Flux and Quality during Ultrafiltration of Mosambi Juice. *Journal of Food Engineering*, 78(2): 561–568. doi:10.1016/j.jfoodeng.2005.10.024
Raisi, Ahmadreza, Abdolreza Aroujalian, and Tahereh Kaghazchi. 2008. Multicomponent Pervaporation Process for Volatile Aroma Compounds Recovery from Pomegranate Juice. *Journal of Membrane Science*, 322(2): 339–348. doi:10.1016/j.memsci.2008.06.001
Ramachandra Rao, H.G. 2002. Mechanisms of Flux Decline during Ultrafiltration of Dairy Products and Influence of pH on Flux Rates of Whey and Buttermilk. *Desalination*, 144(1–3): 319–324. doi:10.1016/S0011-9164(02)00336-3
Ramarethinam, S., G.R. Anitha, and K. Latha. 2006. Standardization of Conditions for Effective Clarification and Concentration of Green Tea Extract by Membrane Filtration. *Journal of Scientific and Industrial Research*, 65(10): 821–825.
Rana, D, and T. Matsuura. 2010. Surface Modifications for Antifouling Membranes. *Chemical Reviews*, 110(4): 2448–2471. doi:10.1021/cr800208y
Rastogi, Navin K. 2017. *Reverse Osmosis and Forward Osmosis for the Concentration of Fruit Juices. Fruit Juices: Extraction, Composition, Quality and Analysis*. Elsevier Inc. doi:10.1016/B978-0-12-802230-6.00013-8
Schroën, Karin, Anna M.C. van Dinther, Solomon Bogale, Martijntje Vollebregt, Gerben Brans, and Remko M. Boom. 2010. Membrane Processes for Dairy Fractionation. In *Membrane Technology*. Wiley Online Books, pp. 25–43. doi:10.1002/9783527631384.ch2
Shalini, H.N., and Chetan A. Nayak. 2016. Forward Osmosis Membrane Concentration of Raw Sugarcane Juice. In: I. Regupathi, K. Vidya Shetty, and M. Thanabalan (eds), *Recent Advances in Chemical Engineering*. Singapore: Springer, pp. 81–88.
She, Manjuan, and Sun T. Hwang. 2006. Recovery of Key Components from Real Flavor Concentrates by Pervaporation. *Journal of Membrane Science*, 279(1–2): 86–93. doi:10.1016/j.memsci.2005.11.034
Shi, Changrong, Darryn W. Rackemann, Lalehvash Moghaddam, Baoyao Wei, Kai Li, Haqin Lu, Caifeng Xie, Fangxue Hang, and William O.S. Doherty. 2019. Ceramic Membrane Filtration of Factory Sugarcane Juice: Effect of Pretreatment on Permeate Flux, Juice Quality and Fouling. *Journal of Food Engineering*, 243 (September 2018): 101–113. doi:10.1016/j.jfoodeng.2018.09.012
Singh, Rajindar. 2005. Introduction to Membrane Technology. In: Rajindar Singh (ed.), *Hybrid Membrane Systems for Water Purification*. Amsterdam: Elsevier Science, pp. 1–56. doi:10.1016/B978-185617442-8/50002-6
Smithers, Geoffrey W. 2008. Whey and Whey Proteins : From ' Gutter-to-Gold'. *International Dairy Journal*, 18: 695–704. doi:10.1016/j.idairyj.2008.03.008
Sohrabvandi, S., S.M. Mousavi, S.H. Razavi, A.M. Mortazavian, and K. Rezaei. 2010. Alcohol-Free Beer: Methods of Production, Sensorial Defects, and Healthful Effects. *Food Reviews International*, 26(4): 335–352. doi:10.1080/87559129.2010.496022
Sondhi, Rishi, and Ramesh Bhave. 2001. Role of Backpulsing in Fouling Minimization in Crossflow Filtration with Ceramic Membranes. *Journal of Membrane Science*, 186(1): 41–52. doi:10.1016/S0376-7388(00)00663-3

Sotofta, L.J., Christensena, K.V., Andrésenb, R., and Norddahl, B. 2012. Full Scale Plant with Membrane Based Concentration of Blackcurrant Juice on the Basis of Laboratory and Pilot Scale Tests. *Chemical Engineering & Processing: Process Intensification*, 54: 12–21. doi:10.1016/j.cep.2012.01.007

Souza, André L.R., Monica M. Pagani, Manuel Dornier, Flávia S. Gomes, Renata V. Tonon, and Lourdes M.C. Cabral. 2013. Concentration of Camu-Camu Juice by the Coupling of Reverse Osmosis and Osmotic Evaporation Processes. *Journal of Food Engineering*, 119(1): 7–12. doi:10.1016/j.jfoodeng.2013.05.004

Todisco, S., P. Tallarico, and B.B. Gupta. 2002. Mass Transfer and Polyphenols Retention in the Clarification of Black Tea with Ceramic Membranes. *Innovative Food Science and Emerging Technologies*, 3(3): 255–262. doi:10.1016/S1466-8564(02)00046-2

Vaillant, F., P. Millan, G. O'Brien, M. Dornier, M. Decloux, and M. Reynes. 1999. Crossflow Microfiltration of Passion Fruit Juice after Partial Enzymatic Liquefaction. *Journal of Food Engineering*, 42(4): 215–224. doi:10.1016/S0260-8774(99)00124-7

Vera, Edwin, Jenny Ruales, Manuel Dornier, Jacqueline Sandeaux, Françoise Persin, Gerald Pourcelly, Fabrice Vaillant, and Reynes Reynes. 2003. Comparison of Different Methods for Deacidification of Clarified Passion Fruit Juice. *Journal of Food Engineering - J Food Eng*, 59(October): 361–367. doi:10.1016/S0260-8774(02)00495-8

Verma, S.K., R. Srikanth, S.K. Das, and G. Venkidachalam. 1996. An Efficient and Novel Approach for Clarification of Sugarcane Juice by Micro-and Ultrafiltration Methods. *Indian Journal of Chemical Technology*, 3(3): 136–139.

Versari, Andrea, R. Ferrarini, Giovanni Battista Tornielli, G.P. Parpinello, Carlo Gostoli, and Emilio Celotti. 2011. Treatment of Grape Juice by Osmotic Evaporation. *Journal of Food Science*, 69(May): E422–E427. doi:10.1111/j.1750-3841.2004.tb18019.x

Vincze, Ivetta, and Gyula Vatai. 2004. Application of Nanofiltration for Coffee Extract Concentration 162: 287–94.

Vyas, H.K., and P.S. Tong. 2010. Process for Calcium Retention During Skim Milk Ultrafiltration. *Journal of Dairy Science*, 86(9): 2761–2766. doi:10.3168/jds.s0022-0302(03)73872-7

Wang, Y.-N., Wang, R., Li, W., and Tang, C.Y. 2017. Whey Recovery Using Forward Osmosis – Evaluating the Factors Limiting the Flux Performance. *Journal of Membrane Science*. doi:10.1016/j.memsci.2017.03.047

Warczok, J., M. Ferrando, F. Lopez, and C. Güell. 2004. Concentration of Apple and Pear Juices by Nanofiltration at Low Pressures, *Journal of Food Engineering*, 63: 63–70. doi:10.1016/S0260-8774(03)00283-8

Weschenfelder, Thiago André, Pedro Lantin, Marcelo Caldeira Viegas, Fernanda De Castilhos, and Agnes De Paula Scheer. 2015. Concentration of Aroma Compounds from an Industrial Solution of Soluble Coffee by Pervaporation Process. *Journal of Food Engineering*, 159: 57–65. doi:10.1016/j.jfoodeng.2015.03.018

Wijmans, J.G., J. Kaschemekat, R.W. Baker, and V.L. Simmons. 1991. Treatment of Organic-Contaminated Wastewater by Pervaporation. *Environmental Progress*, 9(4).

Wilson, D. Ian. 2018. Fouling during Food Processing – Progress in Tackling This Inconvenient Truth. *Current Opinion in Food Science*, 23(January): 105–112. doi:10.1016/j.cofs.2018.10.002

Wrolstad, Ronald E. Mcdaniel, Mina R. Durst, Robert W. Micheals, Nancy Lampi, Keith A. Beaudry, Edward, G. 1993. Composition and Sensory Characterization of Red Raspberry Juice Concentrated by Direct-Osmosis or Evaporation. *Journal of Food Science*, 58(3): 633–637. doi:10.1111/j.1365-2621.1993.tb04344.x

Yan, S.H., C.G. Hill, and C.H. Amundson. 1979. Ultrafiltration of Whole Milk. *Journal of Dairy Science*, 62(1): 23–40. doi:10.3168/jds.s0022-0302(79)83197-5

Youn, Kwang Sup, Joo Heon Hong, Dong Ho Bae, Seok Joong Kim, and Soon Dong Kim. 2004. Effective Clarifying Process of Reconstituted Apple Juice Using Membrane Filtration with Filter-Aid Pretreatment. *Journal of Membrane Science*, 228(2): 179–186. doi:10.1016/j.memsci.2003.10.006

Yousefnezhad, Bahman, Hossein Mirsaeedghazi, and Akbar Arabhosseini. 2016. Pretreatment of Pomegranate and Red Beet Juices by Centrifugation Before Membrane Clarification: A Comparative Study: Red Beet and Pomegranate Juices Pretreatment. *Journal of Food Processing and Preservation*, 41(April). doi:10.1111/jfpp.12765

Zhu, Zhenzhou, Houcine Mhemdi, Lu-Hui Ding, Olivier Bals, Michel Jaffrin, Nabil Grimi, and Eugene Vorobiev. 2015. Dead-End Dynamic Ultrafiltration of Juice Expressed from Electroporated Sugar Beets. *Food and Bioprocess Technology*, 8(March). doi:10.1007/s11947-014-1427-2

7 Irradiation Technology for the Food Industry

Dipika Trivedi and Anil Kumar Dikshit
Indian Institute of Technology, Bombay, Maharashtra, India

CONTENTS

7.1 Introduction 165
7.2 Working Principles 166
7.3 Objectives 168
7.4 Advantages and Disadvantages of Irradiation 168
7.5 Classification 169
7.6 Sources of Irradiation Used to Date 169
7.7 Impacts on Food and Environment 170
7.8 Legislation and Regulation 172
7.9 Industrial Set-Up and Challenges 173
7.9.1 Setting Up a γ-Ray-Based Irradiation Facility 173
7.9.2 Setting Up an X-Ray-Based Irradiation Facility 175
7.9.3 Setting Up an Electron Beam-Based Irradiation Facility 175
7.10 Issues and Challenges 176
7.10.1 Cost of Food 176
7.10.2 Consumer Awareness and Labeling 176
7.10.3 International Trade 177
7.10.4 Security Of Foods Being Irradiated 177
7.11 Economics 178
7.12 Future Trends 179
7.13 Concluding Remarks 179
References 180

7.1 INTRODUCTION

Food irradiation is a process that employs ionizing radiation or electron beams for enhancing the shelf life and safety of food. Foods that undergo irradiation processes retain their flavours and aroma that would be diminished by other processing methods such as heating. The application of irradiation on retention of food quality includes application of gamma irradiation on starch and wheat flour (Bashir et al., 2017), gamma radiation and electron beam to conserve bioactive compounds of wild

Arenaria montana L (Pereira et al., 2016), effect of gamma radiation and electron beam on fresh Agaricus bisporus (Cardoso et al., 2019), effect of irradiation on chemical and antioxidant properties of mushroom (Fernandes et al., 2015; Fernandes et al., 2016). There were no toxicological risks, nutritional or microbial issues when food was subjected to ionizing radiation of less than 10 kGy. Application of high-dose irradiation to microbial decontamination and sterilization of different foodstuffs includes gamma irradiation of sorghum flour (Mukisa et al., 2012), space food sterilized by irradiation (Song et al., 2012), gamma-irradiation on Dakgalbi (Yoon et al., 2012), gamma-ray irradiation on beef meat (Stefanova et al., 2011). The controlled application of energy to foods and related products from ionizing radiation using γ-rays, X-rays, or high-energy electron beams for a specified period of time is called irradiation. During the past few decades, it has proved to be a safe and effective technology for food products, being free of harmful chemicals and achieving a significant reduction in the incidence of foodborne diseases. International statutory bodies have not only accepted and encouraged irradiation processing as the best available option but have also helped to standardize its industrial application, and to approve and certify products for marketing and international trade. The history of irradiation spans more than 100 years of scientific research and testing on a wide spectrum of edible items and raw foods.

From the beginning of civilization, human societies have benefited from the presence of insect and pests in agricultural commodities, the destruction of microbes, pathogens and parasites causing food spoilage, the stoppage of tuber and bulb sprouting, the delay in ripening/senescence of perishable fruits, vegetables, milk, poultry, eggs, fish, meat, spices and other dry ingredients, the in-situ handling of items both with and without packaging, and the extension of storage for long/infinite shelf life with and without refrigeration. There is no need for further chemical treatment and the addition of artificial additives (Farkas, 1998).

When irradiation technology began to be applied in the food industry, several allied disciplines became involved, including microbiology, biotechnology, nutrition science, toxicology, physics, chemistry, food technology, packing and supply chain, and logistics.

The resistance and the quantum (quality and quantity) of microbial and pathogenic loads dictate the intensity and duration of the radiation energy required and, accordingly, the specific process/protocol. Customizing irradiation technology and its associated costs involves various factors, such as the type and composition of product, the moisture conditions, raw or processed form, fresh, refrigerated or frozen state, the operating environment with or without oxygen, and the proposed application.

7.2 WORKING PRINCIPLES

The basic science behind irradiation is that the γ-rays and accelerated electrons differ from other forms of radiation based on their ionizing ability. Once these are absorbed by materials, the chemical bonds are broken and the ionization produces many kinds of electrically charged ions and neutral free radicals. These radicals cause radiolysis of microbes, insects, and parasites, oxidizing them directly or rupturing their cell walls through reaction(s) with the wall materials. A series of such reactions is thought

$$H_2O \rightarrow H_2O^+ + e^-$$

$$e^- + H_2O \rightarrow H_2O^-$$

$$H_2O^+ \rightarrow H^+ + OH.$$

$$H_2O^- \rightarrow H. + OH^-$$

$$H. + H. \rightarrow H_2$$

$$\text{Or } OH. + OH. \rightarrow H_2O_2$$

$$\text{Or } H. + OH. \rightarrow H_2O$$

$$\text{Or } H. + H_2O \rightarrow H_2 + OH.$$

$$\text{Or } OH. + H_2O_2. \rightarrow H_2O + H_2O.$$

$$H. \rightarrow O_2HO_2.$$

FIGURE 7.1 Reactions for water breakdown due to ions and radicals from irradiation

to be the cause of the destruction of parasites during food irradiation (see Figure 7.1 for a typical example). Fortunately, radiation cannot make changes to food materials, including their mineral and nutrient contents. However, the loss of vitamins is a cause for concern.

As seen in Figure 7.1, the water present in foods is ionized by radiation and electrons are released. The electrons trigger various chemical reactions, and the radicals generated cause lysis of chemical bonds. These then recombine to form hydrogen radicals, hydroxyl radicals, hydroperoxyl radicals, hydrogen, hydrogen peroxide, etc. Although, these radicals have a life of less than 10^{-5} s, they destroy bacterial cells and kill microorganisms. Treated food thus not only becomes safe but also has extended shelf life. However, for certain food items and products, appropriate pretreatment may be required.

Farkas (1998) has documented the effect of irradiation both on the microbial quality of various foods and on various microbes, fungi, algae, parasites, insects etc. Irradiation is very successful against food-contaminating living organisms that contain DNA and/or RNA. It does not cause significant loss of macronutrients, micronutrients, vitamins and minerals present in food, although unfortunately some sensory changes in treated food have been reported. Carbohydrates, lipids, and proteins undergo very little modification in nutritional value even with irradiation doses over 10 kGy. Essential amino acids, fatty acids, minerals, and trace elements are also unchanged. There can be a decrease in certain vitamins (mostly thiamin) of the same order of magnitude as occurs in other food preservation techniques like drying or canning (IAEA, 1996; WHO, 2005). Moderate irradiation with radiation doses of 1–5 kGy is regarded as the best available technology for controlling foodborne diseases and spoilage.

7.3 OBJECTIVES

The major objectives of irradiation of food products and items are:

- disinfestation of insects and pests in agricultural commodities;
- elimination of microbes responsible for food spoilage;
- destruction of food pathogens and parasites causing foodborne diseases;
- inhibition of sprouting of tubers and bulbs;
- extension of fruit and vegetable ripening and senescence;
- decontamination of poultry, eggs, fish, meats, and other perishable items to make them safe for consumption whether raw or cooked;
- extension of shelf life with and without refrigeration;
- elimination of chemical and artificial additives in food;
- treatment of items without or with packaging;
- significant reduction in foodborne diseases among consumers;
- ease of both processing and preservation.

7.4 ADVANTAGES AND DISADVANTAGES OF IRRADIATION

The major advantages are:

- chemical-free substitute;
- ease of implementation and day-to-day operation;
- enhancement of food safety;
- reduction in food loss;
- reduction in foodborne diseases;
- achievement of desired effects, such as killing pathogens, controlling sprouting, extending shelf life, replacing chemical fumigation, addition of chemicals or artificial additives, etc.;
- no change in texture with partial or complete cooking due to implementation of technology in a common environment;
- low to moderate operating costs compared to most heat-based technologies.

The disadvantages are:

- high capital cost;
- treatment with γ-rays requires approval, clearance, and strict inspection from the regulatory authority;
- public perception and fear of radioactivity in the irradiated food;
- concern and misconceptions about loss in nutritional values, particularly for vitamins;
- economic and logistical complexities;
- opposition due to public perceptions and lack of knowledge on wholesomeness of irradiated food;
- noticeable sensory changes limit dose for certain items;
- not appropriate for foods that cannot be frozen as irradiation at sub-freezing temperatures can reduce off-flavor formation in certain foods;

- mandatory post-irradiation packaging increases costs;
- not intended to provide universal solution for all food preservation issues.

7.5 CLASSIFICATION

Irradiation has many applications and a very versatile role in the processing and conservation of foods. It can be classified according to intensity of irradiation or type. The types of irradiation process can be categorized by the intention of processing and the dose used (Kalyani and Manjula, 2014).

The radiation response in microbes, expressed by the decimal reduction dose (D_{10}-value), is a measure of the degree of effectiveness of radiation. The D_{10}-value is defined as the radiation dose required to kill 90% of the total viable number of microorganisms. The radiation dose, measured in Gray (Gy), is the amount of ionizing energy that is absorbed by food during the duration of irradiation. One Gy is one joule energy absorbed by one kilogram of food. 1 kilo Gray (kGy) is equal to 1,000 Gy. The absorbed radiation dose used to be measured in rad, with 1 Gy being equal to 100 rad. The radiation dose to be applied during the irradiation of a specific food item is determined according to two factors: the resistance of the organisms present, and the objective of the treatment. The World Health Organization (WHO), in its guideline (1981) has recommended that the average dose should not exceed 10 kGy and the maximum dose must be below 15 kGy.

The classification of applications based on radiation intensity is shown in Table 7.1, while that based on type of radiation is given in Table 7.2. The tables also show irradiation dose (D_{10}-value) suggested to eliminate some non-spore-forming pathogenic bacteria in non-frozen and frozen foods, according to the guidelines provided by various agencies.

7.6 SOURCES OF IRRADIATION USED TO DATE

At present, sources of radiation in irradiation equipment are either a machine radiating a high-energy isotope such as cobalt-60 (^{60}Co) or cesium-137 (^{137}Cs) to produce γ-rays, a machine using electricity to produce rays such as X-rays, or a high-energy electron beam. The ionizing energy of any type of ray is the same and is measured by the same unit of Gy. The degree of radiation or goals achieved depends upon the dose applied and several other factors, including the power rating of the source/machine. The sources and their associated characteristics for three kinds of irradiators are given in Table 7.3. More innovative and safe sources of irradiation are yet to be invented.

γ-rays: electromagnetic radiation of very short wavelength, emitted by radioactive isotopes of cobalt-60 (^{60}Co) and cesium-137 (^{137}Cs).

X-rays: ionizing electromagnetic radiation of a wide variety of short wavelengths, usually produced by a machine using electrical power in which a high-energy electron beam bombards a metallic target in high vacuum and generates X-rays.

High-energy electrons: the streams or beams of electrons accelerated by a machine to energies of up to 10 million electron volts (MeV). Here, the electrons, commonly called β-rays, may also be emitted by some radioactive materials.

TABLE 7.1
Classification Based on Intensity of Radiation

Intensity of Dose	Purpose	Typical Dose (kGy)	Typical Example
Low (0–1 kGy)	Inhibition or prevention of sprouting and delay in senescence	0.01–0.15	Potatoes, onions, garlic, root ginger, yam
	Disinfection to quarantine	0.1–0.7	Fruits
	Insect disinfection	0.2–0.7	Grains, pulses, cereals, flour, dried fruits and vegetables, dried nuts, dried fishery products, coffee beans, spices, and other dried food products
	Delay in ripening	0.2–1	Fruits and mushrooms
	Inactivation of pathogenic parasites	0.3–0.5	Meats
Medium (1–10 kGy)	Increasing shelf life and/or enhancing preserving quality	1–2	Meat, poultry, and seafood
	Improving hygienic quality	1–3	Refrigerated foods
	Improving technological properties	2–7	Grapes (for increasing juice yield) and dehydrated vegetables (for reducing cooking time)
High (10–100 kGy)	Microbial decontamination	10–30	Dried spices, herbs and other dried vegetable seasonings
	Decontamination	10–50	Food additives, enzyme preparations, natural gum and ingredients
	Achieving long shelf life without refrigeration	20–70	Food products in hermetically sealed containers
	Industrial sterilization (achieved in conjunction with mild heat)	30–50	Prepared foods, poultry, meat, seafood, sterilized hospital diets, meals for astronauts

7.7 IMPACTS ON FOOD AND ENVIRONMENT

The impacts of irradiation on bacteria, yeasts, molds, viruses, and insects are discussed in detail by Mostafavi et al. (2009). It has been reported by several researchers and testing agencies that, even at doses higher than 10 kGy, very few changes are caused by irradiation to carbohydrates, proteins, and lipids, while essential amino acids, minerals, trace elements, and most vitamins are not altered at all (Mostafavi et al., 2009).

Vitamins such as riboflavin (B_2), niacin (B_3), pyridoxine (B_6), and D are not affected by irradiation, but vitamins A, B_1 (thiamine), C (ascorbic acid), E, and K are negatively impacted as they are sensitive. However, their sensitivity is not same for all kinds of food. The complexity of the food system, the solubility in water or fat, and the environment dictate the extent of loss (Marcotte, 2005; Miller, 2005).

The nutritional quality of foods is not subject to change during irradiation treatment. For example, the nutritional value of foods is insignificantly affected at low doses up to 1 kGy, has minor losses at medium doses of 1–10 kGy, and major losses at high doses above 10 kGy. Hence, carrying out irradiation at low temperature with

TABLE 7.2
Classification Based on Different Forms of Radiation Applied in Food

Method of Radiation	Purpose	Typical Dose (kGy)	Typical Example
Photo-sanitization	Protecting plant health/ ensuring their insured control	0–1	Potatoes, onions, garlic, root ginger, yam, grains, pulses, cereals, flour, fruits and vegetables, meat, mushrooms, dried nuts, dried fishery products, coffee beans, spices, other dried food products
Randomization/ pasteurization by irradiation	Extending shelf life	1–10	Potatoes, onions, garlic, root ginger and yam
Rededication/ radio-pasteurization	Improving the hygienic quality	7–10	Fruits
Radappertization	Extending extra-long shelf life under ambient conditions	25–70	Food additives, enzyme preparations, natural gum, ingredients and food products in hermetically sealed containers
Radiation sterilization	Storing foods at room temperature/preventing recontamination	30–50	Prepared foods, poultry, meat, seafood, sterilized hospital diets and meals for astronauts

TABLE 7.3
Characteristics of Radiation Sources

Source	Half-life (years)	Energy (MeV)
Cobalt-60 to produce γ-rays	5.3	1.17, 1.33
Cesium-137 to produce γ-rays	30	0.66
Electricity to produce X-rays	–	7.5
Electricity to accelerate electron beams	–	10

a low dose and in the absence of oxygen (vacuum inside packing) for longer duration is recommended as more practicable.

Contrary to popular belief, irradiation does not make food radioactive or create any new radioactive material. The types of radiation source approved for the treatment of foods cannot make food become radioactive. Moreover, the food never comes into contact with the source. The radiation source is generally kept at a lower level than the treatment platform or conveyor. During use, the source of γ-rays is raised and the radiation falls upon the items only for a specified time, much like luggage passing through an airport X-ray scanner or teeth being X-rayed. When the radioactive source, such as cobalt or cesium, reaches its expiry date, it is returned to the supplier or recycler in accordance with regulatory requirements.

When the source of irradiation is X-ray or electron-beam generating machines that use electricity as their power source, there is no concern either about radioactivity or

radioactive materials. There have been several cases of botulism poisoning due to canned foods. The same possibility also exists with irradiation; hence, food under low- and medium-dose irradiation must be exposed to extended time so that spoilage and pathogenic bacteria, including *Clostridium Botulinum* bacteria, are killed. The spoiling impact of yeasts and molds on most foods can be controlled using irradiation at a higher dose of at least 3 kGy to inactivate them.

Viral contamination cannot be eliminated by irradiation as viruses are highly resistant to radiation and require a very high dose of 20–50 kGy to inactivate them. Since their absence cannot be absolutely guaranteed, irradiation is normally not considered suitable for dealing with viral contamination of foods.

There is a misconception that harmful mutant strains of pathogenic microorganisms might flourish during or even after irradiation. The potential risks associated with irradiation have been studied and it is reported that irradiation inactivates spoilage- and disease-causing bacteria by severely damaging their chromosomes. In other words, any pathogenic bacteria surviving in food post irradiation would be significantly injured and thus unable to reproduce. Normally, irradiation cannot damage the food being treated.

Processing by any preservation method may have an effect on the sensory quality of foods, and irradiation is no different. Hence, an array of food-processing technologies is recommended to render food safe for consumption and to preserve it for an indefinite period.

Milk and dairy products top the list of the most radiation-sensitive foods. 'Off' flavor is imparted to dairy products even at an extremely low dose of 0.1 kGy. That is why milk and dairy products for direct consumption are never irradiated. Certain fresh fruits and vegetables are softened after irradiation as their sensitive cell walls are destroyed or broken down. Many types of meat products are found to have 'off' flavor during high-dose irradiation sterilization. Too high a dose affects the food's sensory qualities, while too low a dose will not be effective. Treatment of foods like meats should therefore be carried out under proper conditions and using the optimum dose for optimum contact time.

As of now, food scientists have developed databases on the suitability of food types and the appropriate environmental and operational conditions to prevent damage or 'off' flavor. A code of practice for various applications, including roots and tubers, cereals and legumes, meat, poultry, fish and seafood, most fruits and vegetables, spices and seasonings, is under development. The aim is to avoid noticeable changes in sensory quality, while sensitive food products such as meat and fish must be irradiated under low temperature within proper atmospheric packaging to avoid 'off' flavor.

7.8 LEGISLATION AND REGULATION

As with any other technology, several agencies are involved in developing proper irradiation techniques and methodologies to meet specific purposes or for specific foods. These include legislation and regulation, granting licenses, approvals, or authorization, and operation and safety regulations. Since irradiation with γ-rays involves the use of radiative source such cobalt-60 or cesium-137, in most countries

proper guidelines are established by atomic research agencies, including inspection and the granting of permission to build and operate.

The safety and effectiveness of irradiation technology as a method of food preservation/processing was recognized by the Codex Alimentarius Commission with the adoption of a Codex General Standard for Irradiated Foods. National clearances on food irradiation are recorded in the database maintained by the Food Preservation Section of the Joint FAO/IAEA (International Atomic Energy Agency) Division in Vienna.

7.9 INDUSTRIAL SET-UP AND CHALLENGES

7.9.1 Setting Up a γ-Ray-Based Irradiation Facility

An important aspect of setting up a radiation processing facility is site selection (Figure 7.2). The geological features of the site should be evaluated to understand the effect on integrity and security of radiation shields. When assessing the suitability of the site, areas with potential subsurface or surface subsidence, uplift, or collapse should be considered.

When locating the irradiator, the following minimum distances should be maintained from the boundary wall of the irradiator:

- At least 30 m from public places and residential areas
- At least 2 km from explosives dumps, ammunition storage, military and civil airfields.

The major components of the facility are:

- A source of γ-radiation (cobalt-60/cesium-137)
- A radiation processing cell (irradiation cell)
- Product conveyors and control mechanisms
- Safety devices and interlocks.

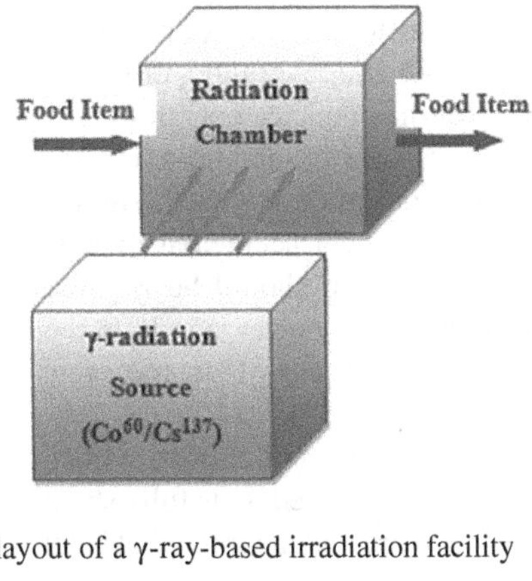

FIGURE 7.2 Typical layout of a γ-ray-based irradiation facility

Source of γ-radiation: The power required for a radiation facility using cobalt-60 or cesium-137 as a source may be estimated by the following equation if the required dose, desired output, and the efficiency of the irradiator are known:

The mechanical strength of the storage area structure should be able to support the source transport container. While the source is stored, forced ventilation or water circulation is provided for continuous removal of decay heat from the radiation source. In forced ventilation, the source coolant is controlled and monitored.

The water pool construction material must be corrosion resistant. To prevent seepage of water, the inner surface of the pool must be lined with either ceramic tiles or stainless steel.

The integral sources are fixed firmly in a rigid frame, which protects them from getting dislodged accidentally or during normal use. The source frame material must be corrosion resistant. Tools used to install and remove the sources should be capable of operation from outside radiation shields.

The safety and smooth working of the process is ensured by the source frames, which should be capable of being moved, by at least two wire ropes, from their shielded position to the irradiation position, which has identical specifications. The wire ropes used to move the source frame should conform to the applicable standard specifications.

The source frame should be able to return to its fully shielded position automatically in the event of any irregularities such as power failure or smoke/fire alarm.

Radiation processing cell: The radiation level outside the walls and above the ceiling of the cell at any occupied location should not exceed 1 μSv/h. The cell must be surrounded by concrete walls 1.5–1.8 m thick to ensure safe levels.

During operation, first the source is raised, then packaged food is loaded onto automatic conveyor belts and passed through the radiation field in a circular path. Thus, the use of the emitted radiation is maximized and a uniform dosage is ensured. A more complex materials-handling system is required for isotope sources than for machine sources. An isotope source is shielded within a pool of water under the process area, as it cannot be switched off and also allows personnel to enter.

Product conveyors and control mechanisms: Processing is carried out either by positioning the product close to the fixed source or by moving the source close to the product via conveyors. In the latter case, the source movement may be either horizontal or vertical. Motive power to the product positioning system and the source movement system may be pneumatic, hydraulic, or electrical. The power should be cut off during any servicing or maintenance operation.

Safety devices and interlocks: Smoke and heat sensors with an audio-visual display are installed in the radiation processing cell. An automatic fire-extinguishing system should be provided to prevent accidental operation. Chemical substances used in the fire-extinguishing system should be such that they do not adversely affect the integrity of the source. The power supply to the cell and the cell ventilation system should be automatically cut off in the event of fire or smoke inside the cell.

Facilities where significant NO_x concentrations are likely should be provided with gas detectors. After the source has returned to its fully shielded position, a time-delay interlock needs to be provided to prevent the immediate entry of personnel to the cell.

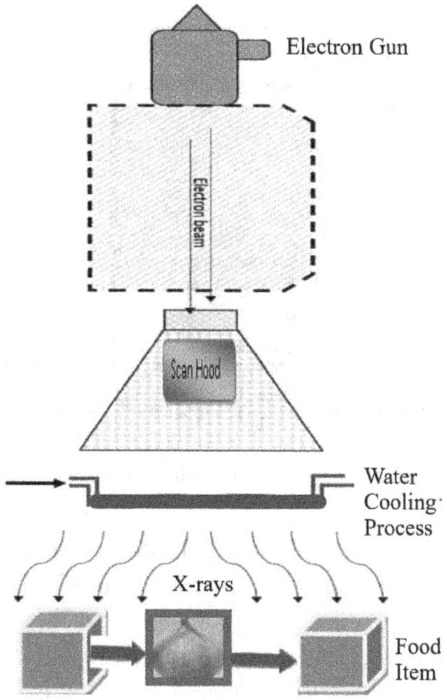

FIGURE 7.3 Typical machine for X-ray-based irradiation facility

7.9.2 Setting Up an X-Ray-Based Irradiation Facility

The machine sources are electron accelerators consisting of a heated cathode which supplies electrons and an evacuated tube in which electrons are accelerated by a high-voltage electrostatic field using electricity. The electrons are bombarded onto a suitable target material to produce X-rays (Figure 7.3).

The advantages of the machine sources are as follows:

- They can be switched off as required.
- To ensure an even dose distribution, the electron beams can be directed over the packaged food.
- Handling equipment is relatively simple.

Even at the lowest commercial doses (0.1 kGy), stringent safety procedures must be followed while working in this environment because just one 5 Gy dose is sufficient to kill an operator. To prevent the source from being raised when personnel are present, entry to all buildings must be prohibited during operations (Aikawa, 2000).

7.9.3 Setting Up an Electron Beam-Based Irradiation Facility

These machine sources are similar to X-ray irradiators except that the electron beams produced are directly accelerated to high kinetic energy (Figure 7.4).

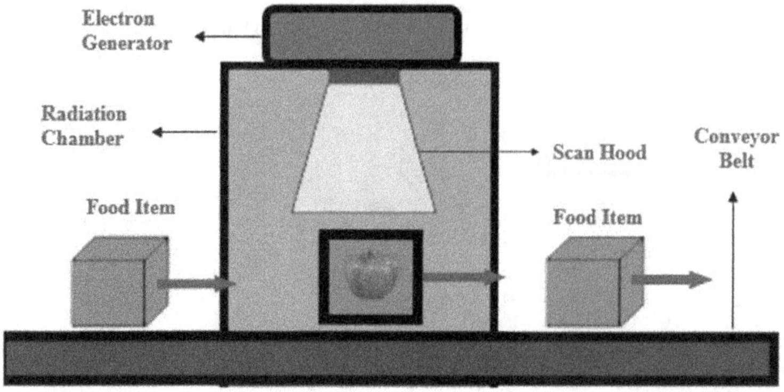

FIGURE 7.4 Typical machine for high-energy electron beam-based irradiation facility

The electrons do not penetrate as far into food as γ-rays or X-rays, and may only be used for treatment of thin packages of food (Figure 7.4). The β-particles that are usually chosen are easy to produce electronically, but they do not have deep penetrating power, and their velocity needs to be enhanced. Electron beams are therefore accelerated and their high energy can also help them penetrate through thick packaging.

7.10 ISSUES AND CHALLENGES

Notwithstanding all these developments, the irradiation of food items is a concern for consumers. The following are a few of the most prominent consumer concerns and issues which need to be alleviated through dissemination of further accredited scientific information.

7.10.1 Cost of Food

The safety and hygiene standards offered by this treatment justify its additional costs. The increased processing costs add to the price of the food but the value added gives the treated food certain competitive advantages. Processes like pasteurization, irradiation, refrigeration, and fumigation increase the end product cost, but consumers benefit in terms of quality, availability, quantity, safety, and convenience.

7.10.2 Consumer Awareness and Labeling

While there is no possibility of radioactivity being induced in food items (Marcotte, 2005), lack of understanding about food irradiation leads to serious misconceptions among consumers who may associate it with nuclear technologies. Consumers may fail to make a distinction between irradiated food and radioactive food. A thorough understanding of the irradiation process would enable them to understand that the food is exposed only to gamma rays (i.e., electron beams of a maximum range of 10 MeV), not to radiation.

FIGURE 7.5 Radura logo for irradiated food

The main aim of labeling is to offer information and choice to consumers. Most national and international authorities that permit the use of food irradiation require food to be treated and labeled properly, with an international food irradiation logo such as the Radura (Figure 7.5) included in the label (WHO, 1981) . The logo, created in the Netherlands, is officially recognized globally by the WHO and the International Consulting Group on Food Irradiation (ICGFI) as the symbol indicating a product processed using irradiation.

7.10.3 International Trade

Although the process and the products of food irradiation have been widely tested, and their high quality has been proven, they account for an insignificant proportion of the global food market. Only a few countries have regulatory approval for irradiation processing, which is also limited to a small number of products and applications.

In general, consumers tend to be reluctant to accept food products treated by any new technology. Some may initially be opposed to irradiation, but if the technology is explained appropriately, they will start to favor it. It is therefore important for professional bodies to launch consumer awareness campaigns that explain the benefits and disadvantages of the technology, to enable informed choice for consumers (Marcotte, 2005). Market-oriented trials could form a part of these campaigns.

There may soon be a significant increase in market share for irradiated food. Irradiation could be used to maintain the strict hygiene standards enabling entry of various food items to international trade. Several irradiated foods, including dried vegetable seasonings, spices, herbs, and mechanically deboned poultry, have successfully entered international markets, and there is growing demand for the international regulation and promotion of worldwide trade in irradiated food.

7.10.4 Security Of Foods Being Irradiated

Scientific studies over a period of 20 years report several methods for checking whether or not foods have been irradiated. Unfortunately, different methods are needed for different foods, and none of them work for all food items.

According to the Codex Alimentarius Commission, the following methods are suggested for analyzing irradiated foods:

1. Gas chromatographic (GC) analysis of hydrocarbons – Irradiated food containing fat
2. GC/mass spectrometric analysis of cyclobutanones – Irradiated food containing fat
3. Electron spin resonance spectroscopy – Irradiated food containing bone
4. Electron spin resonance spectroscopy – Irradiated food containing cellulose
5. Thermo-luminescence analysis – Irradiated food containing silicate materials

Government initiatives and support could be vital in expanding and strengthening collaboration between the various agencies and institutional food suppliers. This will go a long way towards the adoption and integration of irradiated food items into the principal supply chains, helping to promote commercialization and widespread application of the technology.

7.11 ECONOMICS

The economics of radiation processing will depend on different parameters such as dose range, nature and density of product, efficiency of the irradiator, operating hours, annual throughput, operating costs, number of personnel required, and so on. The major capital items include land, building, license and other registration fees, radiation processing cell and shielding, capital cost of irradiation source, warehousing facilities, product handling, transport systems, and auxiliary equipment. Operational items include cost of the irradiation source, salaries, utilities and communications, taxes and interest, inspection costs, and administration.

The total processing cost thus includes the total operating costs, depreciation of capital, and borrowing costs. The cost of unit processing will be determined by equating total processing cost with the annual throughput. A reasonably high profit margin must also be included as it is key to the success of any commercial venture. With appropriate administrative and institutional/logistical arrangements, the break-even point can be achieved in two to three years. Table 7.4 indicates current estimated additional costs of irradiation.

TABLE 7.4
Additional Product Costs Due to Irradiation Treatment

Radiation Treatment	Typical Cost Addition ($/lb)
Low radiation	0.01–0.02
Medium radiation	0.10–0.15
High radiation	0.20–0.30

7.12 FUTURE TRENDS

The use of irradiation as a hygienic treatment will grow with the rise in public awareness levels, increasing consumer health consciousness, and the high risk of disease associated with foods contaminated by a wide range of pathogenic microbes. In a number of countries, especially the USA, the food industry has already embraced the use of irradiation as a cold pasteurization procedure to ensure hygiene quality in solid foods items, similar to the thermal pasteurization successfully applied to liquid foods. It is important to note that irradiation, like other food-processing technologies, can only be applied to foods requiring enhancement for either technical or economic reasons, and there must be minimal nutritional changes after the treatment. There is a need to recognize the fact that irradiation technology cannot and should not be used for all food items.

7.13 CONCLUDING REMARKS

The safety and the effectiveness of radiation have made it a well verified, reliable food process, particularly for the control of foodborne diseases, in addition to its numerous benefits. It has been accepted and adopted by regulatory authorities all over the world. If food safety and security are of paramount importance to a nation, then it is a cheaper as well as an effective option for the food industry. The food industry also finds it attractive on account of several direct benefits; along with reduced chances of illness and deaths caused by accidental marketing of contaminated products goes a reduction in the cost of product recalls and lawsuits, and loss of brand reputation. For organic food such as fresh meat and poultry products, irradiation technology has been the key to quality preservation and standardization, as it ensures microbiological safety.

Although food irradiation may not be the only solution to foodborne diseases, like the heat pasteurization of milk it has definitely prevented many infections caused by solid food items, thereby enhancing the pathogenic safety of important food supply segments at a relatively lower cost than the costs associated with foodborne diseases. Food-exporting countries are finding radiation of treated foods is widely accepted by their main trading partners. It is satisfying to note that WHO and several respected non-profit organizations, including the American Medical Association, the American Dietetic Association, the Institute of Food Technologists, and the US Council for Agricultural Science and Technology, incline positively towards processing of food by irradiation for safety and have included the application of irradiation technology as part of their overall health programs and intervention strategies.

More than 10 decades of research have seen the safe and successful use of irradiation treatment for food safety. In today's food industry its safety and efficacy have been continuously assessed and judged acceptable, and it is now used more than any other technology, even canning. International regulatory bodies, including WHO, FAO, IAEA, and Codex Alimentarius have not only accepted its efficacy and safety but have also approved the process for worldwide applications. Irradiation ability is becoming a decisive factor in food security and consumer safety and the use of irradiated food will gain worldwide acceptance over the next few years. Irradiation should never, however, be accepted as an excuse for poor quality, poor handling, or poor storage of foods, or as a substitute for good manufacturing and hygiene practices.

REFERENCES

Aikawa, Y. 2000. A new facility for X-ray irradiation and its application. *Radiation Physics and Chemistry*, 57: 609–612.

Bashir, K., Swer, T.L., Prakash, K.S., and Aggarwal, M. 2017. Physico-chemical and functional properties of gamma irradiated whole wheat flour and starch. *LWT - Food Science and Technology*, 76: 131–139.

Cardoso, R.V., Fernandes, Â., Barreira, J.C., Verde, S.C., Antonio, A.L., Gonzaléz-Paramás, A.M., Barros, L. and Ferreira, I.C. 2019. Effectiveness of gamma and electron beam irradiation as preserving technologies of fresh Agaricus bisporus Portobello: A comparative study. *Food Chemistry*, 278: 760–766.

Farkas, J. 1998. Irradiation as a method for decontaminating food. *International Journal of Food Microbiology*, 44: 189–204.

Fernandes, Â., Barreira, J.C.M., Antonio, A.L., Oliveira, M.B.P.P., Martins, A., and Ferreira, I.C.F.R. 2016. Extended use of gamma irradiation in wild mushrooms conservation: Validation of 2 k Gy dose to preserve their chemical characteristics. *LWT - Food Science and Technology*, 67: 99–105.

Fernandes, Â., Barreira, J.C.M., Antonio, A.L., Rafalski, A., Oliveira, M.B.P.P., Martins, A., and Ferreira, I.C.F.R. 2015. How does electron beam irradiation dose affect the chemical and antioxidant profiles of wild dried Amanita mushrooms? *Food Chemistry*, 182: 309–315.

IAEA. 1996. *Working Material on the Proceedings of the Regional Seminar on Food Irradiation to Control Food Losses and Food borne Disease and Facilitate Trade*. Rabat, Morocco.

Kalyani, B. and Manjula, K. 2014. Food irradiation – Technology and application. *International Journal of Current Microbiology and Applied Sciences*, 3: 549–555.

Kumar, S., Saxena, S., Verma, J., and Gautam, S. 2016. Development of ambient storable meal for calamity victims and other targets employing radiation processing and evaluation of its nutritional, organoleptic, and safety parameters. *Food Science and Technology*, 69: 409.

Mostafavi, H.A., Fathollahi, H. Motamedi, F., and Mirmajlessi, S.M. 2009. Food irradiation: Applications, public acceptance and global trade. *African Journal of Biotechnology*, 9(20): 2826–2833.

Mukisa, I.M., Muyanja, C.M.B.K., Byaruhanga, Y.B., Schuller, R.B., Langsrud, T., and Narvhus, J.A. 2012. Gamma irradiation of sorghum flour: Effects on microbial inactivation, amylase activity, fermeability, viscosity and starch granule structure. *Radiation Physics and Chemistry*, 81: 345–351.

Pereira, E., Barros, L., Barreira, J.C.M., Carvalho, A.M., Antonio, A.L., and Ferreira, I.C.F.R. 2016. Electron beam and gamma irradiation as feasible conservation technologies for wild Arenaria montana L.: Effects on chemical and antioxidant parameters. *Innovative Food Science and Emerging Technologies*, 36: 269–276.

Song, B., Park, J., Kim, J., Choi, J., Ahn, D., Hao, C., Lee J. 2012. Development of freeze dried miyeokguk, Korean seaweed soup, as space food sterilized by irradiation. *Radiation Physics and Chemistry*, 81(8): 1111–1114.

Stefanova, R., Toshkov, S., Vasilev, N.V., Vassilev, N.G., and Marekov, I.N. 2011. Effect of gamma-ray irradiation on the fatty acid profile of irradiated beef meat. *Food Chemistry*, 127: 461–466.

WHO. 1981. Wholesomeness of Irradiated Foods. Technical Report Series 659, WHO.

WHO. 2005. www.who.Int/mediacentre/fsctsheets.

Yoon, Y.M., Park, J., Lee, J., Park, J., Park, J., Sung, N., Song, B., Kim, J., Yoon, Y., Gao, M., Yook, H., and Lee, J. 2012. Effects of gamma-irradiation before and after cooking on bacterial population and sensory quality of Dakgalbi. *Radiation Physics and Chemistry*, 81: 1121–1124.

8 Cryogenic Freezing

Himani Singh, Murlidhar Meghwal, and Pramod K. Prabhakar
National Institute of Food Technology Entrepreneurship and Management, Sonipat, Haryana, India

CONTENTS

8.1 Introduction 182
8.1.1 Liquid Cryogenic Agent Storage 182
8.1.2 Superconductivity 183
8.1.3 Superfluidity 183
8.1.4 Cryobiology 183
8.2 Food Freezing 183
8.2.1 Slow Freezing 184
8.2.2 Fast Freezing 184
8.3 Cryogenic Freezing 184
8.3.1 Advantages of Cryogenic Freezing of Food 185
8.4 Principles of Freezing 185
8.4.1 Freezing Time 186
8.4.2 Freezing Rate 186
8.4.3 Thermo-Mechanical Effects During Freezing and Heat Transfer During Cryogenic Freezing 187
8.5 Properties of Cryogenic Fluids 188
8.5.1 Liquid Nitrogen 188
8.6 Cryogenic Freezers 188
8.6.1 Tunnel and Spiral Freezers 188
8.6.2 Immersion Freezers 188
8.6.3 Cryogenic Impingement Freezers 188
8.6.4 Free-Flowing Freezers for Liquid Products 189
8.6.5 Cryo-Mechanical Freezers 189
8.6.6 Process Freezer 189
8.6.7 Individual Quick-Freeze (IQF) Freezers 189
8.6.8 Plate Freezers 189
8.7 Quality of Cryogenically Frozen Products 189
8.7.1 Dehydration and Shrinkage 190
8.7.2 Microbiological Activity 190
8.7.3 Product Adhesion During Freezing 190
8.7.4 Recrystallization 190
8.7.5 Mechanical Damage (Freeze-Cracking) 190
8.8 Costs and Design Aspects 191
8.9 Alternative Cryogenic Technologies in Food Industry 191

8.10 Cryogenic Grinding......191
8.11 Health Hazards of Cryogenic Liquids......192
8.11.1 Safety Precautions......192
8.12 Scope of Cryogenic Science and Technology......193
8.13 Conclusions......193
References......193

8.1 INTRODUCTION

The word cryogenic, derived from the Greek word *krýos*, means production or processing under very low temperature, which can be as cold as below –150°C. The study of cryogenics revolves around very low temperatures and the types of characteristics that materials possess under such extreme conditions. Helium, neon, nitrogen, krypton, argon, methane, hydrogen, and liquefied natural gas are mainly used as cryogens in their liquid forms. The cryogen most commonly used in industry is liquid nitrogen. Nitrogen is an inert gas in plentiful supply in nature, making production cheaper. Cryogens, which are stored in insulated containers known as Dewar flasks, show fluid-like behavior at various temperature and pressure combinations, but all are extremely cold and even a small amount of liquid expands in large volumes of gas.

The French engineer Louis P. Cailletet and the Swiss physicist Rasul Pictet are known as the fathers of cryogenics. The thermodynamic process (adiabatic expansion) was used by Cailletet to liquify oxygen. Pictet used the Joule–Thomson effect which states that when a liquid is made to expand below a certain pressure and temperature, its temperature decreases. Two Polish engineers, Karol S. Olszewski and Zygmut von Wroblewski, developed the cascading method. They reported the liquification of nitrogen at 77K and oxygen at 90K. These developments started a race between engineers to achieve 0K temperature or absolute zero temperature. Thereafter, James Dewar was able to liquify hydrogen at a temperature of 20K. But 20K is also the boiling temperature of hydrogen which led to problems with its storage.

8.1.1 Liquid Cryogenic Agent Storage

Storing liquefied gas for long periods presented a great challenge because liquefied gas tries to expand and evaporate. To overcome this problem, Dewar developed a double-walled insulated storage container capable of storing the boiling hydrogen for a few days. Many more advanced versions of these storage containers were developed that remain in use today for the storage of cryogenic liquids. The containers are based on the insulation properties of expanded foam materials and their radiation shielding powers (Venetucci 1980). Further advances in cryogenics were achieved by Dutch physicist Heike Kamerlingh Onnes, who liquified helium at 4.2K and later at 3.2K. The attainment of 0K or absolute zero is ideal in theory but is not achievable as such. The lowest temperature attained to date is 40 millionths of a Kelvin above absolute zero for freezing sodium gas. Superconductivity and superfluidity are the two other fields that are closely related to cryogenics.

8.1.2 Superconductivity

Superconductivity is a phenomenon in which materials lose all electrical resistance when cooled below a certain critical temperature. Onnes discovered that when mercury was cooled to a temperature of 4.15K, the metal lost all electrical resistance. In 1986, K. Alex Müller and J. Gregore Bednorz discovered that oxides of copper, barium, and lanthanum show superconductivity at 30K. Yttrium barium copper oxide (YBCO) becomes superconductive at 95K, which is approximately the same as the boiling point of nitrogen. Compounds containing various mixtures of barium, calcium, thallium, oxygen, and copper are reported to have the highest critical temperature known to date, 125K. Superconductivity is employed in magnetically levitated trains, particle accelerators, magnetic resonance imaging (MRI) devices, fusion energy plants, energy storage, and zero-loss transmission lines.

8.1.3 Superfluidity

Superfluidity is the property of flowing without friction or viscosity, and occurs in liquid helium below about 2.18K. Superfluid helium loses all its viscosity while travelling through a capillary or narrow slit.

8.1.4 Cryobiology

Cryobiology, the study of the effect of low temperature on biological material, has huge potential in the field of medicine. Cryobiology helps to preserve human and animal parts for medical purposes. Cryogenic scalpels are used in surgery to destroy tissue around a wound, so that immediate clotting is achieved to stop blood loss. However, it is not an effective method of saving lives in an accident.

8.2 FOOD FREEZING

Food freezing as a preservation method has been practiced for many decades. In this technique, the temperature of the food product is reduced below freezing point or significantly reduced in such a way that it will not allow microbial growth in food. The two types of freezing used in food processing are slow and fast freezing.

Freezing delays spoilage and keeps foods safe by preventing microorganisms from growing and by slowing down the enzyme activity that causes food to spoil. As the water in the food freezes into ice crystals, it becomes unavailable to those microorganisms that need it for growth. However, most microorganisms (with the exception of parasites) remain alive when frozen so foods must be handled safely both before freezing and once defrosted. Freezing can damage some foods because the formation of ice crystals causes breakage of the cell membranes. This has no adverse effects in terms of safety (indeed some bacterial cells would also be killed); however the food loses its crispness or firmness. Foods that do not tolerate freezing well include salad vegetables, mushrooms, and soft fruits. Foods with higher fat content, such as cream and some sauces, tend to separate when frozen. Commercial freezing freezes foods rapidly so that smaller ice crystals are formed. This causes less damage to cell membranes so that quality is even less affected.

8.2.1 Slow Freezing

Slow freezing, or sharp freezing, occurs when food is directly placed in freezing rooms called sharp freezers where air is circulated by convection through a specially insulated tunnel, either naturally or with the aid of fans. Relatively still air is a poor conductor of heat, which is why a long time is required to freeze the food. The temperature ranges from −15 to −29°C and freezing may take between 3 and 72 hours. Large ice crystals form in extra-cellular spaces, resulting in disruption to the structure of food. Thawed food therefore cannot regain its original water content. Large ice crystals create quality problems such as mushiness in vegetables.

8.2.2 Fast Freezing

Vigorous circulation of cold air enables freezing to proceed moderately fast. The temperature is kept between −32 and −40°C and the food attains the stage of maximum ice crystal formation in 30 minutes or less. Small ice crystals are formed within the cells so the structure of the food is not damaged. On thawing, the structure of the original food is maintained.

8.3 CRYOGENIC FREEZING

Cryogenic freezing has been used by the food industry since the 1960s (Almqvist 2003). It is used to freeze products such as processed fruits and vegetables, processed bakery products, processed red meat, processed sea food and poultry, ice-creams and other desserts, processed potato products (baked and uncooked), pizzas, meat substitutes, and ready-to-eat meals. Cryogenic freezing is carried out using cryogens, which are gaseous refrigerants that expand, sublime, or evaporate at lower temperatures and at pressures equivalent to atmospheric pressure. Commonly used cryo-refrigerants are oxygen, carbon dioxide, nitrogen, argon, and hydrogen. Cryogenic freezing is generally achieved by immersing the product into liquid nitrogen, exposing it to a mixture of gaseous carbon dioxide and solid snow, or spraying with nitrogen. Food manufacturers use cryogens in their (high-pressure) liquid form, which makes the handling of these refrigerants much easier. The liquid refrigerants are delivered to manufacturers in cylinders, with bulk quantities delivered by the tankers straight into their insulated storage systems. The stored cryogen is sprayed into insulated tunnels where the product to be treated is transferred (Estrada-Flores 2016).

Cryogenic freezing is both productive and highly efficient. The quality of products treated with cryogens is superior to those treated by slower freezing methods, due to the faster freezing rate. Food safety risks are reduced with cryo-freezing. Energy consumption is very low, as are set-up and maintenance costs (Estrada-Flores 2012). Among the limitations of cryogenic freezing is that it is most efficient for freezing small or medium-sized products only, since freezing rate is limited by inter-heat transfer in larger products (Cleland and Valentas 1997). Cryogenic freezing incurs higher operational costs than other refrigeration systems (Chourot et al. 2003). It is best suited to seasonal produce, new products, and small-scale production where fast freezing is required (ASHRAE 2006).

8.3.1 Advantages of Cryogenic Freezing of Food

The quality of cryogenically frozen foods is greatly superior to that of food frozen using conventional methods. This is due to the faster freezing rates, whereby smaller ice crystals are formed causing minimum dislocation of water in the structure and minimum structural loss in the product. Singh and Meghwal (2019a) studied the effect of cryo-grinding and conventional grinding on the antioxidants of ajwain seeds. They found that ajwain seeds which were frozen cryogenically prior to grinding retained more antioxidants than those ground conventionally. When unpackaged food is exposed to blast freezers, immense water loss of up to 4% occurs, which is detrimental to high-value foods. In cryogenic freezing there is minimal (up to 0.5%) or no weight loss, which helps to preserve the economic viability of the product. Water content or moisture content contribute significantly to effective and efficient operations. Studies by Singh and Meghwal (2019b) on ajwain seeds, and Meghwal and Goswami (2015) on black pepper powder show the effect of moisture content on processing during cryogenic or conventional operations.

Cryogenic freezers using liquid carbon dioxide and liquid nitrogen require very low capital investment compared with other mechanical freezers of similar capacities. Cryogenic freezers also need much less space than plate or blast freezers of the same capacity. Cryogenic freezing has proved efficient for freezing products with a wide variety of shapes, sizes, and structures (Fennema 1978).

8.4 PRINCIPLES OF FREEZING

Freezing for food preservation works on the principle of the formation of ice from the water present in the food using low temperatures that retard biochemical changes and growth of microorganisms in the product. As water is not available for the growth of microorganisms and for biochemical changes, freezing helps to extend shelf life, though it may not sterilize the product.

During freezing the water present in food converts into ice thermodynamically and via kinetic factors. The thermodynamic factors express the characteristics of the whole system under equilibrium. The rate of attainment of equilibrium is measured by kinetics. The freezing process consists of, first, nucleation or ice crystal formation, and second, the growth of ice crystals. Ice crystal formation is dependent on temperature. Latent heat and sensible heat are removed during freezing.

According to the International Institute of Refrigeration, the three freezing phases are based on the temperature changes that a product undergoes while freezing (Figure 8.1). In the first stage of the freezing curve, the product cools down to a freezing point from its initial temperature. At this point the food system is in equilibrium, i.e., the liquid phase (water from the product) and the solid phase (ice crystals) coexist in the system. During this stage, the water in the food is converted to ice, sensible heat is lost from the food, and as the process of freezing progresses, the concentration of solutes in the unfrozen section of the food increases. In this stage the freezing curve is steep and the rate of cooling is fast. The second stage is marked by an increase in temperature. This is the period of thermal arrest, when all the water is now in a frozen state and the food is free from latent heat. In the third stage, the phase

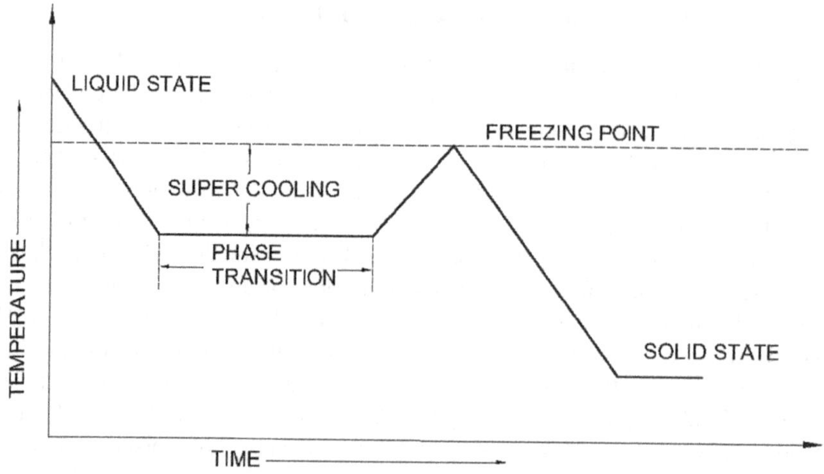

FIGURE 8.1 Different phases of food products during freezing

change temperature is further reduced to the final freezing temperature. Due to ice cooling, a small amount of heat is released in this stage. The remaining water also freezes completely in this stage (Ojha et al. 2016).

The important factors determining the design of any freezing system, including a cryogenic freezing system for foods, are freezing rate and time.

8.4.1 Freezing Time

The freezing time can be calculated using various equations or numerical simulation (Norton and Tiwari 2013). Plank's analytical equation to estimate freezing time (Eq. 8.1) is based on the assumption that initially product properties are independent of temperature and there is a negligible volume change as the entire food system is initially at freezing temperature (Ojha et al. 2016).

$$FT = \frac{\lambda\rho}{\Delta\theta}\left[\frac{ax}{h} + \frac{bx^2}{k}\right] \tag{8.1}$$

where $\Delta\theta$ is the temperature difference between the freezing medium and freezing point of the food; λ is latent heat offood; ρ is density of product; a and b are shape-specific constants.

Among the important parameters in a food system that influence freezing time are refrigeration medium temperature, thermal properties of the food, product shape, size and thickness, and initial and final temperature of the product.

8.4.2 Freezing Rate

Freezing rate is the ratio of temperature change divided by the time required. The freezing rate is lower at the center and higher at the surface of the product as the

surface is in direct contact with the freezing medium. Higher freezing rates give products with better quality and longer shelf life. This is because with higher freezing rates, smaller ice crystals are formed which are evenly distributed throughout the food matrix, whereas with slower freezing rates, larger ice crystals are formed which are not evenly distributed in the food matrix (Bronfenbrener and Rabeea 2015). These larger needle-shaped ice crystals damage food matrix tissue, resulting in food products of lower quality and poorer texture.

8.4.3 Thermo-Mechanical Effects During Freezing and Heat Transfer During Cryogenic Freezing

Among the many factors affecting the rate of heat transfer during freezing are packaging material, physical properties of the food product (such as size, shape, and available surface area), air gaps forming part of the food matrix structure, and thermal properties of a food product (specific heat, thermal diffusivity, and thermal conductivity). While these factors are a crucial part of both cryogenic and mechanical freezing, heat transfer occurs quite differently in the two kinds of freezing.

In cryogenic freezing the product surface is sprayed with liquid CO_2 or N_2, which work in different ways. Liquid carbon dioxide expands as soon as it is released through the spray nozzle; half by weight is converted into solid particles and the other half into vapor. As soon as the CO_2 particles come into contact with the food surface, they cool it by subliming. Particle sublimation provides 85% of the total cooling. The remaining CO_2 particles and vapor travelling through the tunnel provide the remaining cooling/freezing in three ways: by heat transfer through convective currents of CO_2; by conductive transfer of heat from the product to the belt; and by radiative transfer of heat between the surrounding colder surfaces and the food surface.

Liquid nitrogen is generally used in spiral and tunnel-type freezers. Once sprayed into the tunnel the liquid nitrogen is separated into vapor and liquid droplets. When it comes into contact with the surface of the food, the liquid nitrogen undergoes phase change into the gaseous phase, and extracts 50% of the latent heat from the product. The remaining freezing is carried out by the gaseous nitrogen surrounding the product inside the tunnel.

In immersion-type freezing, the product is completely immersed in the liquid cryogen and freezing occurs due to the high heat transfer coefficients, which vary from 200 to 8,000 W/m^2K in the case of liquid nitrogen (Singh and Mannapperuma 1990; Estrada-Flores 2012; Pham 2014). An important aspect of cryo-freezing is the measurement of the freezing times. Models used for cryo-freezing include the finite element method (Zosifescu et al. 2013; Huan et al. 2003) and finite difference method (Shaikh and Prabhu 2007; Agnelli and Mascheroni 2001), and Pham's method (2014), which models product freezing time using the sum of precooling, phase change, and subcooling stages. A modified form of Newton's law of cooling which takes into account the effect of internal resistance of the product to heat flow is used to calculate the cooling time during precooling and subcooling. Plank's equation is used to calculate cooling time during phase change (Plank 1941). Both Pham's method and the finite difference method give results that agree within ±10%.

8.5 PROPERTIES OF CRYOGENIC FLUIDS

8.5.1 Liquid Nitrogen

Liquid nitrogen (LN_2) is considered to be the safest cryo-refrigerant, as it is chemically inactive, non-toxic, and non-explosive. Liquid nitrogen is a clear and colorless fluid. At sea level, liquid nitrogen boils at −195.79°C (77K; −320°F). Nitrogen becomes solid at 63.148K. The density of liquid nitrogen at its normal boiling point is 808.9 kg/m^3 and its latent heat of vaporization is 198.3 kJ/kg (Timmerhaus and Flynn 1989). It is produced industrially by fractional distillation of liquid air.

8.6 CRYOGENIC FREEZERS

8.6.1 Tunnel and Spiral Freezers

Tunnel freezers contain a straight, continuous, or winding conveyer belt. Products are transported along the conveyer belt and sprayed with cryogen to achieve freezing. The angular design of a tunnel freezer achieves optimal use of the cryogen, increasing the cost-effectiveness of the operation (Praxair Technology Inc. 2015). This type of tunnel freezer reduces both access heat loads and air infiltration rates. Spiral freezers (Figure 8.2) work on similar principles to tunnel freezers, differing only in their surface shape.

8.6.2 Immersion Freezers

In immersion-type freezers the product is directly immersed in coolant, generally glycol or brine. It is important to prevent cross-contamination between product and coolant, so the product is generally packaged. Immersion freezers have high freezing rates. The refrigerant liquids are cooled by cooling coils, jackets built into the liquid tank, or by circulation in a heat exchanger.

8.6.3 Cryogenic Impingement Freezers

Impingement freezers have high-velocity jets attached to them that direct cool air onto the surface of the product. It is mostly the top and bottom surfaces of the product

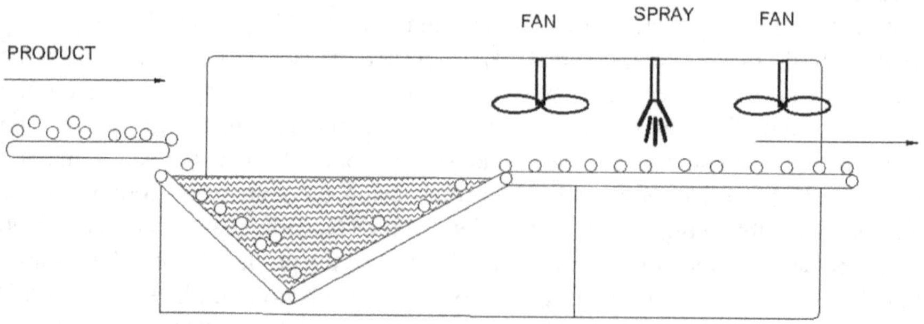

FIGURE 8.2 Spiral freezer

that are exposed to the cool air. The hot layer surrounding the product is blasted away by these jets leading to freezing of the product.

8.6.4 Free-Flowing Freezers for Liquid Products

Free-flowing freezers consist of perforated trays and a freezing chamber containing gaseous and liquid cryogen. Liquid product is poured onto the perforated tray and falls into the freezing chamber as droplets which immediately solidify. The solid beads are collected from the freezing chamber and packaged (Jones 1992; Jones et al. 2001).

8.6.5 Cryo-Mechanical Freezers

The two operations conducted in cryo-mechanical freezers are cryo-immersion followed by blast freezing. The main objective of these freezers is to maintain a balance between loss of product quality and moisture, and operational costs. The initial temperature reduction that occurs when the product is exposed to the mechanical freezer is achieved by cryo-immersion, in which most of the product quality is retained (Cleland and Valentas 1997).

8.6.6 Process Freezer

Process freezers are used towards the packaging end of freezing operations.

8.6.7 Individual Quick-Freeze (IQF) Freezers

IQF freezers are of two types, mechanical and cryogenic, and are used for preparing foods such as pasta or pizza, or for fruits and vegetables. They may consist of a spiral conveyer inside an insulated freezing chamber, or a tunnel through which the product flows in a linear fashion on a conveyer belt.

8.6.8 Plate Freezers

This type of freezer is generally used to freeze flat or box-shaped products. The product is placed between two hollow plates through which refrigerant liquid continuously flows.

8.7 QUALITY OF CRYOGENICALLY FROZEN PRODUCTS

Even the best freezing technology may not be able to prevent product quality loss. Faster freezing rates do not necessarily lead to higher-quality products but may cause rupture of the cell structure or internal strains, or have detrimental effects on the product surface. Cryogenically frozen products are more sensitive to the quite high temperature variations that occur during transportation and storage.

8.7.1 Dehydration and Shrinkage

During freezing, solute concentration in the frozen medium increases due to various factors, including the difference between the water vapor pressure of the product and the surrounding environment, the formation of ice, and the loss of intercellular liquid due to structural damage to the cellular wall (Reid 1993; Goff 2002). Cryogenic freezing incurs slower rates of dehydration and shrinkage because of the higher freezing rate and the rapid reduction of vapor pressure on the product surface. Weight reduction of cryo-frozen products is less than 1% and they have higher porosity, lower bulk density, higher aroma retention, and superior taste and quality as compared with conventionally frozen products.

8.7.2 Microbiological Activity

Freezing increases product shelf life because the microbial growth rate is drastically reduced. Ice formation, dehydration, concentration of solutes, temperature shock, and toxicity of the intercellular solutes may contribute to the negative impact of freezing upon microorganisms (Zaritzky 2000). Bacterial growth increases with quick freezing and thawing processes. In order to compensate for dehydration losses, some manufacturers mist frozen product surfaces with water or other liquids, predisposing them to microbial risks (Everstine et al. 2013). These practices need to be controlled in order to produce safe cryo-frozen products.

8.7.3 Product Adhesion During Freezing

During their passage into the freezers on metallic belts, products tend to build adhesive forces between their surface and the metallic belt. These adhesive forces are temperature dependent and follow non-linear behavior. As the temperature decreases on entry to the freezer, the temperature of the metallic belt also decreases, strengthening the adhesive forces until they become larger than those of the ice. As soon as the temperature of the metal is increased, the strength of the adhesive force decreases and the product is easily removed from the belt (Kennedy 1998).

8.7.4 Recrystallization

Recrystallization means the changes that occur in the ice crystals after the completion of the freezing process and during storage. Crystals having flatter surfaces are more stable than those with sharp edges, as they become smoother with time. During storage smaller ice crystals merge to form larger ice crystals, leading to changes in their shape and orientation. These mechanisms can be minimized by maintaining low temperature constants during storage. The very low temperature to which the product is subjected during cryo-freezing is generally not maintained during storage, leading to more recrystallization (Zaritzky 2000).

8.7.5 Mechanical Damage (Freeze-Cracking)

In addition to the structural damage that may occur during slow freezing, there is another type of damage – mechanical. Frozen and freeze-dried foods become very

fragile and brittle, and need extra care during packaging, storage, and transportation to avoid mechanical damage.

8.8 COSTS AND DESIGN ASPECTS

The major cost incurred in cryo-freezing is the cryogen used to carry out the operation. Liquid carbon dioxide and nitrogen consumption ranges from 0.7 to 1.3 kg and 0.4 to 1.6 kg per kg of the food product (James and James 2003). Further costs are incurred in storage and transportation of the cryogen. While the capital cost of cryo-freezers is low, their operational costs are quite high (Shaikh and Prabhu 2005). Operational costs can be greatly reduced if the freezer is designed according to the production requirements (Lang 2006). In cryo-freezers, 85–95% of the total heat load is represented by the heat load of the product and other components such as heat ingress (via air interchange, storage vessel losses, and insulation) and the heat from fans is minimal in the best cryo-freezers (Cleland and Valentas 1997). The heat load of the product determines the rate at which the cryogen is consumed, which in turn determines the cost-effectiveness of the operation. The cryogen delivery system, including piping and supply, should be designed carefully (Potter and Hotchkins 1998).

8.9 ALTERNATIVE CRYOGENIC TECHNOLOGIES IN FOOD INDUSTRY

To meet the increasing demand for chilled and frozen food, along with traditional liquid nitrogen, dry ice or chilled water systems, mechanical chillers and freezers are needed for mass and continuous production of chilled and frozen food items.

8.10 CRYOGENIC GRINDING

Cryogenic freezing principles are also used for size reduction of various raw materials, including spices and herbs, that have a high concentration of volatile oils. Cryogenic grinding is also known as freezer grinding, freezer milling, or cryo-milling. In cryogenic grinding, the raw material is first cooled down to very low temperatures and then reduced into smaller particles. When the food material comes into contact with the liquid cryogen, the cryogen extracts all its latent heat, rendering it brittle. The embrittled product is easily ruptured with very low force, creating extremely fine particles. In conventional grinding the temperature inside the mill can reach up to 93°C/200°F, which is detrimental to the heat-sensitive components of the food product. In the case of herbs and spices, these high temperatures reduce both quality and quantity of the essential oils. This loss can be prevented by cryo-grinding. The spices are precooled prior to grinding and subjected to liquid cryogen during the grinding process to maintain the low surrounding temperature (Murthy and Bhattacharya 2008). Materials with low melting points are sensitive to oxygen and elastic in nature, and have lower combustion temperatures, making them suitable for grinding under cryogenic conditions. Meghwal and Goswami have studied cryo-treatment of fenugreek and its powder flow characteristics, the energy

efficiency of cryogenic grinding of fenugreek, the effect of different types of mills on the cryogenic processing of black pepper and fenugreek, and the effects of various packaging material on the shelf life of black pepper processed cryogenically (Meghwal and Goswami 2013a, 2013b, 2014a, 2014b). These studies prove the improved efficiency of cryogenic grinding/processing/freezing over conventional processing methods.

8.11 HEALTH HAZARDS OF CRYOGENIC LIQUIDS

Hazards to human health related to cryogenic liquids include frostbite, extremely low temperatures, pressure build-up, fire, explosion, and asphyxiation (*Chemical Health & Safety* 2000).

Frostbite. The fluid in the eyes freezes as soon as it comes into contact with the cryogen, resulting in permanent eye damage.

Extremely low temperatures. The cold boil-off vapor of cryogenic liquids rapidly freezes human tissue. Cold burns and frostbite caused by cryogenic liquids can cause extensive tissue damage.

Pressure build-up and fire. The cryogens have high ratios of conversion from liquid to gaseous form. They are also capable of freezing moisture in the surroundings, so they may block the opening of the vessel, causing an explosion. This explosion may result in fire. It is therefore important to store cryogenic liquids in insulated Dewar vessels.

Asphyxiation. Cryogens have the potential to make the environment deficient in oxygen, which can make a closed chamber lethal. A person may lose consciousness due to lack of oxygen and die from asphyxiation.

8.11.1 Safety Precautions

- Full-face shields and splash-resistant goggles should be worn to protect the eyes and face. Contact lenses should be avoided.
- Cryogen gloves should be worn to protect the hands. Jewelry and watches should be removed.
- Full-coverage shoes and long-sleeved shirts are preferred.
- Respirators should be used when handling cryogens if there is a possibility of the atmosphere becoming oxygen deficient.
- Cryogenic systems should have pressure relief devices in place.
- Cryo-operations should be performed in properly ventilated areas to avoid vapor or gas accumulation.
- All equipment and containers should be free of oil, dirt, and grease to prevent fire hazard.
- No body part should touch insulated pipes or vessels containing cryogenic liquids.
- Containers should not be rolled, dropped, or tipped on their sides.
- All connections should be protected with safety valves.

8.12 SCOPE OF CRYOGENIC SCIENCE AND TECHNOLOGY

Cryogenic freezing can be one of the best options for the separation of waste material, especially from the automobile industry. Another application is the preservation of biological activities, enzymatic activities and the integrity of food products. Cryogenic freezing can also be used to preserve, store, and transport perishable food commodities at low temperature; cryogenic grinding can retain functional and nutraceutical compounds in herbs and spices. Other applications are in the fields of medicine, biological science, space technology, musical instruments, and sports.

8.13 CONCLUSIONS

The role of cryogenic processing for food preservation and quality retention has increased significantly in recent decades. Although these techniques are expensive, they help to retain color, flavor, aroma, taste, and biological activities of the products.

REFERENCES

Agnelli, M.E. and Mascheroni, R.H. 2001. Cryo-mechanical freezing. A model for the heat transfer process. *J. Food Eng.* 263–270.

Almqvist, E. 2003. *History of Industrial Gases*. New York: Kluwer Academic/Plenum Publishers.

ASHRAE. 2006. Industrial food freezing systems. In *ASHRAE Handbook: Refrigeration. The American Society of Heating, Refrigerating and Air-Conditioning Engineers.*

Bronfenbrener, L. and Rabeea, M.A. 2015. Kinetic approach to modeling the freezing porous media: Application to the food freezing. *Chem. Eng. Process. Process Intensif.* 87, 110–123.

Chourot, J.M., Macchi, H., Fournaison, L. and Guilpart, J. 2003. Technical and economical model for the freezing cost comparison of immersion, cryo-mechanical and air blast freezing processes. *Energy Convers. Manag.* 44, 559–571.

Cleland, D.J. and Valentas, K.J. 1997. Prediction of freezing time and design of food freezers. In: Valentas, K.J., Rotstein, E. and Singh, R.P. (eds), *Handbook of Food Engineering Practice*. Boca Raton, FL: CRC Press.

Estrada-Flores, S. 2012. Chilling and freezing by cryogenic gases and liquids (static and continuous equipment). In: Mascheroni, R.H. (ed.), *Operations in Food Refrigeration*. Boca Raton, FL: CRC Press.

Estrada-Flores, S. 2016. *Cryogenic freezing of food*. Melbourne: Plant and Food Research.

Everstine, K., Spink, J. and Kennedy, S. 2013. Economically motivated adulteration (EMA) of food: Common characteristics of EMA incidents. *J. Food Prot.* 76, 723–735.

Fennema, O. 1978. Cryogenic freezing of food. In Timmerhaus, K.D. (ed.), *Advances in Cryogenic Engineering*. New York and London: Plenum Press.

Goff, G.H. 2002. Theoretical Aspects of the Freezing Process (Online). University of Guelph. Available: http://www.foodsci.uoguelph.ca/dairyedu/freeztheor.html. Accessed on 16.10.02).

Huan, Z., He, S. and Ma, Y. 2003. Numerical simulation and analysis for quick-frozen food processing. *J. Food Eng.* 60, 267–273.

James, C. and James, S. 2003. Cryogenic freezing. In Caballero, B. and Trugo, L. (eds), *Encyclopedia of Food Sciences and Nutrition*. Academic Press.

Jones, C. 1992. Method of preparing and storing a free flowing, frozen alimentary dairy product. US Patent Application 07/762, 072.

Jones, M., Jones, C. and Jones, S. 2001. Cryogenic processor for liquid feed preparation of a free-flowing frozen product and method for freezing liquid composition. US Patent Application 6223542.

Kennedy, C.J. 1998. The future of frozen foods. *Food Sci. Technol.* 14, 7–14.

Lang, G. 2006. Cryogenic freezing: Industrial refrigeration consortium research andtechnology forum. University of Wisconsin-Madison, Madison, WI.

Meghwal, M. and Goswami, T.K. 2013b. Evaluation of size reduction and power requirement in conventional and cryogenic ground fenugreek powder. *Adv. Powder Technol.*, *24*(1), 427–435.

Meghwal, M. and Goswami, T.K. 2013a. Ambient and cryogenic grinding of fenugreek and flow characterization of its powder. *J. Food Process Eng.*, *36*(4), 548–557.

Meghwal, M. and Goswami, T.K. 2014a. Comparative study on ambient and cryogenic grinding of fenugreek and black pepper using hammer, pin, rotor and ball mill. *Powder Technol.*, 267, 245–255.

Meghwal, M. and Goswami, T.K. 2014b. Effect of different grinding methods and packaging materials on fenugreek and black pepper powder quality and quantity under normal storage conditions. *Int. J. Agric. Biol. Eng.* MS1271, *7*(4), 106–113.

Meghwal, M., and Goswami, T.K. 2015. Flow characterization of ambiently and cryogenically ground black pepper (piper nigrum) powder as a function of varying moisture content. *J. Food Process Eng.*, *40*(1), E12304.

Murthy, C.T. and Bhattacharya, S. 2008. Cryogenic grinding of black pepper. *J. Food Eng.* *85*(1), 18–28.

Norton, T. and Tiwari, B. 2013. Aiding the understanding of novel freezing technology through numerical modelling with visual basic for applications (VBA). *Comput. Appl. Eng. Educ.* *21*(3), 530–538.

Ojha, K.S., Kerry, J.P., Tiwari, B.K. and O'Donnell, C. 2016. *Freezing for Food Preservation. Reference module in Food Science.* Elsevier.

Pham, T.Q. 2014. Freezing time formulas for foods with low moisture content, low freezing point and for cryogenic freezing. *J. Food Eng.* 127, 85–92.

Plank, R. 1941. Contribute to the calculation and evaluation of the freezing speed of food. *Supplements magazine Gesampte Kalte-Industrie*, *3*(10), 22.

Potter, H.N. and Hotchkins, J.H. 1998. *Food Science.* Springer.

Praxair Technology Inc. 2015. Cold Front Cryo-Saver Tunnel Freezer.

Reid, D. 1993. Physical phenomena in the freezing and thawing of plant and animal tissues. In Mallet, C.P. (ed.), *Frozen Food Technology.* Blackie Academic & Professional.

Shaikh, N.I. and Prabhu, V. 2005. Vision system for model based control of cryogenic tunnel freezers. *Comput. Ind.* 56, 777–786.

Shaikh, N.I. and Prabhu, V. 2007. Mathematical modeling and simulation of cryogenic tunnel freezers. *J. Food Eng. 80*, 701–710.

Singh, H. and Meghwal, M. 2019a. Ajwain a potential source of phytochemical for better health. *The Pharma Innovation Journal*, *8*(6), 599–604.

Singh, H. and Meghwal, M. 2019b. Physical and thermal properties of various Ajwain (Trachyspermum Ammi L.) seed varieties as a function of moisture content. *J. Food Process Eng. 43*(2): E13310.

Singh, R.P. and Mannapperuma, J. 1990. Developments in food freezing. In: Schwartzberg, H.G. and Rao, M.A. (eds), *Bioprocess and Food Process Engineering.* New York: Marcel Dekker, Inc.

Timmerhaus, K.D. and Flynn, T.M. 1989. Properties of cryogenic fluids. In *Cryogenic Process Engineering.* The International Cryogenics Monograph Series. Boston, MA: Springer.

Venetucci, J.M. 1980. Cryogenic storage vessels. In Norman R. Braton (ed.), *Cryogenic Recycling and Processing.* London: CRC Press, Taylor and Francis Group.

Zaritzky, N.E. 2000. Factors affecting the stability of frozen foods. In Kennedy, C.J. (ed.), *Managing Frozen Foods.* Woodhead Publishing Ltd.

Zosifescu, C., Damian, V. and Coman, G. 2013. Cryogenic freezing of berries-assessment of phase change heat transfer. In *ARA Annual Congress Proceedings*, pp. 454–457.

9 Nanofiltration: Principles, Process Modeling, and Applications

Siddhartha Vatsa, Manibhushan Kumar, Neeraj Ghanghas, Pramod K. Prabhakar, and Murlidhar Meghwal
National Institute of Food Technology Entrepreneurship and Management, Sonipat, Haryana, India

CONTENTS

9.1 Introduction 196
9.2 Scope and Opportunities 196
9.3 Nanofiltration Membrane Materials and Preparation 197
9.4 Nanofiltration Modules 199
9.4.1 Flow Geometries 202
9.4.2 Membrane Characterization 202
9.4.3 Performance Parameters 204
9.4.3.1 Morphology Parameters 204
9.4.3.2 Charge Parameters 205
9.4.3.3 Membrane Charge and Species Transport 206
9.5 Nanofiltration Modeling 207
9.5.1 Theory 208
9.5.2 Model Categorization 210
9.5.2.1 Membrane-Dependent Semi-Empirical Model: Solution-Diffusion Model (SDM) 210
9.5.2.2 Phenomenological Models 212
9.6 Applications 213
9.6.1 Nanofiltration in the Food Industry 213
9.6.2 Nanofiltration in Water Treatment 213
9.6.3 NF as a Pre-Treatment for Desalination 216
9.6.4 NF in Trace Contaminant Removal 217
9.7 Conclusions 218
References 218

9.1 INTRODUCTION

Nanofiltration is the most recent of the four pressure-driven membrane technologies and was developed in the 1970s mainly to bridge the gap between ultrafiltration (UF) and reverse osmosis (RO). The earliest membranes were microfiltration (MF), developed in the 1900s, with a wide range of applications in medicine, pharmaceuticals, and microbiology, followed by RO to obtain drinking water from seawater and brackish water, invented by Loeb and Sourirajan in 1959 at the University of California, Los Angeles (Ismail and Matsuura, 2018). Soon afterwards, UF was developed, falling between the principles of microfiltration (passage of salt ions, rejection of particles) and RO (rejection of salt ions). Together, RO and UF set an important standard that worked well for different applications, but there was a growing demand for a membrane with performance characteristics lying between RO and UF. The earliest documented application of nanofiltration membranes is for water softening at a Florida-based company in the late 1970s. First commercialized in 1983 for desalting food-grade dye, the term nanofiltration was coined in 1984 by FilmTec Corporation, based on the estimated size of the pores in an NF membrane.

Nanofiltration has very similar operating principles to RO and is also known as modified RO. As a pressure-driven membrane filtration process, NF also utilizes semi-permeable membrane and cross-flow filtration to separate the feed stream into purified permeate stream and a concentrate stream containing a high percentage of impurities. Ouyang et al. (2008) reported that lower operating pressures are required and the structure is more open in NF than RO, which allows only monovalent ions to enter, not divalent ones.

NF is largely used in water softening, where hardness of water is reduced by the rejection of organics, color, bacteria, and several other impurities found in raw water (Burn et al., 2015). RO is used in desalination of seawater and brackish water containing very high total dissolved solids (TDS), but most water supplies do not require the total removal of impurities (Burn et al., 2015). Thus, NF partially demineralizes water, removing between 10 and 90% of impurities, as compared with RO which removes 99.5%.

9.2 SCOPE AND OPPORTUNITIES

The main objective of nanofiltration is to partially eliminate salts, but not to completely remove them. Nanofiltration membranes allow the passage of monovalent ions, but not divalent or larger ones. The reason is that monovalent ions have smaller hydrated radii, and the power of attracting or repelling them is less (Garcia, 2000). Most of the time NF membranes are negatively charged at their surface at neutral pH, and thus repel negatively charged multivalent ions. There is no net transfer of charge across the membrane. Thus, membrane characteristics, as well as the composition of the feed mixture, dictate which ions will be rejected and which permitted. Fridman-Bishop et al. (2018) termed this rejection behavior Donnan exclusion behavior. Donnan exclusion behavior works on rejection theory: for example, if the membrane is positively charged, the rejection sequence would be $CaCl_2$>NaCl>Na_2SO_4; if it is negatively charged, the same sequence would become Na_2SO_4>NaCl>$CaCl_2$. But

this sequence need not be accurate in all cases because of the large deviations occurring in complex mixtures (Mohammad et al., 2015). There is a distribution of both types of ions at the membrane surface, leading to an additional separation also known as Donnan exclusion. This separation efficiency reduces with increasing salt concentration. Apart from the separation of charged ions, what about the separation of uncharged (organic) solutes? Labban et al. (2017) described a system based on ultrafiltration. The capability of ultrafiltration to hold the uncharged solutes is defined by molecular weight cut-off (MWCO), defined as the solute molecular weight at which 90% of it is retained. Solutes of higher molecular weight than the MWCO are rejected and those of lower molecular weight pass through the filter. However, the molecular weight is only one of several criteria that determine the permeation rates of solutes. Other properties like steric hindrance and polarity also affect permeation rate, where the process happens in sub-nanometer range. The MWCO values for nanofiltration range from 150 to 3,000 Da. Most pesticides and micro-pollutants lie between 150 and 300 Da, hence nanofiltration is largely preferred for the purification of drinking water and water used in other applications. But high rejections were not always observed for small hydrophobic solutes and several micro-pollutants.

9.3 NANOFILTRATION MEMBRANE MATERIALS AND PREPARATION

Nanofiltration membranes are made of a thin barrier layer supported by a porous membrane. The thin barrier layer selectively rejects ions and salts from the parent material passing through it. The bottom support layer provides mechanical strength. The thickness of the membrane is inversely proportional to the permeable flux (Han et al., 2015). NF membranes are classified as loose or tight according to their materials and synthesis. Tight NF membranes are also known as loose RO membranes, while loose NF membranes are very similar to UF membranes. Most of these membranes are polymeric (solvent filters) or ceramic structures, or a combination of both. The characteristics of tight and loose NF membranes are shown in Table 9.1.

Nanofiltration membranes are synthesized in a way similar to that of loose RO membranes (Jye and Ismail, 2016). They are composite structures with a polyamide layer on the top and a polysulfone layer on non-woven support below. The sub-layer is immersed in a solution first of aqueous amine and then of organic acyl chloride, giving a very thin polyamide layer with outstanding separation capacity. The final

TABLE 9.1
Differences between Tight and Loose Nanofiltration Membranes

Basis of Difference	Tight NF Membrane	Loose NF Membrane
Properties close to	Near RO	Near UF
Divalent ions retention	Ca, SO_4>99%	Ca, SO_4 90–99%
Monovalent ions retention	Na, Cl 60–90%	Na, Cl 10–60%
MWCO	MW 200 Da organics	500–1,000 Da organics
Material	PA, PI	P(E)S, ceramics

step is the optimization of the top layer, which makes it more permeable than RO membranes. The parameters affecting membrane performance are reaction time, type of monomer, and monomer concentration (Saha and Joshi, 2009). The nanofiltration membrane is generally designed to have >99% rejection of multivalent ions and 60–90% rejection of monovalent ions, while MWCO is around 200.

Purkait et al. (2018) describe loose NF membranes that are similar to UF membranes, but made through phase inversion instead of interfacial polymerization. Phase inversion is a controlled transformation of the cast polymeric solution from a liquid to a solid state. Phase inversion is completed by any of the following four methods: (a) Immersion precipitation (immersion in a non-solvent bath); (b) controlled evaporation; (c) thermal precipitation (cooling); and (d) precipitation from the vapor phase. Immersion precipitation, which is the most commonly used, requires three components: (i) a polymer; (ii) a solvent for the polymer; and (iii) a non-solvent (Hołda et al., 2013).

The process is simple. The non-solvent component diffuses into the polymer-rich phase and simultaneously the solvent component diffuses into the polymer-lean phase. This leads to complete removal of the solvent part from the polymer-rich phase, which eventually leads to formation of nuclei that grow and coalesce, finally building a membrane structure for nanofiltration (Campbell, 2014).

Unlike others, ceramic membranes have asymmetric structures and three layers with different-sized pores. The three layers are (i) a macro-porous layer providing mechanical strength, (ii) ameso-porous intermediate layer, and (iii) a thin uppermost layer with pores in the sub-nanometer range.

The most common ceramic material is α-Al_2O_3 with a pore size <= 1 μm. The quality of the membrane is highly dependent on the support layer, the quality of which can be increased by reducing the roughness, and minimizing irregularities and defects. Top-layer defects can be eliminated by addition of amesoporous intermediate layer, formed by the sol–gel process. A suitable colloidal suspension can be made in two ways: mixing a precursor with a large amount of water to form a colloid of a few nanometers; and mixing precursors with a small amount of water in an organic solvent, resulting in branched polymeric molecules.

The colloidal route is followed for the intermediate layer using Al, Zr, Ti, and Si oxides. The Ti and Zroxides are much more stable than the rest because they are not corroded even at extreme pH values (Ng et al., 2013). Colloids are generated by hydrolysis of organometallic species:

$$M(OR)_4 + H_2O \rightarrow HO - M(OR)_3 + ROH;\ M = Ti, Zr.$$

The initial hydroxides further condense thus:

$$(OR)_3M - OH + HO - M(OR)_3 \rightarrow (OR)_3M - O - M(OR)_3 + H_2O;\ M = Ti, Zr$$

The risk of aggregation is prevented by addition of HCl or HNO_3 to stabilize the colloids. The intermediate layer is formed by dipping in the resulting dispersion. Dave and Nath (2016) found that below 100°C, the dried product is a gel, leading to

formation of a network structure of particles. Polyvinyl alcohol (PVA) is used to prevent cracks caused by asymmetric drying.

A similar method using polymeric sols forms the thin top layer. The polymeric structures penetrate each other to form three-dimensional structured networks. More and more pores are formed as a result. The calcination temperature should be around 300°C, which will provide for easy separation of the top layer afterwards.

9.4 NANOFILTRATION MODULES

For high efficiency the membrane properties should match the requirements for surface/volume ratio, less flow restriction, reduced polarization, reduced construction cost, etc. The main types of NF module (Figure 9.1) are plate and frame, tubular, spiral wound, and hollow fiber (Sagle and Freeman, 2004). The simplest configuration is plate and frame, but the most popular type in the industry is spiral wound.

The tubular module uses selective tubes having an inner diameter of about 1 cm which are often made of ceramic materials. Many tubes are placed parallel to each other inside a housing, like a shell and tube heat exchanger. To avoid the polarization effect (accumulation of solute inside the tubes), feed should be introduced into the lumen of these tubes with a high axial velocity. Permeate passes through to the shell side whilst retentate exits from the other end of the tubes (Figure 9.2). An advantage of this module type is that high-concentration suspension can be used. Large size and low packing density are disadvantages (Nagy, 2018).

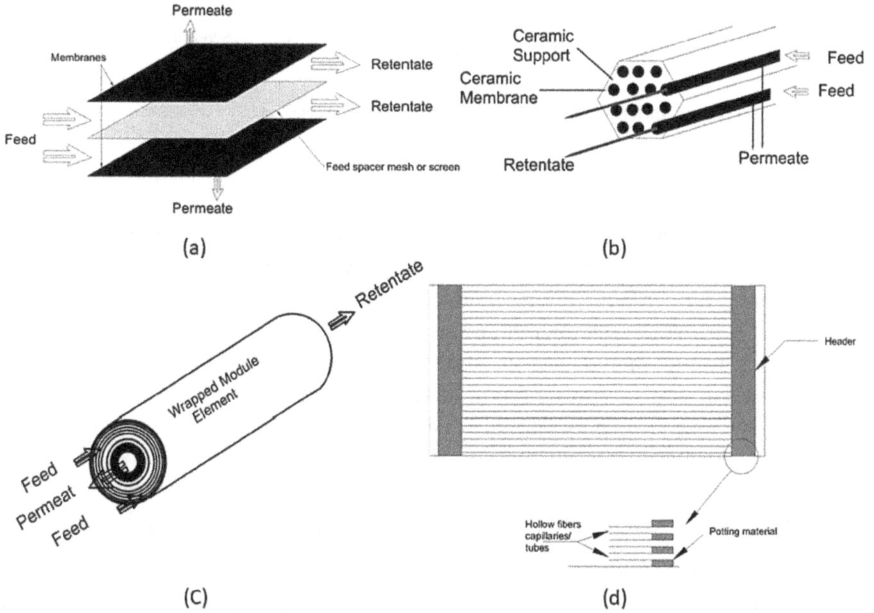

FIGURE 9.1 Nanofiltration modules: a) plate and frame, b) tubular, c) spiral wound, and d) hollow fiber

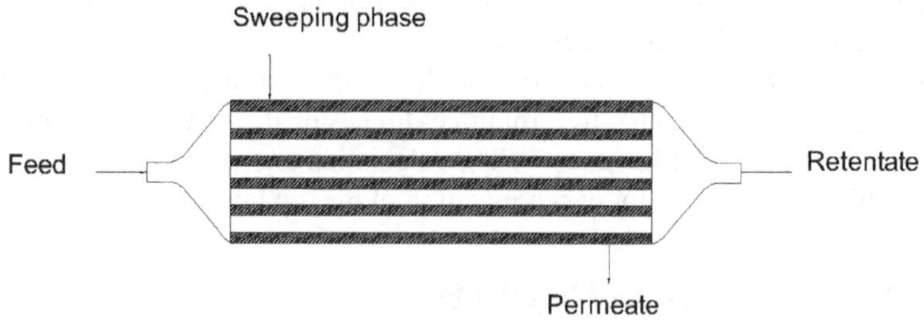

FIGURE 9.2 Tubular membrane module

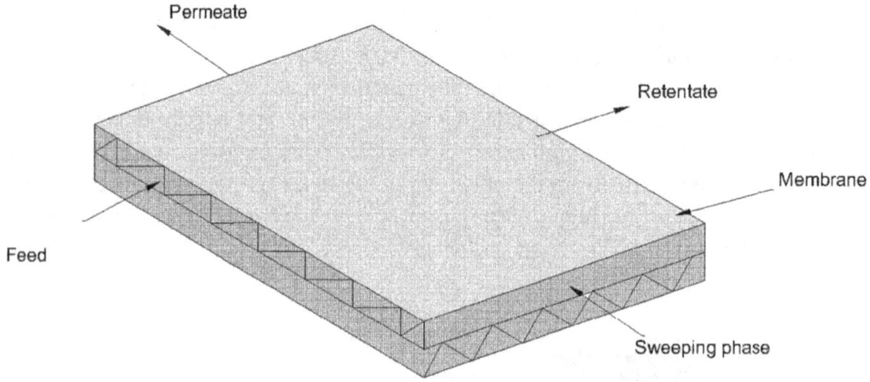

FIGURE 9.3 Plate-and-frame module for membrane separation

The plate-and-frame module has a sandwich-like structure (Figure 9.3), consisting of a selective membrane supported between two plates. This configuration provides channels for permeate to flow on one side of the membrane and retentate on the other. Flow is parallel to the membrane.

In a spiral-wound module a plane sheet membrane forms a spiral around a perforated cylindrical tube (permeate tube). Membrane sheets are separated by polyester or thin plastic spacers. To avoid edge leakage the membrane is sealed with epoxy adhesive (Figure 9.4).

The hollow-fiber module is also known as a capillary membrane module. Unlike the tubular module it has a high packing density with values ranging between 600 and 1,200 $\frac{m^2}{m^3}$; the membrane thickness ranges between 200 and 600 μm, depending on the application. The hollow fibers have two layers, an active layer and a support layer. The active layer should be as thin as possible (<1 μm), whereas the thickness of the support layer depends on the mechanical demands made of it. There are two module configurations: inside-out (Figure 9.5a) and outside-in (Figure 9.5b).

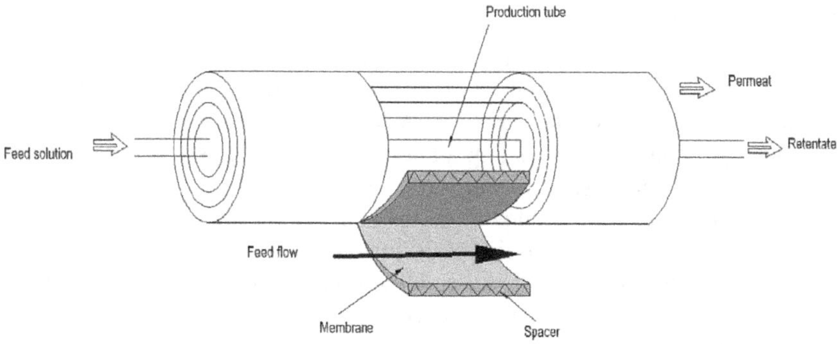

FIGURE 9.4 Spiral-wound module

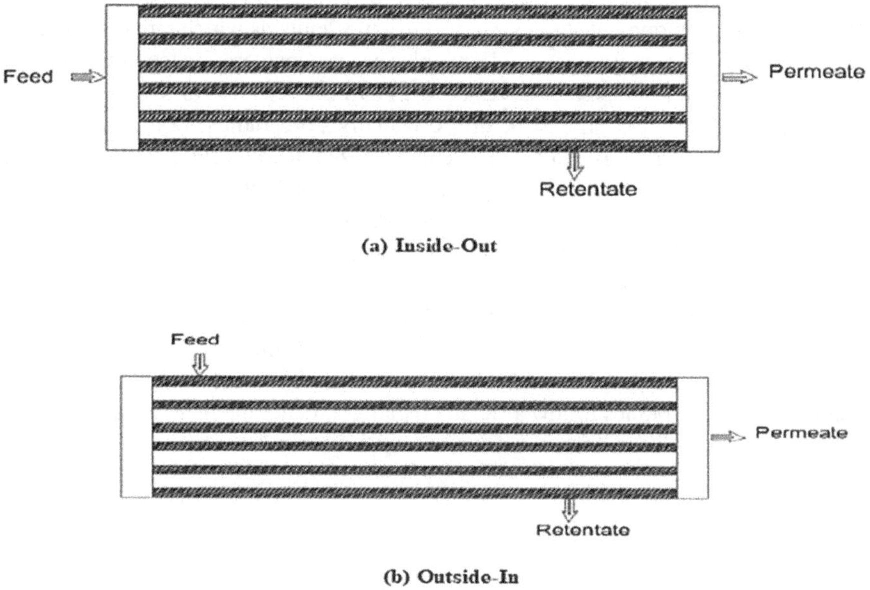

FIGURE 9.5 Capillary hollow-fiber module

The active layer has greater protection in the inside-out module, as the skin layer is on the lumen side. The disadvantage of this flow pattern is that the small tube diameter for the feed encourages plug formation, so it requires a clean feed. The pretreatment required for a clean feed solution makes production costs high. In the outside-in module, the skin layer is on the shell side, which gives a higher membrane surface area. However, channeling on the shell side is a disadvantage of this configuration (Nagy, 2018).

9.4.1 Flow Geometries

Membrane processes have two main flow configurations, dead-end and cross-flow filtration (Figure 9.6). In dead-end filtration, the fluid flow is only perpendicular to the membrane surface, so retentate collects on it, forming a filtration cake. This cake over the filter provides an additional filtration effect, improving separation efficiency. At the same time, an increase in pressure drop reduces overall filtration performance. The feed solution flow in cross-flow filtration is parallel to the membrane surface. Retentate is collected on the same side of the membrane but at its end, whilst permeate flows through the membrane. Dead-end filtration is used at laboratory scale as a batch-type process and has low operating costs but membranes eventually foul. In cross-flow filtration the sweeping effect caused by the flow of the feed solution over the membrane surface reduces fouling (Nagy, 2018).

9.4.2 Membrane Characterization

Membrane characterization is used to select membranes for particular applications. Research and development studies can identify mechanisms which reduce separation effectiveness. The performance of NF membranes is considerably degraded by filtration of concentrated salt solutions, resulting in reduced rejection and decreased volumetric flux. Membrane pore structure, especially pore radius, pore density, pore shape, pore length and tortuosity can be characterized. Results can aid the analysis of solute transport, membrane fouling, etc. Instrumental characterization methods include contact angle measurement, atomic force microscopy (AFM), Fourier transform infrared (FT-IR) spectroscopy, and scanning electron microscopy (SEM). Membrane performance can also be directly observed experimentally.

Passage of uncharged organic molecules through a nanofiltration membrane is determined by their molecular weight. Other physicochemical effects such as dipole interaction may also be present. All models of organic molecules' passage through nanofiltration membranes follow a sieving paradigm. Other interactions are neglected

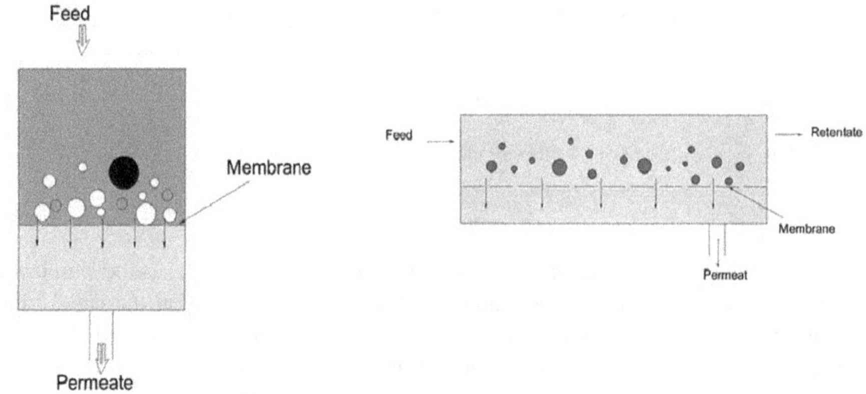

FIGURE 9.6 Two main types of filtration a) dead-end; b) cross-flow

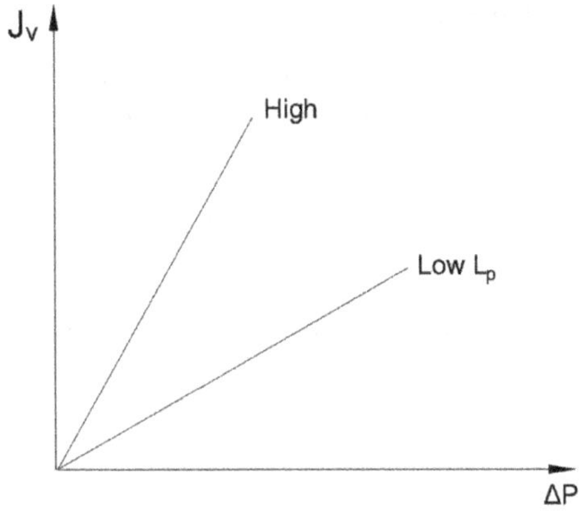

FIGURE 9.7 Permeability: Volume flux as a function of the operating pressure

as per Košutić and Kunst (2002) and Van der Bruggen et al. (2000). Hydrodynamic permeability (L_P), solute permeability (ω), and reflection coefficient (σ) determine solute transport through the membrane. For pure water ($\Delta\Pi = 0$) the slope of the volume flux vs. operating pressure graph gives the permeability value, as shown in Figure 9.7.

The solute permeability and reflection coefficient of a membrane can be calculated from experimental data (Figure 9.8). An expression for solute flux is below.

$$\frac{Js}{\Delta C} = \omega + (1-\sigma) J_V \frac{C'}{\Delta C} \tag{9.1}$$

Artuğ (2007) used three groups to characterize membranes: 1) performance parameters, 2) morphology parameters, and 3) charge parameters.

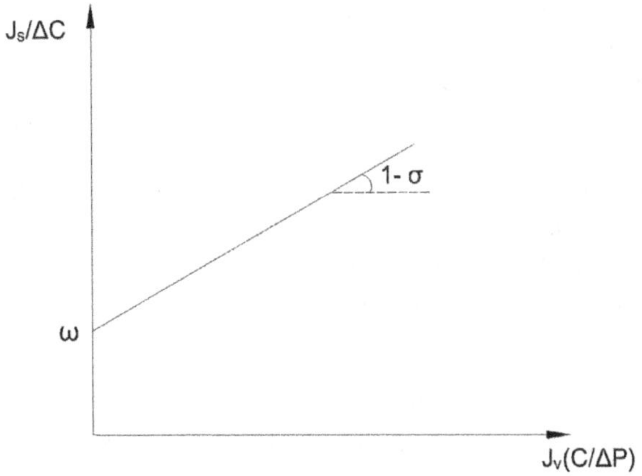

FIGURE 9.8 Relationship for obtaining ω and σ for a nanofiltration membrane

9.4.3 Performance Parameters

Rejection and permeate flux are two major parameters usually used to characterize membrane performance. Observed rejection is the extent to which a component is retained.

$$R_o = \left(1 - \frac{c_p}{c_F}\right) \cdot 100 \tag{9.2}$$

The real rejection is based on the concentration on the feed side membrane surface

$$R_{real} = \left(1 - \frac{c_p}{c_m}\right) \cdot 100 \tag{9.3}$$

where C_m = solute concentration at the feed-membrane interface, C_p = permeate solute concentration, C_F = solute concentration in feed.

The difference between the C_m and C_F terms becomes more important for complete modules due to the accumulation of solute over the membrane surface, which is known as concentration polarization. There is no way to determine membrane surface concentration directly, but it can be estimated from measurements made at low feed concentration, low ΔP, and high Reynolds number. Throughout the process, feed chamber concentration will be the same.

$$R_{real} \approx R_0 = \left(1 - \frac{c_p}{c_F}\right) \cdot 100\, AsC_F \approx C_m \tag{9.4}$$

The boundary layer thickness is determined entirely by the system hydrodynamics and decreases when turbulence increases. Consequently, rejection of a component changes with the operating conditions.

Permeate flux is determined by the driving force and the total resistance of the membrane and the interfacial region adjacent to it. Fouling and cleaning procedures therefore affect the permeate flux. Prediction of solute and solvent fluxes for NF membranes is discussed in the modeling Section 9.5.

Rejection and flux are directly correlated, but there is no fixed relationship between them, especially for NF membranes. Increasing flux through pressure or cross-flow velocity increases may increase or decrease rejection, depending upon concentration polarization effects and the module design. Slightly different systems may yield markedly improved performance for the same NF membrane. Consequently, empirical measurements are commonly used to test delivered membranes against specifications and to select a membrane for a particular application. This procedure may be called membrane screening.

9.4.3.1 Morphology Parameters

Methods for the determination of membrane morphology are summarized in Table 9.2.

AFM is a newer technique in surface characterization than SEM. Hilal et al. (2003) have quantified NF membrane properties (pore size distribution, thickness,

TABLE 9.2
Morphology Characterization Methods for NF Membranes

Method	Characteristic(s)
Gas adsorption/desorption	Pore size, surface area
Permporometry	Pore size/porosity
Microscopy	Pore size, porosity
Field emission microscopy (FEM)	Surface roughness, pore size, porosity
Scanning electron microscopy (SEM)	
Atomic force microscopy (AFM)	
Spectroscopy, ATR-FTIR, ESR/NMR	Chemical composition
Raman spectroscopy, XPS	
Contact angle, captive bubble method	Hydrophobicity
Sessile drop method	

Abbreviations: ATR-FTIR, Attenuated total reflection-Fourier transform infrared; ESR, Electron spin resonance; NMR, nuclear magnetic resonance; XPC, X-ray photoelectron spectroscopy

and surface morphology) by AFM. SEM is widely used to obtain membrane images. Samples must be electrically conductive, and sample preparation of polymeric membranes may modify their surface structure. This technique is used in membrane fouling studies, surface structure or cross-section investigations.

The hydrophilic/hydrophobic character of the membrane influences membrane wettability and may give indications of the fouling tendency and membrane chemical and mechanical stabilities. Hydrophobic membranes have good chemical and mechanical stabilities but show greater fouling tendency than hydrophilic membranes. Hydrophilic membranes usually have more functional groups that can dissociate and generate a charge on the membrane, and they generally show higher fluxes. In experiments on filtration of wastewater containing non-ionic surfactants, Van der Bruggen et al. (2005) observed effect related to membrane hydrophilic character and MWCO. Three mechanisms affects flux. First, flux declined when the molecular weight was less than the MWCO of the membrane, due to narrowing of membrane pores through the adsorption of surfactant monomers. Second, improved wettability of the membrane surface through adsorption of monomers onto hydrophobic groups caused the flux to increase above that of pure water. Third, decreased wettability through adsorption of monomers onto hydrophilic groups caused flux decline. Thus hydrophilic/hydrophobic character was considered to exert a decisive effect on membrane flux. Membrane chemical structure also plays an important role. Improvement of post-treatment in membrane manufacture can introduce functional groups onto the surface or change the membrane hydrophilicity. ATR-FTIR is generally used to determine the functional chemistry of membranes.

9.4.3.2 Charge Parameters

Ionic charge is an important physicochemical parameter in respect of the rejection mechanisms and fouling tendency of a membrane. Matsuura (2001), Nghiem et al. (2005), Wang and Chung (2005) and Bouchoux et al. (2005) have studied the effects

of solute physicochemical properties and solution chemistry on their interaction with membranes. Since the hydrophobic character of the membrane may affect fouling, this property has been explored by the pharmaceutical, pulp, and paper industries. Increasing solution pH can improve the rejection of negatively charged components by negatively charged membranes. Charge interactions between the membrane and charged solute components can be used to separate charged from neutral and oppositely charged components (Artuğ, 2007). According to Bouchoux et al. (2005), the rejection performance of an NF membrane for glucose (neutral solute) in the absence of sodium lactate (charged) was feasible. The separation was not possible in the presence of sodium lactate. This shows that the transport mechanism for a neutral solute can be altered by the presence of charged components.

9.4.3.3 Membrane Charge and Species Transport

Because membrane charge is strongly correlated with species transport across the boundary layer and through the membrane itself, it is very important to understand the charging behavior of the NF membrane to maximize process performance. Typical NF membrane functional groups are carboxylic acid ($-CO_2H$), sulphonic acid ($-SO_3H$) and primary and secondary amine ($-NH_2$, $-NH$) groups. A hypothetical NF membrane with dissociated carboxylic groups in an electrolyte aqueous solution is presented in Figure 9.9. Dissociated carboxylic groups lead to a negative membrane charge, which repels divalent sulfate ions strongly but allows monovalent ions to pass.

Membrane surface charge influences the distribution of ions in solution due to the electro neutrality condition. Counter ions concentrate near the membrane surface. Figure 9.10 illustrates the concentration of ions inside a negatively charged membrane and in the bulk.

An electrochemical double layer, divided into the immobile Stern layer and the mobile/diffuse layer, exists at the surface/solution interface (Figure 9.11). A fixed double layer is assumed adjacent to the surface, whilst beyond is an outer mobile layer in which ion concentration decreases with increasing distance from the membrane surface. The ion distribution in this layer is described by the Poisson–Boltzmann equation.

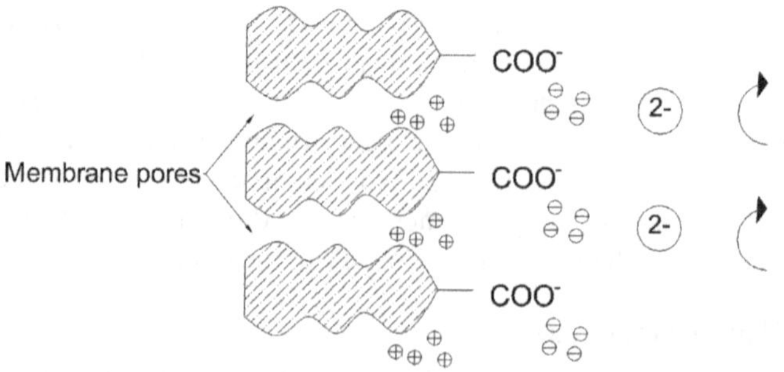

FIGURE 9.9 Hypothetical NF membrane with dissociated carboxylic groups

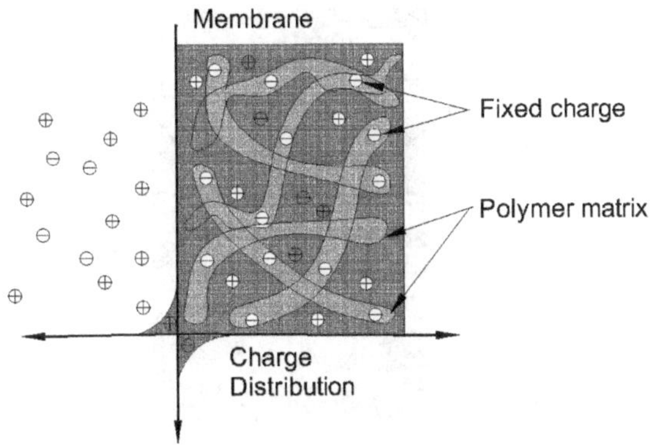

FIGURE 9.10 A schematic illustration of ions inside a negatively charged membrane and in the bulk

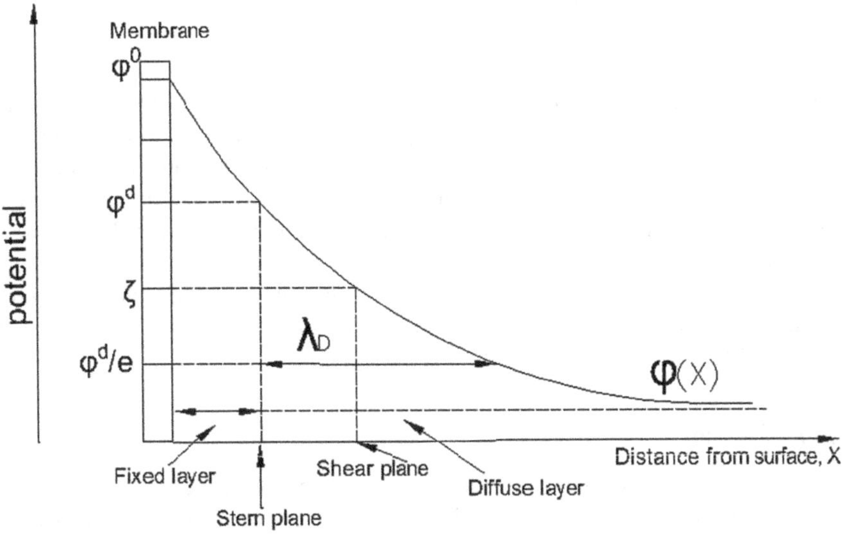

FIGURE 9.11 A schematic drawing of the electrochemical double layer at a membrane solution interface

9.5 NANOFILTRATION MODELING

The design and operation of membrane processes require quantitative methods for the description and prediction of filtration performance. The separation of solutes in NF depends on the micro-hydrodynamics and interfacial events occurring at the membrane surface and inside the membrane. Generally, four layers of transport are considered (Figure 9.12).

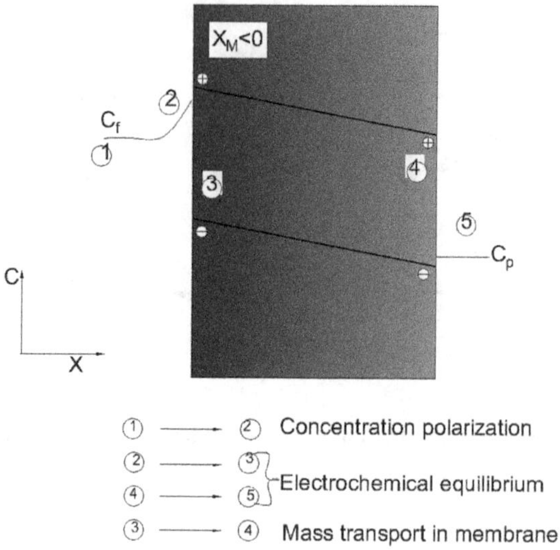

FIGURE 9.12 Mass transport in NF systems

The feed solution is transported to the membrane surface (1→2). Due to ion/membrane interactions, some solutes will be rejected whilst others, and the solvent, will pass through the membrane. The rejected components will accumulate on the membrane surface and diffuse back to the bulk. In this boundary layer, the diffusive and convective fluxes will at equilibrium be equal. With high cross-flow velocities mass transport in the boundary layer can be neglected, as when a turbulence promoter (feed spacer) is used and the yield of the system is low. At interfaces, there is electrochemical equilibrium (2→3 and 4→5). Mass transport occurs inside the membrane (3→4) by diffusion, convection, and electro-migration.

9.5.1 Theory

In RO, NF, and UF, movement of solute is best modeled using irreversible thermodynamics where the membrane is considered as a black box. Spiegler and Kedem (1966) used the following equations to relate the volumetric flux J_v with the solute flux J_s through a membrane:

$$J_V = L_P\left(\Delta P - \sigma\Delta\Pi\right) \tag{9.5}$$

$$J_S = C'\left(1-\sigma\right)J_V + P_S\Delta C \tag{9.6}$$

$$C' = \frac{c_f - c_p}{\ln\left(\frac{c_f}{c_p}\right)} \tag{9.7}$$

where J_V = volume flux, L_P = water permeability, ΔP = trans-membrane pressure, ΔΠ = osmotic pressure difference, ΔP_e = (ΔP −σΔΠ) effective pressure driving force, P_s = solute permeability, σ = reflection coefficient, C_p = permeate concentration, C_f = feed concentration, C = logarithmic concentration, and ΔC = concentration difference between the feed and the permeate.

Several researchers have related the hydraulic permeability (L_P) to the solvent viscosity (μ) and the effective membrane thickness Δx by a Hagen-Poiseuille type relationship.

$$L_p = \frac{r_p^2}{8\mu\Delta x} \tag{9.8}$$

where r_p = average membrane pore radius.

Equation 9.6 shows that the solute flux includes both convective and diffusive effects. Solute is transported convectively due to the applied pressure gradient across the membrane, whereas diffusive solute transport occurs due to the concentration difference across the membrane. When high concentration differences exist between the reject and the permeate, Spiegler and Kedem (1966) used the above equations and obtained the following expression for the rejection rate of the solute as a function of the permeation flux:

$$R = 1 \quad \frac{C_p}{C_m} = \frac{\sigma(1-F)}{1-\sigma F} \tag{9.9}$$

where R = Rejection, C_m = solute concentration at the feed-membrane interface and C_p = permeate solute concentration.

This equation describes salt rejection, R, at different flow rates (J_V) as a function of F (permeation flux)

$$\mathrm{F} = e^{-J_v A.};\mathrm{A} = \frac{1-\sigma}{P_S} \tag{9.10}$$

Equation 9.9 indicates that rejection increases with increasing water flux. The parameters σ and P_s can be determined from experimental measurements of rejection (R) as a function of volume flux (J_V) using the best-fit method. The reflection coefficient (σ) measures the semi-permeability of the membrane, reflecting its ability to pass the solvent in preference to the solute. When an osmotic pressure difference (ΔΠ) across an imperfectly semi-permeable membrane is compensated by an applied pressure (ΔP), so that the volumetric flow is zero (ΔP is smaller than ΔΠ), the ratio between the two is defined (σ), as shown in Eq. 9.11.

$$\sigma = \left(\frac{\Delta P}{\Delta \Pi}\right)_{J_V=0} \tag{9.11}$$

When $\sigma = 1$, convective solute transport through the membrane is absent. When $\sigma < 1$ convective transport takes place, there flection coefficient will be $\sigma < 1$ and will be greater when the solute is small in comparison with the membrane pores.

9.5.2 Model Categorization

Mass transport models can be categorized into three groups: pure empirical, semi-empirical, and physically founded. Purely empirical models require multiple experimental measurements at the operating condition. Regression analysis of experimental data can establish the relationship between the modeling function and observed phenomena (Artuğ, 2007). Another limitation is that separation behavior can only be described by these functions within the range of experimental measurements. Semi-empirical models require fewer experimental measurements, and by extrapolation separation beyond their range can be at least qualitatively described. Semi-empirical models can be divided into two groups: membrane-dependent models where the membrane characteristics are taken into account and membrane-independent models, also called phenomenological models. The solution-diffusion model (SDM) is membrane dependent. The pore model, which is also a membrane-dependent semi-empirical model, is applied to UF and microfiltration, but is not suitable for NF membranes since selectivity occurs only through steric factors and mass transport inside the membrane is purely convective. In Weber (2001), NF membranes are idealized as nonporous or micro-porous, a modification to SDM that can successfully account for the separation behavior of NF membranes. In membrane-independent models the system is considered as a black box. The membrane cannot therefore be characterized in terms of its charge and structural properties. One well-known approach is the Spiegler–Kedem model (Spiegler and Kedem, 1966).

Physical or transport-mechanistic models have focused mainly on three parameters: structure of the membrane; physicochemical properties of the solution; and interactions between membrane, solvent, and solutes. Physical models fall into two categories: those based on a space-charge (SC) model proposed by Morrison and Osterle (1965), and the Meyer–Sievers (1953) (TMS) model. In SC models, the membrane is considered as porous and there are radial and axial distributions of electric potential and ion concentrations across the pores. The radial distributions are described by the Poisson–Boltzmann equation. Ions are treated as point charges, transport of which are described by an extended Nernst–Planck (ENP) equation and the volumetric flow by the Navier–Stokes equation. Application of this model is limited by the numerical complexity of the calculations. The TMS model is a simplified SC model in which the radial distribution is neglected. Bowen and Welfoot (2002) considered that this neglect is valid under normal NF conditions in which the surface charge density is reasonably small and the pores are sufficiently narrow. A uniform membrane and uniform distribution of fixed charge density inside the membrane are assumed. It has been shown that the SC and TMS models give similar results when the pore diameter is less than 2 nm since the radial variation of the potential distribution is quite small in the SC model (Wang et al., 1995).

9.5.2.1 Membrane-Dependent Semi-Empirical Model: Solution-Diffusion Model (SDM)

The assumptions made in this model are:

- The membrane is a nonporous, homogeneous continuum. The components dissolve into the membrane and diffuse through it due to the chemical potential gradient.

- There is chemical equilibrium at phase interfaces.
- Solute and solvent fluxes are not coupled.

The flux of a component in a membrane is modeled thus by Teorell (1935):
Flux = Concentration × Mobility × Driving force
Mobility is given by the Nernst–Einstein equation

$$D_i = R_g \cdot T \cdot u_i \tag{9.12}$$

The driving force for non-electrolyte components is the chemical potential gradient. Accordingly, the flux of component i per unit of membrane area is:

$$\acute{n}_l = -c_i \cdot \frac{D_i}{R_g \cdot T} \cdot \frac{d\mu_i}{dx} \tag{9.13}$$

where $\acute{n}$ = Flux, D = Diffusion, μ = Chemical potential, c = concentration, R_g = Gas constant, T = temperature, u = mobility, and v = velocity.

Integration of Eq. 9.13 across the membrane (0–Δx), gives the volumetric water flux by assuming constant mobility through the membrane. The final equation for water flux is

$$J_w = \frac{C_w \cdot D_w \, \acute{V}_{w}\square}{\Delta x \cdot R_g \cdot T} \cdot \frac{M_w}{\rho_w} \cdot \left[P^F - P^P - \prod_w^F + \prod_w^P \right] \tag{9.14}$$

After simplification the resultant equation for water flux according to SDM is

$$J_w = L_P \left(\Delta \text{P} - \Delta \Pi \right) \tag{9.15}$$

Equation 9.15 indicates the dependency of the solvent flux on the applied pressure and the concentration difference between the feed and the permeate sides.

In the SDM, the chemical potential gradient is the only driving force for salt flux and electrical interactions are not taken into account. The latter play an important role in the case of NF membranes. When the concentration of the rejected component in the permeate is very low, the salt flux can be neglected in comparison to the water flux and the following approximation is obtained:

$$C_p = \frac{J_c}{J_c + J_w} \approx \frac{J_c}{J_w} \tag{9.16}$$

When this relation is substituted for real rejection in Eq. 9.9

$$R = \left(1 \frac{J_i}{C_m \cdot J_w} \right) \cdot 100 \tag{9.17}$$

Solute flux is defined by

$$J_i = \text{B} \cdot \Delta \text{C} \tag{9.18}$$

where B is the membrane constant for solute flux. By putting the value of J_i in Eq. 9.16, the final equation is

$$R = \left(\frac{1}{1 + \frac{B}{J_w}} \right) \cdot 100 \tag{9.19}$$

Equation 9.19 shows the rejection increase with increasing trans-membrane flux approximates to the limiting value (100% rejection), since the water and salt fluxes are not coupled. A limitation of the SDM model is that rejection is independent of feed concentration, which makes it inappropriate for modeling nanofiltration processes, where concentration has a significant effect on separation performance.

9.5.2.2 Phenomenological Models

Initial descriptions of ion transport in NF membranes were based on phenomenological equations, which are based on irreversible thermodynamics. In the following, the well-known Spiegler–Kedem model will be introduced.

9.5.2.2.1 Spiegler–Kedem Model

According to Spiegler and Kedem (1966), partial fluxes of solute and solvent in the SDM are connected by a convective transport term. For that purpose, the Staverman reflection coefficient, σ, which represents the selectivity of a membrane for the solute, is used.

$$J_w = -L_w \left(\frac{dp}{dx} - \sigma \frac{d\Pi}{dx} \right) \tag{9.20}$$

$$J_S = -L_i \cdot \frac{dC_i}{dx} + (1 - \sigma) \cdot C_i \cdot J_w \tag{9.21}$$

σ can be determined when the osmotic pressure is balanced by hydraulic pressure so that no permeate flux is observed:

$$\sigma = \frac{\Delta P}{\Delta \Pi} \Big|_{J_{p=0}} \tag{9.22}$$

For an unselective membrane ($\sigma = 0$) the osmotic pressure effect in Eq. 9.20 disappears and solute convective transport is predominant. Whereas, for an ideal semipermeable membrane ($\sigma = 1$), water flux definition turn to the one in SDM and convective salt transport vanishes. By integrating Eqs. 9.20 and 9.21 over the membrane thickness Δx and implementing in the rejection equation gives.

$$R = \frac{\sigma(1 - F)}{1 - \sigma F} \tag{9.23}$$

where

$$F = e^{-J_v \left(\frac{1-\sigma}{L_S} \right)} \tag{9.24}$$

L_s and σ can be determined from the experimental data of rejection as a function of water flux. Rejection is independent of salt concentration since the solute permeability coefficient L_s is constant. However, it is not realistic for NF membranes since such coefficients depend on the feed concentration of electrolyte solutions. Perry and Linder (1989) extended the Spiegler–Kedem model in order to include electrostatic exclusion in three ion systems, in which one ion is totally rejected due to the electrostatic exclusion and influences the charge balance of the whole system.

9.6 APPLICATIONS

9.6.1 Nanofiltration in the Food Industry

Membrane separation provides an alternative to the conventional processes used for concentration or fractionation. Nanofiltration (NF) is a pressure-driven separation process in which the filtration efficiency depends on steric (sieving) and charge (Donnan) effects (Warczok et al., 2004). Nanofiltration has been addressed by researchers for application in the food and allied industries due to its advantages over conventional methods. Aqueous solutions of organic solutes with a molecular weight of 100–1,000 Da can be concentrated, fractionated, or purified using nanofiltration with an applied pressure of 1–4 MPa (Salehi, 2014). NF has been extensively studied in the past few decades for applications in the concentration of juices (Warczok et al., 2004), in winemaking (García-Martín et al., 2010), in the dairy industry (Atra et al., 2005), and for separation of various bioactive compounds (Tundis et al., 2018).

In the beverage industry nanofiltration is mainly used for the concentration of juices (Warczok et al., 2004), reduction of alcohol content in alcoholic beverages (Goncalves and Maria, 2007) and the concentration of phenolic compounds (Murakami et al., 2011). NF is also used for the separation of bioactive compounds from juices (Arriola et al., 2014; Tundis et al., 2018). In the dairy industry a major challenge is the disposal of acid whey because it contains lactic acid and minerals (Talebi et al., 2020). Nanofiltration is effective in whey processing (Chandrapala et al., 2016), demineralization of whey and milk (Suárez et al., 2006), recovery of lactose (Magueijo et al., 2005) and lactic acid (Li and Shahbazi, 2006; Li et al., 2007). Nanofiltration allows by-product utilization, and reduces waste water organic load and the environmental impact of dairy industry waste. Applications of nanofiltration in food and allied industries are shown in Table 9.3.

9.6.2 Nanofiltration in Water Treatment

Nanofiltration can reduce particulate contamination, remove hardness, remove organic matter, and reduce ionic strength. Sombekke et al. (1997) compared the effectiveness of nanofiltration for water softening with that of pellet softening and granular activated carbon. Nanofiltration is generally beneficial with respect to health aspects and investment costs. Jacangelo et al. (1997) reported that NF and RO have been regularly used to remove natural organic matter (NOM). Nanofiltration has also been used to treat groundwater having a high total

TABLE 9.3
Application of Nanofiltration in Food and Allied Industries

Application in		Type of Membrane	Results	References
Fruits processing	• Apple Juice • Pear Juice	Tubular membrane	• NF can be used for the concentration of Apple and Pear juice. • Permeate flux depends on the composition of juice. • Membrane roughness doesn't depend on structure density. • Membrane roughness is linked to yield of irreversible fouling	Warczok et al. (2004)
	• Strawberry Juice	–	• Effective for Phenolic compound maintenance and prevents the loss of bioactive compounds. • A good alternative for processing of strawberry juice even in the presence of light and oxygen.	Arend et al. (2019)
	• Grape Juice	Spiral model of NF membrane	• NF is an effective technique for the concentration of procyanidins from grape juice. • Fast separation. • Low energy consumption. • Zero oxidation loss	Li et al. (2019a)
	• *Citrus aurantium* Juice	–	• NF can be used for the separation of synephrine from citrus juice. • A linear relation between operation pressure and membrane flux.	Li et al. (2019b)
	• Pequi Juice	Flat membrane	• The direct NF process resulted in almost 100% retention of phenolic compounds • In the direct NF process, the permeate flux is quite low. • In the sequential NF process (combined with micro and ultrafiltration processes as pre-treatments), the permeate flux is twice the permeate flux in the direct NF process.	de Santana Magalhães et al. (2019)
	• Blackberry Juice	Flat sheet membrane	• A higher trans-membrane pressures increases permeate flux and increase of retention of total anthocyanins (>94%) and total soluble solids (44–97%) • Retention of 100% of total ellagitannin.	Acosta et al. (2017)
	• Grapes must	–	• NF membranes provided a high rejection for sugars (range 77–97%) and low malic acid retention (range 2–14%). • A two-stage NF process allows the concentration of the sugars, i.e., the potential alcohol. • The wine obtained from grape must be processed using the NF membrane contains high total polyphenols and optical density at 420 nm.	Versari et al. (2003)

(Continued)

Dairy	• Whey	Flat sheet membrane	• NF can be used to separate lactate acid from lactose in acid whey streams. • NF can retain over 90% of the lactose.	Chandrapala et al. (2016)
	• Second cheese whey	Flat sheet membrane	• NF is effective in the separation of lactose from SCW. • Reduces wastewater organic load. • This leads to the valorization of the by-product of cheese and curd cheese manufacturing.	Magueijo et al. (2005)
	• Cheese whey	–	• NF membrane with a molecular weight cutoff of 100–400 can be used to separate lactic acid from lactose and microbial cells.	Li et al. (2007)
Biomolecules	• Amino acids solution	–	• Neutral amino acids can be readily separated according to their size using NF with multilayer polyelectrolyte membranes.	Hong and Bruening (2006)

hardness, high color, and high loading of disinfection by-product (DBP) precursors. This makes nanofiltration a reliable method although the main focus is to remove organics rather than water softening.

Yeh et al. (2000) studied the softening of lake water by GAC and pellet softening, a conventional method followed by ozone treatment and by an integrated UF/NF membrane process. It was found that the integrated membrane process yielded the best water quality, with a turbidity of 0.03 NTU. Total hardness rejection was 90% and dissolved organic rejection was 75%. Nanofiltration also removes organic pollutants effectively. Koyuncu and Yazgan (2001) reported good results obtained from Kucukecmece Lake water present in Istanbul. The salty and polluted water was treated using the TFC-S NF membrane. NF easily removes viruses and bacteria. *Bacillus subtilis* spores were reduced from 5.4 to 10.7 log by Reiss et al. (1999) using an integrated microfiltration and nanofiltration membrane structure. Studies have also shown that nanofiltration is quite efficient in the removal of arsenic. As (V) rejection is as high as 90% but that of As (III) is only 30%.

Wastewater can be remediated using NF. Afonso and Yan (2001) treated fish meal wastewater using NF. The total organic load was reduced and the water made reusable by partial desalination. Rautenbach et al. (2000) studied different combinations of membranes in order to achieve 100% water recovery. They developed an integrated RO/NF/high-pressure RO membrane which recovered more than 95% of wastewater from dumpsite leachate. Hafiane et al. (2000) studied the removal of chromate (VI) from aqueous solution using the TCF-S NF membrane. BNF was more effective than conventional methods being highly dependent on pH and ionic strength. At a pH of 8, retention was around 80%. Tang and Chen (2002) treated textile wastewater using NF at an operating pressure of 500 kPa. 90% of dye was removed with 14% removal of NaCl.

9.6.3 NF as a Pre-Treatment for Desalination

Feed pre-treatment plays a major role in determining the effectiveness of desalination. Pre-treatment processes previously used include media and cartridge filters supported by chemical treatment, coagulation, flocculation, acid treatment, etc. These pre-treatments can cause corrosion of equipment surfaces making them more susceptible to scale deposits. They are also more complex and time-consuming. Al-Sofi et al. (1998) used nanofiltration in the pre-treatment of seawater prior to reverse osmosis (SWRO), multistage flash (MSF), and SWRO-rejected multistage flash (SWRO-MSF) desalination. Changes were observed in hardness, turbidity, and microbial load.

Sulfate, magnesium, calcium, and bicarbonate ions were reduced by 97.8, 94.0, 89.6, and 76.6% respectively. Separation of monovalent ions was also good at 40.3% for chloride, potassium, and sodium ions. The TDS observed was 57.7% and total hardness was 93.3%. Advantages of NF pre-treatment include:

1) Prevention of membrane fouling due to the removal of bacteria and turbidity.
2) Prevention of scale build-up.
3) Reduced seawater TDS reduced the required feed pressure.

Drioli et al. (1999) demonstrated that NF membranes improved pre-treatment performance at the same energy requirement. Drioli et al. (1999) used an integrated nanofiltration, reverse osmosis, and membrane crystallizer membrane structure to achieve total recovery of desalted water. The introduction of NF increased the recovery of the RO unit by 50% and MC increased it to 100%. The brine disposal problem was eliminated and crystalline solids were obtained.

9.6.4 NF in Trace Contaminant Removal

Sorption diffusion plays an important role in nanofiltration besides size exclusion and charge repulsion during the separation process. Trace contamination removal is generally discussed in terms of organics and inorganics removal. Saini et al. (2014) stated that the separation process is physico-chemical, meaning that both physical and chemical processes contribute to selectivity. The molecular weights of trace organic contaminants generally range from less than 100 to a few hundred Daltons. The trace organic contamination is further subdivided into four groups: (a) pesticides; (b) trihalomethanes (THMs); (c) polychlorinated biphenols (PCBs); and (d) polycyclic aromatic hydrocarbons (PAHs).

Maximum contamination levels for many compounds are issued by the World Health Organisation, but many compounds have no set level because of difficulties in their analysis at trace levels. A wide variety of synthetic organic compounds are in widespread use such as pesticides, pigments, dye carriers, preservatives, pharmaceuticals, refrigerants, propellants, heat transfer media, dielectric fluids, degreasers, and lubricants. By-products from the production of these chemicals, and their metabolites, may pose a greater risk to human health and the environment than the parent compounds. Nghiem et al. (2005) divided these compounds into different groups, including persistent organic compounds (POCs), pesticides, pharmaceutically active compounds (PhACs), and endocrine-disrupting chemicals (EDCs). Kristiana et al. (2013) stated that NF reduces DBP concentrations in finished water quite effectively. Natural organic matter is generally of high molecular weight. Therefore, NF can effectively eliminate DBPs by removing naturally occurring organic matter before disinfection. Hence, the effectiveness of DBP removal is increased by reducing the DBP formation potential. NF can also directly remove DBPs following disinfection. Size exclusion, charge interaction and adsorption are important mechanisms in trace organic contamination removal. Removal of inorganic contamination is mainly achieved by two NF processes:

Convection: The solvent stream passes through the membrane and the larger solutes are retained (based on physical selectivity).

Sorption diffusion: Solutes are transported across the membrane by diffusion, driven by a chemical potential gradient (based on ion-exchange selectivity and diffusion coefficient).

Bodzek et al. (2011) explained that NF is mainly used to eliminate boron, bitrate, and fluoride ions where these are found as unwanted trace contaminants, especially in drinking water.

The hydrated radius of the nitrate ion is slightly higher than that of chloride and both have a similar charge. However, Wang et al. (2014) stated that chloride is usually better retained by the membrane. This cannot be explained solely by a size exclusion mechanism or the extended Nernst–Planck equation. In this case 'solute membrane affinity' appears to work synergistically with ion-exchange selectivity. It thus seems that sorption diffusion plays an important role in the transport of nitrate ions in the nanofiltration membrane.

9.7 CONCLUSIONS

Although nanofiltration was initially used only to process drinking water, it is now effectively used in wastewater treatment and has applications in other industries. Current nanofiltration materials are increasingly inefficient and are reducing the cost of installation. Different models of nanofiltration are a source for the development of future applications. Nanofiltration has the potential for wider application in fruit juice processing, dairy processing, and biomolecule separation. Nanofiltration has also emerged as a game-changer in desalination and wastewater treatment. Finally, trace contaminants which threaten humanity, the environment, and the ecosystem can be effectively controlled using nanofiltration, and much research is ongoing in this field.

REFERENCES

Acosta, O., Vaillant, F., Pérez, A.M., and Dornier, M. 2017. Concentration of polyphenolic compounds in blackberry (rubusadenotrichosschltdl.) juice by nanofiltration. *Journal of Food Process Engineering*, *40*(1): E12343.

Afonso, M.D., and Yan, R.B. 2001. Nanofiltration of wastewater from the fishmeal industry. *Desalination*, *139*(1–3), 429.

Al-Sofi, M.A., Hassan, A.M., Mustafa, G.M., Dalvi, A.G.I., and Kither, M.N. 1998. Nanofiltration as a means of achieving higher TBT of≥ 120 C in MSF. *Desalination*, *118*(1–3), 123–129.

Arend, G.D., Rezzadori, K., Soares, L.S., and Petrus, J.C.C. 2019. Performance of nanofiltration process during concentration of strawberry juice. *Journal of Food Science and Technology*, *56*(4), 2312–2319.

Arriola, N.A., dos Santos, G.D., Prudêncio, E.S., Vitali, L., Petrus, J.C.C., and Castanho Amboni, R.D. 2014. Potential of nanofiltration for the concentration of bioactive compounds from watermelon juice. *International Journal of Food Science &Technology*, *49*(9): 2052–2060.

Artuğ, G. 2007. *Modelling and Simulation of Nanofiltration Membranes*. Cuvillier Verlag, pp. 22–50.

Atra, R., Vatai, G., Bekassy-Molnar, E., and Balint, A. 2005. Investigation of ultra-and nanofiltration for utilization of whey protein and lactose. *Journal of Food Engineering*, *67*(3), 325–332.

Bessarabov, D., and Twardowski, Z. 2002. Industrial application of nanofiltration-new perspectives. *Membrane Technology*, *9*(6).

Bodzek, M., Konieczny, K., and Kwiecińska, A. 2011. Application of membrane processes in drinking water treatment–state of art. *Desalination and Water Treatment*, *35*(1–3), 164–184.

Bouchoux, A., Roux-de Balmann, H., and Lutin, F. 2005. Nanofiltration of glucose and sodium lactate solutions: Variations of retention between single-and mixed-solute solutions. *Journal of Membrane Science*, *258*(1–2), 123–132.

Bowen, W.R., and Welfoot, J.S. 2002. Modelling the performance of membrane nanofiltration: Critical assessment and model development. *Chemical Engineering Sscience*, *57*(7), 1121–1137.

Bruggen, B.V.D. 2013. Nanofiltration. In *Encyclopedia of Membrane Science and Technology*. John Wiley & Sons.

Burn, S., Hoang, M., Zarzo, D., Olewniak, F., Campos, E., Bolto, B., and Barron, O. 2015. Desalination techniques: A review of the opportunities for desalination in agriculture. *Desalination*, *364*, 2–16.

Campbell, J. 2014. The development of hybrid polymer-metal organic framework membranes for organic solvent nanofiltration. doi:10.25560/33219

Chandrapala, J., Chen, G.Q., Kezia, K., Bowman, E.G., Vasiljevic, T., and Kentish, S.E. 2016. Removal of lactate from acid whey using nanofiltration. *Journal of Food Engineering*, *177*, 59–64.

Dave, H.K., and Nath, K. 2016. Graphene oxide incorporated novel polyvinyl alcohol composite membrane for pervaporative recovery of acetic acid from vinegar wastewater. *Journal of Water Process Engineering*, *14*, 124–134.

de Santana Magalhães, F., Sá, M.D.S.M., Cardoso, V.L., and Reis, M.H.M. 2019. Recovery of phenolic compounds from pequi (Caryocarbrasiliense Camb.) fruit extract by membrane filtrations: Comparison of direct and sequential processes. *Journal of Food Engineering*, *257*, 26–33.

Drioli, E., Lagana, F., Criscuoli, A., and Barbieri, G. 1999. Integrated membrane operations in desalination processes. *Desalination*, *122*(2–3), 141–145.

Fridman-Bishop, N., Tankus, K.A., and Freger, V. 2018. Permeation mechanism and interplay between ions in nanofiltration. *Journal of Membrane Science*, *548*, 449–458.

Garcia, C.M. 2000. Ion separation from dilute electrolyte solutions by nanofiltration. *INIS*. http://inis.iaea.org/search/search.aspx?orig_q=RN:32046478.

García-Martín, N., Perez-Magariño, S., Ortega-Heras, M., González-Huerta, C., Mihnea, M., González-Sanjosé, M.L., and Hernández, A. 2010. Sugar reduction in musts with nanofiltration membranes to obtain low alcohol-content wines. *Separation and Purification Technology*, *76*(2), 158–170.

Goncalves, F.D.S., and Maria, N.C.D.P. 2007. U.S. Patent Application No. 10/561, 540.

Hafiane, A., Lemordant, D., and Dhahbi, M. 2000. Removal of hexavalent chromium by nanofiltration. *Desalination*, *130*(3), 305–312.

Han, Y., Jiang, Y., and Gao, C. 2015. High-flux graphene oxide nanofiltration membrane intercalated by carbon nanotubes. *ACS applied materials & interfaces*, *7*(15), 8147–8155.

Hilal, N., Mohammad, A.W., Atkin, B., and Darwish, N.A. 2003. Using atomic force microscopy towards improvement in nanofiltration membranes properties for desalination pretreatment: A review. *Desalination*, *157*(1–3), 137–144.

Hołda, A.K., Aernouts, B., Saeys, W., and Vankelecom, I.F. 2013. Study of polymer concentration and evaporation time as phase inversion parameters for polysulfone-based SRNF membranes. *Journal of Membrane Science*, *442*, 196–205.

Hong, S.U., and Bruening, M.L. 2006. Separation of amino acid mixtures using multilayer polyelectrolyte nanofiltration membranes. *Journal of Membrane Science*, *280*(1–2), 1–5.

Ismail, A.F., and Matsuura, T. 2018. Progress in transport theory and characterization method of Reverse Osmosis (RO) membrane in past fifty years. *Desalination*, *434*, 2–11.

Jacangelo, J.G., Trussell, R.R., and Watson, M. 1997. Role of membrane technology in drinking water treatment in the United States. *Desalination*, *113*(2–3), 119–127.

Jye, L.W., and Ismail, A.F. 2016. *Nanofiltration Membranes: Synthesis, Characterization, and Applications*. CRC Press.

Košutić, K., and Kunst, B. 2002. Removal of organics from aqueous solutions by commercial RO and NF membranes of characterized porosities. *Desalination*, *142*(1), 47–56.

Koyuncu, I., and Yazgan, M. 2001. Application of nanofiltration and reverse osmosis membranes to the salty and polluted surface water. *Journal of Environmental Science and Health, Part A*, *36*(7), 1321–1333.

Kristiana, I., Tan, J., Joll, C.A., Heitz, A., Von Gunten, U., and Charrois, J.W. 2013. Formation of N-nitrosamines from chlorination and chloramination of molecular weight fractions of natural organic matter. *Water research*, *47*(2), 535–546.

Labban, O., Liu, C., and Chong, T.H. 2017. Fundamentals of low-pressure nanofiltration: Membrane characterization, modeling, and understanding the multi-ionic interactions in water softening. *Journal of Membrane Science*, *521*, 18–32.

Li, C., Ma, Y., Li, H., and Peng, G. 2019a. Exploring the nanofiltration mass transfer characteristic and concentrate process of procyanidins from grape juice. *Food Science &Nutrition*, *7*(5), 1884–1890.

Li, C., Ma, Y., Gu, J., Zhi, X., Li, H., and Peng, G. 2019b. A green separation mode of synephrine from Citrus aurantium L. (Rutaceae) by nanofiltration technology. *Food Science & Nutrition*, *7*(12), 4014–4020.

Li, Y., and Shahbazi, A. 2006. Lactic acid recovery from cheese whey fermentation broth using combined ultrafiltration and nanofiltration membranes. In: *Twenty-Seventh Symposium on Biotechnology for Fuels and Chemicals*. Humana Press, pp. 985–996.

Li, Y., Shahbazi, A., Williams, K., and Wan, C. 2007. Separate and concentrate lactic acid using combination of nanofiltration and reverse osmosis membranes. In: *Biotechnology for Fuels and Chemicals*. Humana Press, pp. 369–377.

Magueijo, V., Minhalma, M., Queiroz, D., Geraldes, V., Macedo, A., and Pinho, M.D. 2005. Reduction of wastewaters and valorisation of by-products from "Serpa" cheese manufacture using nanofiltration. *Water Science and Technology*, *52*(10–11), 393–399.

Matsuura, T. 2001. Progress in membrane science and technology for seawater desalination: A review. *Desalination*, *134*(1–3), 47–54.

Mohammad, A.W., Teow, Y.H., Ang, W.L., Chung, Y.T., Oatley-Radcliffe, D.L., and Hilal, N. 2015. Nanofiltration membranes review: Recent advances and future prospects. *Desalination*, *356*, 226–254.

Morrison Jr, F.A., and Osterle, J.F. 1965. Electrokinetic energy conversion in ultrafine capillaries. *The Journal of Chemical Physics*, *43*(6), 2111–2115.

Mulder, M. 1996. *Basic Principles of Membrane Technology*, 2nd ed. Dordrecht: Springer Netherlands.

Murakami, A.N.N., Amboni, R.D.D.M.C., Prudêncio, E.S., Amante, E.R., de Moraes Zanotta, L., Maraschin, M., and Teófilo, R.F. 2011. Concentration of phenolic compounds in aqueous mate (Ilex paraguariensis A. St. Hil) extract through nanofiltration. *LWT - Food Science and Technology*, *44*(10), 2211–2216.

Nagy, E. 2018. *Basic Equations of Mass Transport through a Membrane Layer*. Elsevier.

Ng, L.Y., Mohammad, A.W., Leo, C.P., and Hilal, N. 2013. Polymeric membranes incorporated with metal/metal oxide nanoparticles: A comprehensive review. *Desalination*, *308*, 15–33.

Nghiem, L. 2005. Removal of emerging trace organic contaminants by nanofiltration and reverse osmosis. PhD thesis, Univeristy of Wollongong.

Nghiem, L.D., Schäfer, A.I., and Elimelech, M. 2005. Pharmaceutical retention mechanisms by nanofiltration membranes. *Environmental Science &Technology*, *39*(19), 7698–7705.

Ouyang, L., Malaisamy, R., and Bruening, M.L. 2008. Multilayer polyelectrolyte films as nanofiltration membranes for separating monovalent and divalent cations. *Journal of membrane science*, *310*(1–2), 76–84.

Peeters, J.M.M., Mulder, M.H.V., and Strathmann, H. 1999. Streaming potential measurements as a characterization method for nanofiltration membranes. *Colloids and Surfaces A: Physicochemical and Engineering Aspects*, *150*(1–3), 247–259.

Perry, M., and Linder, C. 1989. Intermediate reverse osmosis ultrafiltration (RO UF) membranes for concentration and desalting of low molecular weight organic solutes. *Desalination*, *71*(3), 233–245.

Purkait, M.K., Sinha, M.K., Mondal, P., and Singh, R. 2018. *Stimuli Responsive Polymeric Membranes: Smart Polymeric Membranes*. Academic Press.

Rautenbach, R., Linn, T., and Eilers, L. 2000. Treatment of severely contaminated waste water by a combination of RO, high-pressure RO and NF: Potential and limits of the process. *Journal of Membrane Science*, *174*(2), 231–241.

Reiss, C.R., Taylor, J.S., and Robert, C. 1999. Surface water treatment using nanofiltration: Pilot testing results and design considerations. *Desalination*, *125*(1–3), 97–112.

Sagle, A., and Freeman, B. 2004. Fundamentals of membranes for water treatment. *The Future of Desalination in Texas*, *2*(363), 137.

Saha, N.K., and Joshi, S.V. 2009. Performance evaluation of thin film composite polyamide nanofiltration membrane with variation in monomer type. *Journal of Membrane Science*, *342*(1–2), 60–69.

Saini, P., Reddy, A.S., and Bulasara, V.K.G. 2014. *Pre-Treatment of Textile Industry Wastewater Using Ceramic Membranes (Doctoral dissertation)*.

Salehi, F. 2014. Current and future applications for nanofiltration technology in the food processing. *Food and Bioproducts Processing*, *92(2)*, 161–177.

Sombekke, H.D.M., Voorhoeve, D.K., and Hiemstra, P. 1997. Environmental impact assessment of groundwater treatment with nanofiltration. *Desalination*, *113*(2–3), 293–296.

Spiegler, K.S., and Kedem, O. 1966. Thermodynamics of hyperfiltration (reverse osmosis): Criteria for efficient membranes. *Desalination*, *1*(4), 311–326.

Suárez, E., Lobo, A., Álvarez, S., Riera, F.A., and Álvarez, R. 2006. Partial demineralization of whey and milk ultrafiltration permeate by nanofiltration at pilot-plant scale. *Desalination*, *198*(1–3), 274–281.

Talebi, S., Suarez, F., Chen, G.Q., Chen, X., Bathurst, K., and Kentish, S.E. 2020. Pilot study on the removal of lactic acid and minerals from Acid whey using membrane technology. *ACS Sustainable Chemistry & Engineering*, *8*(7), 2742–2752.

Tang, C., and Chen, V. 2002. Nanofiltration of textile wastewater for water reuse. *Desalination*, *143*(1), 11–20.

Teorell, T. 1935. An attempt to formulate a quantitative theory of membrane permeability. *Proceedings of the Society for Experimental Biology and Medicine*, *33*(2), 282–285.

Teorell, T. 1953. Transport processes in ionic membranes. *Progress in Biophysics & Molecular Biology*, *3*, 305.

Tundis, R., Loizzo, M.R., Bonesi, M., Sicari, V., Ursino, C., Manfredi, I., and Cassano, A. 2018. Concentration of bioactive compounds from elderberry (Sambucusnigra L.) juice by nanofiltration membranes. *Plant Foods for Human Nutrition*, *73*(4), 336–343.

Van der Bruggen, B., Cornelis, G., Vandecasteele, C., and Devreese, I. 2005. Fouling of nanofiltration and ultrafiltration membranes applied for wastewater regeneration in the textile industry. *Desalination*, *175*(1), 111–119.

Van der Bruggen, B., Schaep, J., Wilms, D., and Vandecasteele, C. 2000. A comparison of models to describe the maximal retention of organic molecules in nanofiltration. *Separation Science and Technology*, *35*(2), 169–182.

Versari, A., Ferrarini, R., Parpinello, G.P., and Galassi, S. 2003. Concentration of grape must by nanofiltration membranes. *Food and Bioproducts Processing*, *81*(3), 275–278.

Wang, J., Dlamini, D.S., Mishra, A.K., Pendergast, M.T.M., Wong, M.C., Mamba, B.B., and Hoek, E.M. 2014. A critical review of transport through osmotic membranes. *Journal of Membrane Science*, *454*, 516–537.

Wang, K.Y., and Chung, T.S. 2005. The characterization of flat composite nanofiltration membranes and their applications in the separation of Cephalexin. *Journal of Membrane Science*, *247*(1–2), 37–50.

Wang, X.L., Tsuru, T., Nakao, S.I., and Kimura, S. 1995. Electrolyte transport through nanofiltration membranes by the space-charge model and the comparison with Teorell-Meyer-Sievers model. *Journal of Membrane Science*, *103*(1–2), 117–133.

Warczok, J., Ferrando, M., Lopez, F., and Güell, C. 2004. Concentration of apple and pear juices by nanofiltration at low pressures. *Journal of Food Engineering*, *63*(1), 63–70.

Weber, R. 2001. *Charakterisierung, Stofftransport und Einsatzkeramischer Nanofiltrationsmembranen*. Mensch-und-Buch-Verlag.

Yeh, H.H., Tseng, I.C., Kao, S.J., Lai, W.L., Chen, J.J., Wang, G.T., and Lin, S.H. 2000. Comparison of the finished water quality among an integrated membrane process, conventional and other advanced treatment processes. *Desalination*, *131*(1–3), 237–244.

10 Atmospheric Pressure Non-Thermal Plasma in Food Processing

Mahreen
Centre for Energy Studies, Indian Institute of Technology Delhi, New Delhi, India

Priyanka Prasad
Centre for Rural Development and Technology, Indian Institute of Technology Delhi, New Delhi, India

Satyananda Kar
Centre for Energy Studies, Indian Institute of Technology Delhi, New Delhi, India

Jatindra K. Sahu
Centre for Rural Development and Technology, Indian Institute of Technology Delhi, New Delhi, India

CONTENTS

10.1 Background 224
10.2 Atmospheric Pressure Cold Plasma 225
10.3 Methods of APNTP Generation 228
10.4 Functionality of APNTP 231
10.4.1 Mechanism of Microbial Cells Inactivation Using APNTP 231
10.4.2 The Impact of APNTP on Physical Qualities of Food 233
10.4.2.1 Color 233
10.4.2.2 Texture 233
10.4.2.3 Chemical Qualities 233
10.4.3 Effect of Plasma Parameters on Food 234
10.4.3.1 Effect of Processing Time 234
10.4.3.2 Effect of Applied Voltage and Frequency 234
10.4.3.3 Effect of Working Gas 234
10.4.3.4 Effect of Humidity 235
10.4.3.5 Effect of Gas Flow Rate 235
10.4.3.6 Effect of Treatment Method 235
10.4.3.7 Effect of Type of Microorganism 236
10.4.3.8 Effect of Type of Food 236

10.5 Applications in Food Processing....237
10.5.1 Destruction of Pathogens Related to Foodborne Illness....237
10.5.2 Shelf-Life Extension of Food Products....241
10.5.3 Endogenous Enzymes....241
10.5.4 Modification of Starch and Protein....242
10.5.5 Food Quality and Functional Components....243
10.5.6 Food Packaging....243
10.5.7 Modification of Food Packaging Polymers....244
10.5.8 In-Package Plasma Technology....245
10.6 Challenges....245
10.6.1 Process Control....245
10.6.2 Design of a Plasma Source....246
10.7 Future Scope and Conclusion....247
References....247

10.1 BACKGROUND

Food safety has become a major concern for food producers. Food processing techniques should meet quality criteria such as highly nourishing, long-lasting, chemical-free, good appearance, and freshness. For commercial use a processing technology must be economical and eco-friendly. Treatments must effectively destroy foodborne pathogens and spoilage microorganisms without seriously affecting the essential properties of the food. Thermal treatments have commonly been used for several decades to enhance food shelf life and food safety, but they cause loss of nutrients, and degrade the sensory and functional properties of treated foods. New technologies have the potential to inactivate microorganisms without significantly changing the nutritional and functional characteristics of food. Atmospheric pressure non-thermal plasma (APNTP) is an emerging method for food processing with insignificant effects on food quality. In APNTP a mixture of reactive oxygen and nitrogen species (RONS), including hydroxyl radicals, ozone, superoxide, atomic oxygen, singlet oxygen, nitric oxide, and nitrogen ions, is formed by ionizing the working gas at atmospheric pressure. Reactive species are generated through the interactions of highly excited plasma components with the surrounding atmosphere. It is well documented in the literature that these reactant scan inactivate bacteria, moulds, and yeasts (Tseng et al., 2012; Takamatsu et al., 2015). APNTP offers various advantages over other contemporary technologies making this a very favorable food processing technology. It is fast (between a few seconds and a few minutes), and highly effective. 5 log reductions are possible for microorganisms such as *Salmonella typhimurium*, *S. enteritidis*, *Escherichia coli*, and sporulated microorganisms such as *Bacillus cereus* and *Clostridium botulinum*. The low gas temperature employed makes it suitable for thermally sensitive substrates. It is particularly beneficial for packaged foods. Excessive consumption of water and chemical agents can be reduced by the use of plasma technology. Wastes are significantly reduced, which is both economically and environmentally beneficial. Despite these potential advantages, the interactions of plasma species generated by APNTP with food stuffs are incompletely understood. A detailed understanding would facilitate optimization and scale-up to

commercial levels. This chapter discusses the major aspects of APNTP, methods of generation, and applications in the area of microbial inactivation of food. The effect of different plasma parameters on food quality is also discussed. It is hoped that the current review will encourage greater understanding of this novel eco-friendly technology and its use in food processing operations.

10.2 ATMOSPHERIC PRESSURE COLD PLASMA

In 1928, the term 'plasma' was first coined by Irving Langmuir. He presented plasma as the fourth state of matter. When energy is supplied to a gas above a certain level an ionized gas, consisting of ions, free electrons, neutral atoms and/or molecules will form, which is called a plasma. Approximately 99% of the matter in the universe is in the plasma state. For example, the sun and other stars, the Aurora Borealis (Northern Lights) and the ionosphere are all natural plasmas. Artificial plasmas can be generated both for research and commercial purposes. They are used in television screens, fluorescent tubes, and neon signs. Plasmas are used in the semiconductor industry for surface treatments to modify some material properties (likewise fabrics or electronic elements). Different energy sources can be used to generate plasmas, as shown in Figure 10.1. Electrical energy is most commonly used to generate plasma by ionizing neutral gas (Conrads and Schmidt, 2000). The density and temperature of the electrons depends upon the form and extent of the energy delivered to the plasma. This can produce thermal (high-temperature) plasma, or non-thermal (low-temperature) plasma (Figure 10.2). When the electrons, ions and neutral species are all at the

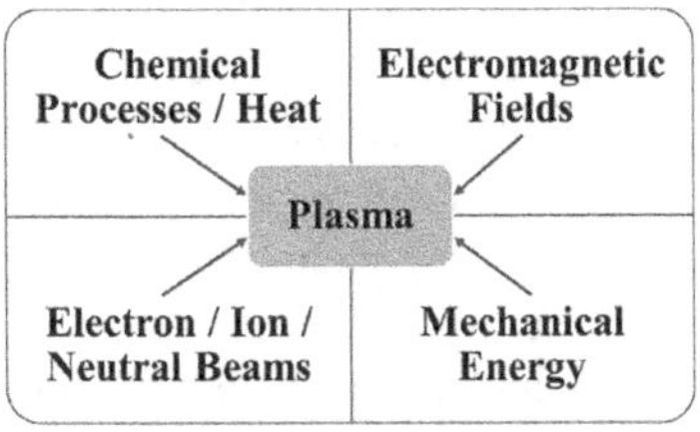

FIGURE 10.1 Different methods of plasma generation

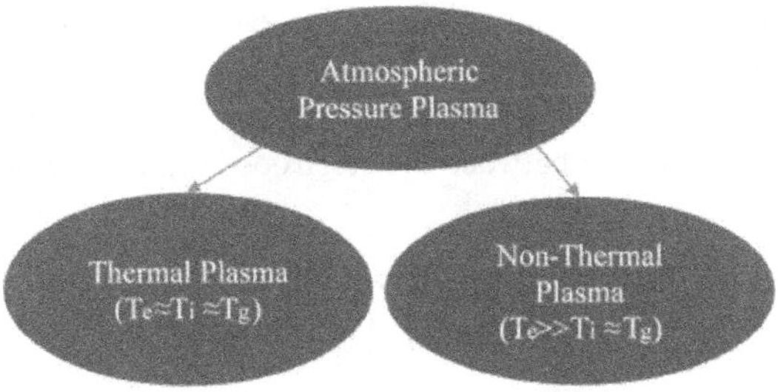

(T_e= electron temperature, T_i= ion temperature, and T_g= gas temperature)

FIGURE 10.2 Classification of atmospheric pressure plasma

TABLE 10.1
Temperature Difference of Atmospheric Pressure Thermal and Non-Thermal Plasmas XTR 4

Plasma	Species Temperature	Heat Sensitive Surfaces
Thermal plasma	$T_e \approx T_i \approx T_g \approx 10{,}000 - 20{,}000K$	Incompatible for heat-sensitive surfaces
Non-thermal plasma	$T_e >> T_i \approx T_g = 300 - 1{,}000K$	Compatible for heat-sensitive surfaces

T_e, electron temperature; T_i, ion temperature; T_g, gas temperature

same temperature the plasma is in thermal equilibrium. In low-temperature plasmas the species are not at thermal equilibrium. The electron temperature is much higher than that of the ions and neutral species. The principal differences between atmospheric pressure thermal and non-thermal plasmas are given in Table 10.1. In one form of non-thermal plasma, referred to as cold plasma, the heavier species remain at room temperature. This is more suitable for the treatment of heat-sensitive surfaces such as human skin. Plasma can also be classified as high pressure (atmospheric pressure) or low pressure. Low-pressure plasma, which requires vacuum equipment, is well established and commercially available. Atmospheric pressure plasma does not require expensive plasma reactor chambers and vacuum pumps. Advantages of atmospheric pressure plasma include easier in-line processing, treatment of complex structures, and simpler sample handling. Dielectric barrier discharge and non-thermal plasma jet sources are suitable for current food industry equipment. Atmospheric pressure non-thermal plasma is attracting attention for food industry applications because it avoids high-temperature treatments. APNTP is generated by the application of an electric field to a noble or molecular gas. The electric field accelerates free electrons which ionize the gas. At atmospheric pressure, momentum transfer through fast collisions between the plasma components increases ionization. When plasma is generated at atmospheric pressure, ionization requires a higher voltage. The widely known Paschen curves for argon and air are displayed in Figure 10.3. At lower pressure (longer mean free path), collisions are fewer and thus ionization requires a higher voltage. At higher pressure (smaller mean free path), collisions are more frequent and a higher voltage is required. This behavior is common to all gases. The Paschen curve gives the minimum voltage (Paschen minima) (Fridman et al., 2005; Schutze et al., 1998).

Breakdown potential is the lowest voltage V_b that is required to ignite the plasma. It is a function of the distance between the electrodes (d) and the pressure (p)

$$V_b = \frac{B(pd)}{\ln\left[A(pd)\right] - \left[\left(1 + 1/\gamma_{\text{sec}}\right)\right]} \tag{10.1}$$

where iuation A and B and are constants and γ_{sec} is the secondary electron emission coefficient that depends on the material of electrodes (values of these constants are given in Table 10.2). The electrode spacing must be chosen to ignite the plasma at a feasible applied potential. Atmospheric pressure plasma sources are therefore small.

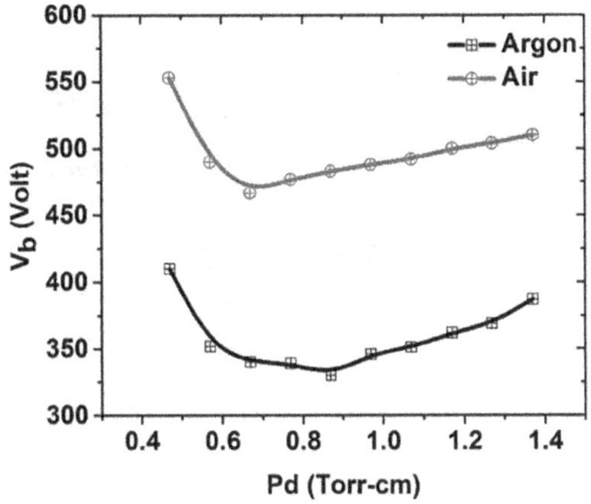

FIGURE 10.3 Breakdown potential of some gases as a function of potential difference for $d = 10$ cm

TABLE 10.2
A and B Values for the Determination of V_b (Fridman et al., 2005)

Gas	Constant A (cm^{-1} $Torr^{-1}$)	Constant B (Vcm^{-1} $Torr^{-1}$)
Air	15	365
N_2	10	310
CO_2	20	466
H_2	5	130
He	3	34
Ne	4	100
Ar	12	180

TABLE 10.3
Required Electrical Field for Gas Breakdown at Atmospheric Pressure

Gas	Air	O_2	N_2	H_2	He	Ne
E/p (kV/cm)	32	30	35	20	10	1.4

Reduced electric field (*E/p*) depends on the product *pd*. A convenient formula for the reduced electric field is directly deduced from Eq. 10.1.

$$\frac{E}{p} = \frac{B}{C + \ln(pd)} \quad (10.2)$$

where B is the constant given in Table 10.2 and $C = \text{In}\, A - \text{In}\, [1 + \text{In}\, (1/\gamma_{sec})]$. Electric field strengths required for breakdown at atmospheric pressure for different gases (Schutze et al., 1998) are shown in Table 10.3.

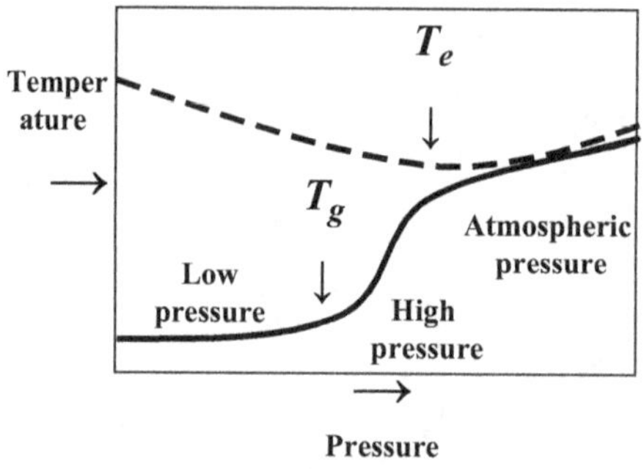

FIGURE 10.4 Electron and gas temperature as a function of pressure

At high pressure, the temperatures of electrons and heavy particles (ions and gas molecules) move towards equilibrium, creating a thermal plasma more prone to arcing (Conrads and Schmidt, 2000; Schutze et al., 1998; Raizer and Allen, 2017) (Figure 10.4). If the current is high enough to heat one of the electrodes it creates a thermal plasma called an arc. If the temperature of the electrodes is sufficiently high for thermionic emission to occur, very high currents (hundreds of amperes) can flow at low voltages (tens of volts) and at high gas temperatures. Due to the high current, arcs should be avoided in many applications. Special techniques are thus used to generate non-thermal plasmas at atmospheric pressure. High currents are prevented by using a high voltage (few kV to tens of kV) and low current (μA to mA) power source with dielectric material(s) placed between the electrodes.

10.3 METHODS OF APNTP GENERATION

Corona discharges, gliding arc discharges, atmospheric uniform glow discharge, dielectric barrier discharge (DBD), and atmospheric pressure plasma jets (APPJs) are among the special devices required for non-thermal plasma generation at atmospheric pressure; the details of these systems can be found elsewhere (Misra et al., 2014). Of these, corona discharge, dielectric barrier discharge, and plasma jets are suitable for food processing. Corona discharge is greatest where an intense electric field surrounds a sharp-edged or pointed metallic electrode (Figure 10.5). This is a

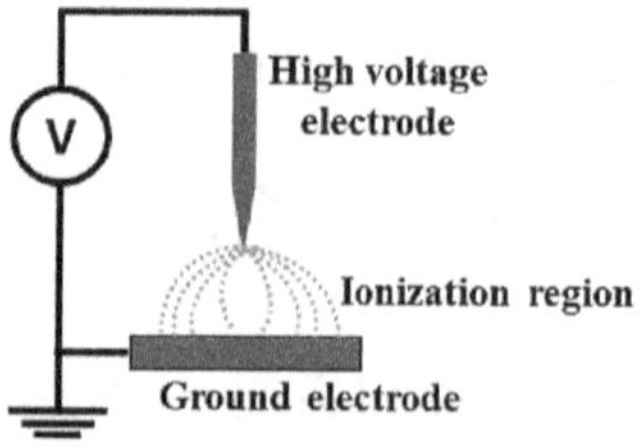

FIGURE 10.5 Schematic of corona discharge

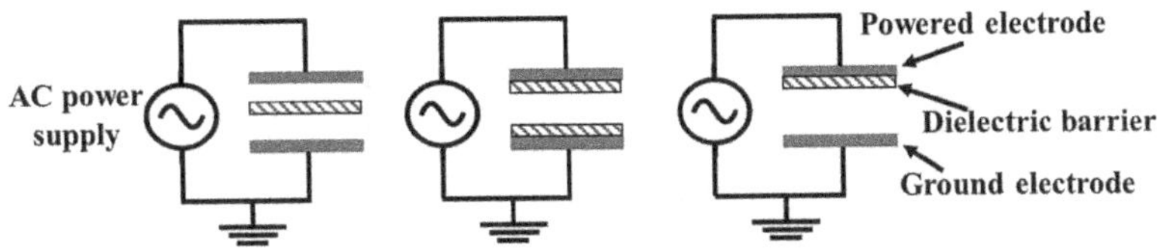

FIGURE 10.6 General dielectric barrier discharge (DBD) configurations

Townsend discharge (low discharge current), and takes place before the electrical breakdown of the gas. The ionization region is limited and only the area around the electrode becomes ionized. Corona discharge can be generated by applying a DC or AC/RF voltage in the kV range.

DBD devices consist of dielectric material between two metal electrodes that work as a ballast resistor to resist the high current formation between two electrodes (Figure 10.6). Dielectric barriers act as a relaxing material that avoids any high current formation between the electrodes and assists homogeneous treatment of substrates by generating a large number of micro-discharges in the gap between the electrodes. DBD involves electrode configuration, diverse gases and uniform discharges for quite a few meters. However, as the process requires high-voltage supplies, there are health and safety implications.

Atmospheric pressure non-thermal plasma jet (APNTPJ) is a device that generates plasma in the surrounding atmosphere by making use of the adequate flow rate of the working gas, which exits through the end of the plasma reactor, giving the plasma a 'jet-like' exterior (Ahmed et al., 2007; Winter et al., 2015; Hofmann, 2013). A two-electrode design plasma jet design is more common, with one electrode typically coupled to the external power supply and another connected to the ground. Different plasma jet arrangements are proposed by several researchers, and can be classified globally according to the following parameters (Winter et al., 2015:).

- Electrode arrangement in a plasma reactor
- Operating temperature of the plasma
- Applied source frequency (pulsed DC/AC/RF/MW) for the plasma generation
- Selection of carrier gas or combination of gases

The two most common electrode configurations for plasma jets are linear and cross field (Figure 10.7). In the linear configuration the paths of the generated electric field and gas flow in the same direction. When the paths of the electric field and gas flow are perpendicular, it is called cross-field configuration. Plasma jets can generate plasma with a wide range of processing parameters. A variety of plasma jet designs with more compound geometries are in development. Details of these systems and the suitability of plasma treatment for a number of objects and goods are available elsewhere (Winter et al., 2015).

Plasma jets have several advantages over DBD discharges (Hofmann, 2013; Kogelschatz, 2002.) For example, DBD discharges, especially in air, are in most

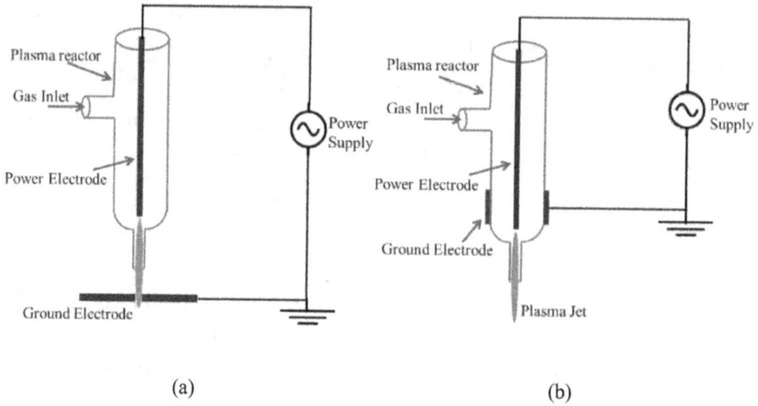

FIGURE 10.7 Plasma jet configurations: (a) linear field; (b) cross-field

cases filamentary and are produced randomly over the surface, while a plasma jet forms in most cases a stable diffuse-looking discharge without needing a surface as a second electrode. This makes it possible to treat rough surfaces homogeneously either in direct contact (Figure 10.7a) or in indirect mode (Figure 10.7b), which is not possible using a DBD geometry. A disadvantage of plasma jets is their relatively small dimensions, causing them to produce small-scale plasma, so the treatment area becomes the major limitation. The ranges of electron and neutral gas temperatures for different plasma discharges are given in Table 10.4. A sufficient density of energetic electrons is generated to dissociate enough O_2 and N_2, which are the main source of different RONS. These reactive species are beneficial in terms of microbial inactivation. The average densities of oxygen by-products – singlet oxygen, ozone, and oxygen atoms – in various atmospheric-pressure plasma jets are given in Table 10.5, which also shows the breakdown voltages and the plasma densities of corresponding discharges. Ozone is the core by-product in dielectric barrier and corona discharges, while an ample amount of oxygen atoms is produced by other plasmas. Considering all the properties of the plasma, it seems that the atmospheric-pressure plasma jet most resembles a low-pressure glow discharge. Thus, it has the potential for use in numerous applications that are restricted to vacuum.

TABLE 10.4
Electron and Neutral Gas Temperatures for Different Plasma Discharges

Plasma Discharge	Electron Temperature T_e (eV)	Gas Temperature T_g (°C)
Arc discharge	2–7	5,000–15,000
Corona discharge	6–8	70–100
DBD	4–6	100–500
Plasma jet	1–5	20–400

TABLE 10.5
Plasma Parameters for Different Plasma Sources

Plasma Source	Breakdown Voltage (kV)	Plasma Density (cm^{-3})	Densities of Oxygen Species (cm^{-3}) O^+, O_2^+, O^-	O	O_3
Low pressure discharge	0.2–0.8	10^8–10^{13}	10^{10}	10^{14}	$< 10^{10}$
Corona discharge	10–50	10^9–10^{13}	10^{10}	10^{12}	10^{18}
Dielectric barrier discharge	5–25	10^{12}–10^{15}	10^{10}	10^{12}	10^{18}
Non-thermal plasma jet	0.05–0.2	10^{11}–10^{12}	10^{12}	10^{16}	10^{16}

10.4 FUNCTIONALITY OF APNTP

To understand the effect of APNTP on food, one first needs to know how this plasma inactivates the microbes on the surface. Here we discuss how APNTP affects the quality of treated food. In the last section, the outcome of plasma parameters on food quality is briefly described.

10.4.1 Mechanism of Microbial Cells Inactivation Using APNTP

Plasma ions and reactive species have important bactericidal effects on microbial cells. The reactive species in plasma are responsible for the direct oxidative effects on the outer surface of the microbial cells. Deoxyribonucleic acid (DNA), which is present in the chromosomes of the microbes, is destroyed when the interaction of microbes takes place with the reactive species (RONS) of the plasma. This destruction of DNA is the prime cause of the bactericidal inactivation in the cells. Studies have shown that the interaction of plasma with a cell is possible through the formation of reactive species directly in the locality of DNA inside a cell nucleus (Wiseman and Halliwell, 1996). When the plasma interacts with microbial cells, malondialdehyde (MDA) is formed in the microbial cells. The reactive species-induced peroxidation of the lipid layer present in the bacterial cell membrane leads to the formation of MDA. The reactive species remove allylic hydrogen of unsaturated lipids to form lipid radicals. The lipid radicals quickly react with oxygen and form lipid peroxyl radicals which further extract hydrogen from another unsaturated lipid to generate a new lipid radical and lipid hydroperoxide. The lipid peroxyl radicals decompose to form MDA (Figure 10.8). MDA further participates in the formation of DNA, adducts namely deoxyguanosine and deoxyadenosine, which cause damage to microbial cells (Dobrynin et al., 2009). Reactive species of plasma interact with water, which leads to the formation of OH* ions. OH* ions are the most reactive and harmful to cells (Zou et al., 2004). OH* radicals are very important in the whole process. This formation of the OH* radicals in the hydration layer around the DNA molecule causes 90% of the damage to the DNA. OH* radicals then react with the surrounding organics and produce the chain oxidation that destroys DNA molecules, cellular membranes, and other cell components (Dobrynin et al., 2009). Despite the likely reaction of several active species with cells, it has been reported that reactive oxygen species

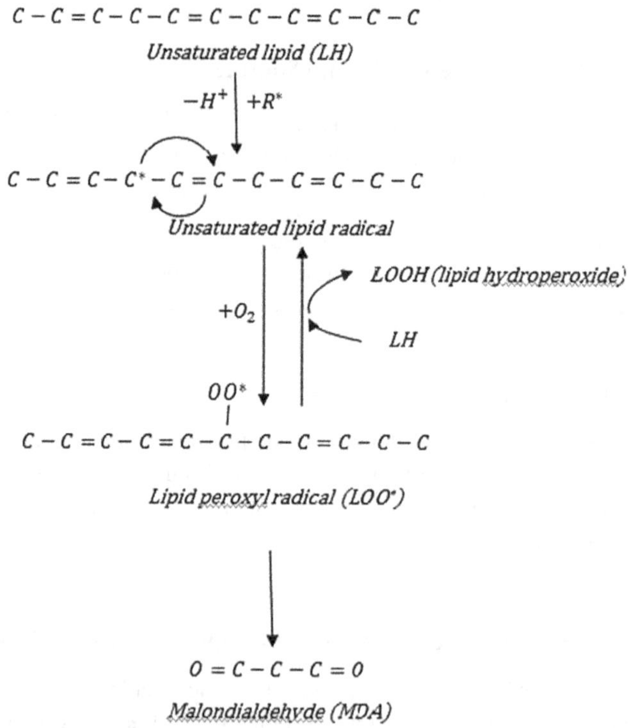

FIGURE 10.8 Formation of malondialdehyde (MDA)

such as oxygen radicals can produce profound effects on cells through reactions with the various macromolecules. The microorganisms are more susceptible to singlet state oxygen leading to deformation of cells. The lipid bilayer of the microbial cell is more susceptible to atomic oxygen as the reactivity of atomic oxygen is much higher than that of the molecular oxygen, which can degrade lipids, proteins, and DNA of cells. Damage of the double bonds in the lipid bilayer causes impaired transportation of molecules in and out of the cell. The showering of reactive species upon the exterior of the bacterial cell also upsets the membrane lipids. Microorganisms are exposed to intense bombardment by the radicals during the application of plasma, provoking the surface lesions which the living cell cannot repair sufficiently fast. This process is called 'etching'. Plasma etching is the foundation of the interplay between energetic ions and RONS with the molecules of the object. Electro-permeabilization is the process that causes cell wall rupture due to the accumulation of electric charges on the exterior of the cell membrane. During the plasma treatment process, where initiation, catalyzation, or continuation of a complex biological response is done by the plasma, the compromised membrane structure (e.g., peroxidation) or change in membrane-bound proteins and/or enzymes leads to complex cell responses as the affected cell signals to other cells (Dobrynin et al., 2009).

10.4.2 The Impact of APNTP on Physical Qualities of Food

10.4.2.1 Color

The visual aspect of food is an important factor in consumer perception and product market growth. Two factors determine whether color is present: pigments (synthetic or natural) and chemical reactions (enzymatic or non-enzymatic). For a product to exhibit continuous market growth, it should not undergo unwanted variation in color. It is reported in the literature that the severity of treatment conditions is mainly responsible for the color of fresh fruits. After APNTP treatments, researchers found that no noteworthy loss of color of strawberry, cherry tomatoes, kiwifruit, lettuce, apples, and carrots is seen (Misra et al., 2014; Ramazzina et al., 2015; Lacombe et al., 2015). Sarangapani et al. (2017) reported that blueberries lose color with higher durations of plasma treatment durations up to 5 minutes' exposure. Change of color after APNTP treatment was slight in fruit juices and not visible with the naked eye (Xu et al., 2017). Some literature reports that pigments like chlorophyll and anthocyanin are fractionally degraded by the treatment (Ramazzina et al., 2015). Generally, it has been clearly observed that shorter APNTP processing times have only a minute effect on the color of the food product.

10.4.2.2 Texture

Results show that the texture of food products is preserved after APNTP processing. APNTP treatment of apple, melons, strawberries, and cherry tomatoes does not reveal any drastic changes (Lacombe et al., 2015). However, diminution in firmness was found after APNTP treatment of blueberries (Sarangapani et al., 2017). During plasma treatment, the temperature rises, and airflow is high, which results in mechanical damage to blueberries. When grains and legumes were treated with APNTP they showed a reduction in firmness and stringiness (Thirumdas et al., 2016). Including drenching or cooking time, the properties of these groups are reduced when treated with plasma for marketing reasons.

10.4.2.3 Chemical Qualities

The study of plasma chemistry is complex, in that it involves several species in innumerable chemical reactions happening at different times. For instance, plasma generated by using air contains nearly a hundred types of chemical species producing several hundred chemical reactions. It is plasma-reactive species that are the main reason for all the variations in chemical quality detected. Plasma-reactive species are chiefly reliant on the types of gases used for plasma generation, and most processed food products display a close relation with pH and acidity, making this the most significant factor in chemical change. Any radical change in the balance can lead to unwanted effects on the texture, taste, and life of the food. Moisture on the food products causes pH and acidity to vary when they encounter plasma-reactive gases. Due to the surface water present on solid food products, acidic compounds will form only on the outside surface, while in the case of liquid products the results are more noticeable due to higher reactive species generation.

10.4.3 Effect of Plasma Parameters on Food

The specific characteristics of plasma depend on the input parameters that have been chosen for the plasma generation. Whatever the equipment used, the energy required to generate the plasma is determined by the power, voltage, applied frequency, flow rate, and admixture of working gases. Direct comparison of the various research studies that have been undertaken is not possible due to the different operating conditions and equipment used. However, some general conclusions can be drawn as to how processing parameters affect the state of microbial inactivation attained through APNTP. Some of these relate to the positions used for plasma production and application, while others are linked with the specific characteristics of the micro be affected and the features of the processing substrate.

10.4.3.1 Effect of Processing Time

Processing time for treatment of food with plasma interaction varies from seconds to minutes. Studies show that microbial inactivation and treatment time of food are linked, and that treatment time may be linear or exponential (Lee et al., 2011; Calvo et al., 2017).

10.4.3.2 Effect of Applied Voltage and Frequency

External energy applied for plasma generation has a powerful antimicrobial effect. The amount of external energy provided to the discharge, the extent (or concentration) of the reactive species produced by this external energy, and the energy absorbed by these reactive species can be determined using a mass spectrometer. An upsurge in voltage alleviates microbial inactivation. By increasing the voltage for the production of helium plasmas (Xiaohu et al., 2013), that the extent of reactive species particularly, N^{+2}, OH, He, and O, and the level of inactivation achieved for *S. aureus*. If the applied voltage remains steady and the excitation frequency or applied power is increased, fatal effects on various microbial species are seen. High voltages and high frequencies are recommended for plasma generation. However, due to the high costs and excessive temperature rise, the use of these treatment conditions at industrial scale is difficult.

10.4.3.3 Effect of Working Gas

Various gases are used to produce the plasma, including argon, nitrogen, helium, oxygen, air, or a mixture. The type of gas used also affects the effectiveness of plasma application. The working gas is selected according to the reactive species required to treat the substrate. All these gases can achieve a certain level of microbial inactivation, but there is no general agreement in the literature as to which is the most effective. Nitrogen plasmas are found most effective against *L. monocytogenes* compared to helium plasma in the study performed by Lee et al. (2011). Other researchers suggest that air plasmas are more effective than nitrogen plasmas for the inactivation of *E. coli*, *B. cereus*, *S. aureus*, and various other bacteria (Calvo et al., 2017; Xiaohu et al., 2013). Conversely, Takamatsu et al. (2015) found a greater bactericidal effect against *S. aureus* and *P. aeruginosa* with nitrogen and CO_2 plasmas than with air or O_2 plasmas. The addition of small amounts of oxygen to rare gases, such as helium

(Lee et al., 2011), argon (Surowsk et al., 2014), or nitrogen (Lee et al., 2011) increases the antimicrobial effectiveness of APNTP against vegetative cells and spore-forming bacteria, due to the presence of reactive species such as hydroxyl and hydroperoxyl radicals, hydrogen peroxide, atomic oxygen, ozone, and singlet oxygen.

10.4.3.4 Effect of Humidity

Gases with high moisture content have shown more effective antimicrobial results than dry gases; for example, fully dry gases are ineffective for *E. coli* inactivation. Ragni et al. (2010) state that as the relative air humidity increases from 35 to 65%, inactivation of *S. Enteritidis* and *S. Typhimurium* also increases from 2.5 to 4.5 log cycles, and this effect is ascribed to a higher concentration of hydroxyl radicals in the plasma. These researchers used plasmas with different moisture content and found a 5–6 log reduction at a humidity of 3 and 10%, and at higher humidity (35 and 65%), there was complete inactivation. With more reactive species, antimicrobial activity is higher, such as N_2O_5, H_2O_2, HNO_4, or hydroxyl radicals, leading particularly to the decay of ozone in water, with the resulting formation of highly oxidizing species, such as hydroxyl and hydroperoxyl radicals, superoxide anion and H_2O_2. On the other hand, a study by Lai et al. (2016) showed that when the relative air humidity increased from 62 to 81% and from 52 to 81%, the inactivation efficiency of APNTP against *Staphylococcus epidermidis* and *E. coli* decreased by 58 and 87%, respectively, which was linked to a reduction in the number of negative ions in the plasma. It is clear from these contradictory results that there should be an ideal moisture value by which maximum antimicrobial activity can be achieved.

10.4.3.5 Effect of Gas Flow Rate

Lai et al. (2016) reported a continuous increase in *S. epidermidis*, *E. coli,* and *P. alcaligenes* inactivation when airflow rates rose from 2 to 7 liters per minute, which means the gas flow rate is one of the key parameters deciding the efficiency of APNTP treatments. It is connected to a linear rise in the number of negative ions in the air plasma. An acute effect on microbial inactivation is observed when the gas flow rate increases to 5–10 liters per minute when air is used. However, when nitrogen is used to generate plasma, the gas flow rate is hardly affected by APNTP treatment and similar behavior is also observed for *S. Enteritidis* and *S. Typhimurium.* There is an ideal flow rate (60 cm^3/min) for the inactivation of *E. coli* when oxygen is used as a working gas. Miao and Jierong (2009) reported that to reduce the lethal effect the optimal flow rate should be exceeded. It may be recognized that when the flow rates are low, there are fewer reactive species, but the higher the average energy, the higher the antimicrobial effectiveness. Hitherto, when flow rates are higher the number of reactive species has been greater, but their average energy will be low, therefore antimicrobial action will be less prominent. Thus, an optimal gas flow rate should be chosen for a specific application.

10.4.3.6 Effect of Treatment Method

There are two APNTP treatment methods: direct or indirect. In direct treatments, the product or subject (working as one of the electrodes) comes into direct contact with reactive species as they are part of the plasma generation system (Figure 10.7a). In

indirect treatments, the product is kept at a distance and with the substrate will be exposed to the outflowing plasma (Figure 10.7b). Direct treatments are more effective than indirect in microbial inactivation, due to high thermal effect. A comparative study of direct and indirect treatments shows that the direct exposure of *B. atrophaeus* spores to plasmas produced with gases of dissimilar composition (air; a mixture of 90% nitrogen and 10% oxygen; a mixture of 65% oxygen, 30% $CO_{2,}$ and 5% nitrogen) initiated the inactivation of at least 6 log cycles, while for indirect exposures inactivation rates fluctuated from 2.1 to 6.3 log units, subject to the type of gas used (Patil et al., 2014). Nonetheless, Han et al. (2016) found that indirect treatments were more effective against *E. coli* and *L. monocyte genes* than direct treatments, because of the short recombination time of reactive radicals in indirect treatments before reaching the microorganisms. However, Ziuzina et al. (2015) found that the effectiveness of direct and indirect treatments is identical when treating *L. monocyte* genes, *E. coli,* and *S. aureus* biofilm. The APNTP treatments are affected by the distance between the point of plasma generation and the type of sample (substrate) used. Thus, taking the results obtained by several authors into account, the antimicrobial effectiveness can be said in general to decrease as distance increases. This may be due to a smaller number of ROS interactions with the microbes.

10.4.3.7 Effect of Type of Microorganism

The structure and sensitivity of microorganisms differ, so the efficiency of APNTP treatment varies from one to another. For example, the exterior properties of lipopolysaccharide membranes make gram-negative bacteria such as *Escherichia coli* and *Salmonella* more susceptible to plasma than gram-positive bacteria such as *Staphylococcus aureus* and *Bacillus subtilis* (Lunov et al., 2016), as the cell thickness of gram-negative bacteria is thinner than that of gram-positive bacteria. In identical inactivation procedures, sporulated microorganisms show more resilience to the plasma action than vegetative cells, as well as high absorption of bacteria clusters, which lowers the penetration ability of reactive species. The cell walls of fungi consist mainly of chitin, which is also more resistant to plasma treatment.

10.4.3.8 Effect of Type of Food

All foods have properties upon which the resistance of microbes against APNTP depend. Liquid media are more affected by APNTP because of the generation of secondary reactive species produced by the interface of reactive species of plasma either with one another or with the water present in the liquid media. Reactive species, namely OH, H_2O_2 and O^+, are mainly responsible for microbial inactivation in liquid media. Likewise, a highly reactive compound with the capability to diffuse through cell membranes is created by the contact of nitric oxide and superoxide producing peroxynitrite $(N)O + O_2^{-\rightarrow ONOO^-}$ (Ercan et al., 2016).

APNTP treatment of food products like fruits and vegetables for surface decontamination is very effective, mainly due to factors like coarseness, porosity, and topography. More even and refined surfaces improve anti-microbial efficiency of APNTP, whereas uneven, permeable, and irregular surfaces, such as those of some foods, provide room for microorganisms to hide, avoiding plasma activity.

10.5 APPLICATIONS IN FOOD PROCESSING

10.5.1 Destruction of Pathogens Related to Foodborne Illness

Foodborne illness causes approximately half lakh deaths worldwide, one-third of which occur among children under 5 years. The loss of productivity and economic activities in developing countries with the majority of the population below the poverty line costs those countries about 100 billion dollars each year (Jaffee et al., 2018). However, these losses can be significantly reduced by adopting farm-to-table food safety measures. As discussed above, cold plasma has remarkable antimicrobial potential against several pathogenic bacteria, fungal species, and bacterial spores (*Salmonella*, *Escherichia coli*, *Staphylococcus aureus*, *Listeria monocytogenes*, *Aeromonas hydrophila*, *Aspergillus*, *Cladosporium*, *Bacillus*, *Clostridium*, *Geobacillus stearothermophilus*).

The effect of cold plasma on the inactivation of microorganisms is summarized in Table 10.6. The antimicrobial effectiveness of cold plasma differs according to the species of microorganism. Generally, the cold plasma tolerance of molds, yeasts, and bacterial spores is higher than vegetative cells. Klämpfl et al. (2012) observed that 30 seconds of APNTP treatment reduced the concentration of bacterial vegetative cells upto 6 times and bacterial spores were reduced up to 4 times (G. *stearothermophilus* and *B. subtilis*) after 1 minute plasma treatment. Maximum cold plasma resistance was exhibited by *Candida albicans*. Similarly, 50% inactivation of vegetative cells of *E. coli*, *P. aeruginosa* and *S. aureus* was achieved after 1 minute of plasma treatment, while the same reduction of *A. niger* and *B. cereus* spores was obtained after plasma treatment of 5–15 minutes. In another study, the decimal reduction time (D-value) of plasma treatment for vegetative cells (*E. coli* and *B. subtilis*) was less than 0.5 min and for spores (*Bacillus*, *Clostridium*) was 2–8 minutes. The D-value is the time required to kill 1 log of microorganisms. Moreover, no significant difference has been observed between the D-values of plasma treatments for *B. stearothermophilus* spore (regarded as the most heat-resistant microorganism) and *B. subtilis* spore, suggesting higher efficacy of plasma treatment in food sterilization (Tseng et al., 2012). In the case of thermal treatment, *B. stearothermophilus* spores can withstand 121°C for 12 minutes with a D-value of 2.4 minutes, whereas *the D*-value at 100°C for *B. subtilis* spores is reportedly 0.5 minutes.

The treatment medium or food matrix plays an important role in the potency of cold plasma treatment. Liquid media are more effective due to their relatively longer shelf life, stability, and rapid dispersion of plasma-generated primary (ozone, atomic oxygen, nitric oxide) and secondary (hydroperoxides, hydroxyl radicals) reactive species, resulting in greater interaction with food. Moreover, nitrogen gas-induced plasma treatment of an aqueous medium also results in its acidification due to the formation of nitric and nitrous acid which contribute to the higher antimicrobial efficacy of cold plasma in a liquid medium. The pH of the plasma-treated liquid medium decreases to pH_3 and reportedly the antimicrobial potential of such a low pH medium is higher than that of a medium which has been acidified using acids (Yost and Joshi, 2015). In the case of a solid medium, cold plasma has been used effectively for sterilization of diverse types of fruits, vegetables, cereals, meat, poultry, seafood, and

TABLE 10.6

Impact of Atmospheric Pressure Non-Thermal Cold Plasma on Microbial Inactivation, Shelf-Life Extension, and Quality Enhancement of Food Products

Plasma Treatment Parameters	Food Products	Microbial Reduction and Effect on Shelf Life	Effect on Quality Attributes	References
Fruit, vegetables and cereals (based products)				
Plasma jet activated PAW, Ar/O2, 5 litre/min	Strawberry	*Staphylococcus aureus*–3.4 up to 4 days storage, no fungal growth up to 6 days storage	No significant change in color, firmness, pH	Ma et al. (2015)
Air, diffuse coplanar surface barrier discharge CP, 10–600 seconds	Wheat seeds	Significant reduction in fungus population	Significant increase in germination rate, dry weight, and water uptake of seedlings 10–50 seconds	Zoharanova et al. (2016)
Microcorona discharge on single dielectric barrier, Air	Rice seed with husk	Inactivation of pathogenic fungi	Increased germination (98%) and water inhibition	Khamsen et al. (2016)
Air, DBD, 90 kV, 30 seconds	Orange juice	*Salmonella enterica*–5 log	No significant change turbidity and pH. Pectin methylesterase activity reduced by 74%.	Xu et al. (2017)
Air, 60 kV, 5–30 minutes, direct exposure	Tomato	*Escherichia coli*–6 log in 15 minutes, stable up to 48 hours at 4°C	Not assessed	Prasad et al. (2017)
N2, Benchtop plasma system, 5–15 minutes	Cashew apple juice	Not assessed	Increased polyphenol, vitamin C, flavonoid content & antioxidant activity	Rodriguez et al. (2017)
Air DBD, 20 kV, direct, 5–35 minutes	Wheat germ flakes	Not assessed	Lipase activity reduced to 25% in 25 minutes, insignificant change in total phenol and DPPH free radical scavenging activity. Storage stability up to 30 days	Tolouie et al. (2018)

Air, plasma jet, 3,000 litre per hour, 30–120 seconds	Orange, tomtao, apple juice, cherry nectar	*Escherichia coli* – apple juice – 4 log, cherry nectar – 3.3 log, orange juice – 1.6 log, tomato juice – 1.4 log	*Insignificant* change in colour, pH. Phenolic content increased up to 15%	Dasan and Boyaci (2018)
Air, DBD, 80 kV, 20 minutes	Wheat and barley	Bacteria and fungi – Wheat – 1.5, 2.5 log Barley – 2.4, 2.1 log	No effect on germination and quality parameters	Los et al. (2018)
Microwave CP, 900 W, 40 minutes	Potato slices	Not assessed	PPO enzyme activity decreased to 50%. Surface-to-volume ratio of slices increased. Browning delayed	Kang, Roh, and Min (2019)
Surface micro-discharge CAP, 7 mW/cm², 6.8 kV, 7 minutes	Corn starch powder	*Bacillus subtilis* spores – 4 log	Not assessed	Pina-perez et al. (2020)
Air, DBD, 60–100 Kv, 1–5 minutes	Fresh cut carrots	Total aerobic mesophiles, yeast and mold – 2 log in 5 minutes	Insignificant changes in color, texture, pH, carotenoids	Mahnot et al. (2020)
Meat and meat products, dairy products, and egg				
Air, DBD, 10 min	Fresh pork and beef	*Listeria monocytogenes– 2 log, Escherichia coli O157: H7–2.5 log, Salmonella Typhimurium– 2.6 log*	Non-significant effect on texture and taste	Jayasena et al. (2015)
Air, 80 kV, 15 minutes, direct and indirect	Egg	*Salmonella Enteritidis– 5 log*	Non significant effect on quality	Wan et al. (2017)
Air, DBD, 2 minutes	Vacuum packaged Beef	*Staphylococcus aureus, Listeria monocytogenes, Escherichia coli* – >2 log. Shelf stability of 10 in vacuum and 3 days in aerobic storage	No effect on color, lipid peroxidation, and protein denaturation	Bauer et al. (2017)
O2/Ar CAP, 180 and 300 seconds	Dry cured beef	*S. aureus, L. monocytogenes*–4 Log. Yeast & mold count – 1.4 and 1.6 log	Insignificant change in sensory	Gök et al. (2019)

(*Continued*)

TABLE 10.6
Continued

Plasma Treatment Parameters	Food Products	Microbial Reduction and Effect on Shelf Life	Effect on Quality Attributes	References
Nitrogen plasma, 600 W, 2 minutes	Egg shell	*Salmonella Typhimurium* – 82%	Insignificant effect on sensory	Lin et al. (2020)
Air, in package DBD, 100 kV, 1–5 minutes	Chicken	Mesophylls, psychrotrophs, Enterobacteriaceae – 2 log	Not assessed	Moutiq et al. (2020)
Air, Resistive barrier discharge, 15 kV, 10–90 minutes	Egg shell	*Salmonella Enteritidis* –4.5 log in 90 minutes	No effect on egg quality	Ragni et al. (2010)
Fish and Seafood				
Air, DBD, 70 kV, 5 minutes	Packaged herring	Significant reduction in fungus population aerobic psychotropics, pseudomonas, lactic acid bacteria, Enterobacteriaceae. Shelf stable for 11 days at 4°C	Insignificant color change and oxidation	Albertos et al. (2019)
Air, 90 s/air, 10 minutes	White shrimp	Shelf life up to 12–14 days in refrigeration	Non significant effect on color, pH, TBRAS, FFA, PV, PPO activity reduced by 50%	Zouelm et al. (2019), de Souza Silva et al. (2019)
Gas phase surface discharge plasma, 6.5 lite per minute, 12.8 kV, 300 seconds	Fish balls	*Psychrobacterglacincola*– up to 6.8 log, *Brochothrixthermosphacta* up to 4.8 log, *Pseudomonas fragi* up to 3.3 log	Not assessed	Zhang et al. (2019)
Air, DBD, 60 kV, 60 seconds	Chub mackerel	Shelf life increased to 14 days. Reduction in count of microorganisms	Insignificant lipid oxidation, low PV & TBA value	Chen et al. (2019b)
O2/Ar DBD HVACP, 2.5–10 minutes	Asian sea bass	Viable bacteria count up to 6 log. Shelf life of 12 days	Minimal lipid oxidation at treatment >5 min	Olatunde et al. (2019)

non-food surfaces. Varied resistance of vegetative cells and spores of microorganisms has been observed for similar treatment conditions of cold plasma on different solid surfaces. Higher log reduction in vegetative cells and spores of bacteria were observed on PET and polycarbonate than in solid foods surfaces like lettuce and strawberry (Miao and Yun, 2011; Fernandez et al., 2013). These differences in the antimicrobial activity of cold plasma have shown the higher efficiency of cold plasma on smooth and polished surfaces (tomato, carrot, berries, cheese slices, sliced ham, smooth packaging material) compared with irregular, rough, and porous surfaces, and this is because such surfaces offer abundant hiding places for microorganisms to attach and avoid plasma interaction.

10.5.2 Shelf-Life Extension of Food Products

Extending the shelf life of fresh food products is imperative to minimize wastage and enhance food security and safety, thus delivering financial benefits to food producers and handlers. Cold plasma treatment has been shown to significantly improve the shelf life of a variety of fruits, vegetables, cereals, meat products, poultry, and seafood (Table 10.6). The efficiency of cold plasma in reducing biofilm formation by *Aeromonas hydrophila* and the population of planktonic cells on lettuce is reported by Jahid et al. (2014). The treatment of cold oxygen plasma for 5 min decreased biofilm formation by 5 log units, while the population of planktonic cells on lettuce and *E. coli, L. monocytogenes, and S. Typhimurium* on cherry tomatoes and strawberry *reduced by 6 log units after cold plasma treatment of 10–15 seconds* (Bauer et al., 2017) Furthermore, the effect of cold plasma on *Aspergillus flavus* and *Bacillus cereus* spores in red pepper powder showed 2.5 and 3.4 log reduction respectively in 20 minutes' treatment time (Kim et al., 2014). Another new application of cold plasma is when a plasma jet is applied to an aqueous solution to produce an acidified (pH2–6) medium with anti-microbial properties called plasma-activated water (PAW). PAW solution contains reactive species like hydrogen peroxide, nitrates, and nitrites (half-life of several days), which subsequently react into hydroxyl radical, nitric oxide, superoxide radical, peroxy nitrous acid (half-life of a few seconds and less than 1 second) (Zhou et al., 2020), which inactivate or kill microbial cells. It has been reported that the presence of lignin, cellulose, hemicelluloses, and resistant starch in the surface (upper peel portion) of fruits and vegetables prevents the penetration of reactive species from PAW into them (Perinban et al., 2019). Reactive species have been shown to exist in PAW for up to 30 minutes (Zhai et al., 2019), 3 days (at a storage temperature of 4°C), and 30 days (at a storage temperature of −80°C) (Shen et al., 2016). Moreover, studies have reported no significant effect of PAW treatment on firmness, pH, sensory, and nutritional attributes of fruits and vegetables (Ma et al., 2015). So, during the time the reactive species are active in PAW, it can be used as a decontaminant wash for fruits and vegetables.

10.5.3 Endogenous Enzymes

Freshly cut fruits and vegetables like lychee, sugarcane, pear, apple, peach, banana, potato, and mushroom develop an undesirable flavor and color due to enzymatic

browning, leading to decay. Studies have demonstrated the effect of cold plasma on the inactivation of polyphenol oxidase (PPO), peroxidase (POD), pectin methylesterase (PME), superoxide dismutase (SOD), lysozyme, and catalase in various foods and model food systems (Han et al., 2019). PPO is commonly present in many fruits and vegetables and is related to off-flavor and off-color in raw cut fruits and vegetables. PPO is a gold standard enzyme for estimating the extent of thermal treatment in food processing that is generally inactivated at a higher temperature (70–80°C). Thermal processing techniques for enzyme inactivation result in loss of nutritional and sensory attributes of food, whereas plasma treatment effective at room temperature has an insignificant effect on the quality of food. Plasma treatment causes bombardment of reactive species on enzymes (etching) which leads to breakage of C–H, C–N, and N–H bonds and subsequent changes in the secondary and tertiary structures of enzyme proteins (Pankaj et al., 2013; Choi et al., 2017). Moreover, Takai et al. (2012) observed that interaction of plasma-generated reactive species (OH^*, O_2, peroxide radical, NO^*) with enzymes results in the denaturation of the reactive side chain of cysteine, and aromatic rings of phenylalanine, tyrosine, and tryptophan, causing loss of their activity. Similarly, an increase in β-sheet regions and the consequent reduction in α-helix content and activity of PPO and POD in model food systems after plasma treatment was reported by Surowsky et al. (2013). Furthermore, reduction in POD and PPO activity of upto 90 and 85% was found after plasma treatment for 3 and 4 minutes respectively.

10.5.4 Modification of Starch and Protein

Cold plasma induces surface functionalization in polymers, which alters their rheological properties, molecular structure, surface morphology, and thermal stability. Modification of starch is mainly caused by the interaction between starch and plasma-generated reactive species, resulting in cross-linkage, depolymerization, and addition of new functional groups such as carbonyl and carboxylic (Muhammad et al., 2018). These changes convert smooth hydrophobic surfaces of starch to rough hydrophilic surfaces, decrease the molecular size, and change the gelatinization and retrogradation properties of starch. The plasma-influenced cross-linking in starch is explained by Zou et al. (2004). In plasma-modified starch, water molecules are removed due to cross-linking between glucose units, and the C=O bond is not formed (Eq. 10.3). The cross-linking is a result of plasma electric field-induced polarization of a single bond between the hydroxyl group and the development of ionic bonds. The most active position for cross-linking is the second carbon of the glucose ring.

$$Starch - OH + HO - Starch \underrightarrow{CP} Stach - O - Starch + H_2O \tag{10.3}$$

Proteins are used as texturizing and thickening agents in the food industry, so improved functionality can enhance their targeted performance, increasing their usefulness and commercial value in the food industry. The possible mechanisms through which cold plasma-generated reactive species affect protein (etching effect) include alteration in secondary and tertiary structures, increase in carbonyl group, surface hydrophobicity, and reduction of thiol group. Held et al. (2019) observed an increase

from 15.2 to 27.9% in β-turns and a decrease from 59.4 to 47.9% in β-sheets of the secondary structure of soft wheat protein after cold plasma treatment, indicative of a superior hydrated gluten network and viscoelasticity. Viscoelasticity and gelation are important properties of the protein in the context of food-processing industries. Cold plasma treatment improved gelation, water, and fat-binding properties and emulsion stability of pea-protein isolate and improved the foaming and emulsifying properties of whey protein isolate. So in the case of protein and starch, cold plasma treatment exposes the hidden hydrophobic groups and cut hydrophilic ends in starch backbone, respectively, while the structure of the macromolecules is slightly altered due to breakage and formation of bonds, thereby improving their interfacial performance.

10.5.5 Food Quality and Functional Components

Quality food with optimum nutritional benefits is a driver of high consumer demand. The food industry is constantly seeking novel processing technologies. The application of cold plasma for decontamination, packaging, and food functionalization involves the generation of reactive species that may affect color, acidity, taste, and physicochemical (vitamins, polyphenols, flavonoids content) attributes of fruits, vegetables, and food products. Color is the most important aspect and has a significant effect on consumer perception of quality food products. Studies have observed that cold plasma treatment of 45 seconds significantly reduced $b*$ values; an increase in treatment time (1–2 minutes) showed significant loss of both $L*$ and $a*$ values of berries while significantly enhancing the brightness and whiteness index of brown rice (Sarangapani et al., 2015) and red color of pork. There was insignificant loss of color and texture in fresh fruits and vegetables, according to Yong et al. (2019). The effect of DBD cold plasma mist treatment for 5 minutes on baby kale leaves showed insignificant changes in color, browning index, and color stability during refrigerated storage for 12 days. However, higher plasma treatment (>5 minutes) caused leaf damage and visible color changes in kale leaves (Shah et al., 2019). Other studies observed slight insignificant changes in the flavor and taste of milk (Kim et al., 2015), overall acceptability of cheese (Lee et al., 2012), and pork (Ulbin-Figlewicz et al., 2015).

Functional food components like polyphenol compounds, carotenoids, phytosterols, glucosinolates, lycopene, isoflavones, β-glucan, and lignans, have no nutritional benefits, but strengthen the body's antioxidant mechanism along with fiber, minerals, and vitamins, improving human health. The effect of cold plasma on bioactive compounds depends upon factors like treatment gas, treatment time and voltage, and the distance between the plasma discharge and the target food product. The effects of non-thermal plasma on quality attributes and functional components of food products are summarized in Table 10.6.

10.5.6 Food Packaging

Cold plasma is widely used in food packaging. Compared with traditional thermal processes like pasteurization and sterilization, cold plasma processing is an energy-efficient, green, and fast process applicable to a wide range of packaging materials.

TABLE 10.7
Research Findings Evaluating the Effect of Cold Plasma on Food Packaging Material

Plasma Conditions	Packaging Material	Research Findings	References
O2, 900 W, 40 minutes	Polylactic acid films	Thermal, antimicrobial properties, printability, tensile strength increased. Storage stability of film up to 56 days.	Song et al. (2016)
ACP	Carboxymethyl–cellulose coated PP film	Surface hydrophilicity increased, contact angle decreased, lower WVP	Honarvar et al. (2017)
N2 plasma, 300 W, 300 seconds	LDPE	Surface hydrophillicity and antimicrobial properties increased	Karam et al. (2016)
Ar CP, 400 W, 15 minutes	Defatted soybean meal film	Elongation, water vapor permeability, and tensile strength increased	Oh et al. (2016)
Air, 60 seconds	Zein film	Significant increase in elongation, flexibility, barrier properties, thermal stability	Chen et al. (2019a)
Glow discharge CP, 2–5 minutes	Fish protein film	Tensile strength, barrier properties increased significantly	Romani et al. (2019)
O2 CP, 4–8 minutes	Starch chitosan film	Significant increase in tensile strength and hydrophillicity	Farhoodi and Beikzadeh (2020)

Moreover, the secondary features of cold plasma include its dry nature and suitability for the treatment of heat-sensitive polymeric packaging materials. Plasma only alters the surface morphology and chemistry of packaging polymers (to a depth depending on plasma ion energy) without changing the nature of the native polymer. A summary of research studies evaluating the effect of cold plasma on food packaging materials is presented in Table 10.7.

10.5.7 Modification of Food Packaging Polymers

Cold plasma surface treatment applications in food packaging materials involve functionalization, surface cleaning or etching, sterilization, and surface deposition. Surface functionalization is the introduction of specific chemical functional groups onto the surface packaging polymer. Etching is performed to eliminate unwanted materials and contaminants from the surface of the polymer, and deposition deals with the coating of thin layers of chemicals on the polymer surface to impart anti-microbial properties. These processes improve surface wettability, sealability, printability, dye uptake, glaze resistance, barrier properties, and attachment to other polymers without compromising the bulk properties of the original polymer. Moreover, barrier characteristics like water vapor and oxygen permeability, along with surface hydrophobicity of treated polymers, are enhanced. For food-based applications, packaging materials should conform to

the microbiological requirements specified in regulatory guidelines. Cold plasma treatment increases the surface contact angle of labeling polymers, providing resistance against abrasion and improved printability and ink adhesion. Plasticizer migration, where plasticizers leave the packaging polymer and move to the surrounding foodstuffs results in chemical contamination of packaged food products and a reduction in the physical properties of the packaging material. Cold plasma treatment of packaging material has been shown to reduce the migration of low molecular weight plasticizers into food (Audic et al., 2001).

10.5.8 In-Package Plasma Technology

Dielectric barrier discharge-based non-thermal plasma treatment of a sealed package containing food products is currently subject to research to explore its potential use in extending the shelf life of packaged food products without affecting their quality. The in-package decontamination system involves simultaneous cold plasma treatment of both food and packaging material. Food is placed inside the packaging material during plasma application and the polymeric package (LDPE, HDPE, PS) acts as the dielectric. In a DBD system, the reactive species produced inside the food packet sterilize both the food and the inside of the packaging material and avoid recontamination of the food product. This approach can be successfully scaled up to continuous industrial processing, resulting in the output of good-quality food. Ziuzina et al. (2020) investigated the efficacy of a DBD-based pilot-scale, non-thermal atmospheric cold plasma generated in batch or continuous mode for decontamination of fresh strawberries and spinach. Post-plasma discharge concentration of ozone and H_2 inside food packages after storage duration of 5–27 minutes was within the range of 1.500–400 and 4,000–2,000 ppm by volume respectively. Static cold plasma treatment for 2.5 minutes showed significant reductions in the population of bacterial cells inoculated on strawberries and spinach with insignificant changes in color, pH, and firmness (Ziuzina et al., 2020). The effect of in-package DBD cold plasma treatment on microbiological quality and color of *Campylobacter*- and *Salmonella*-inoculated chicken breast was studied by Zhuang et al. (2019). The plasma treatment not only reduced the growth of pathogenic bacteria inoculated in the meat samples but also retarded the growth of other meat-based psychrophilic microorganisms. The antimicrobial activity of DBD plasma treatment was highest at treatment time of 1 minute, but increasing the treatment time to 5 minutes did not result in further reduction of the microbial population. Moreover, the in-package plasma treatment had an insignificant effect on the color of chicken meat samples (Zhuang et al., 2019).

10.6 CHALLENGES

10.6.1 Process Control

In view of knowledge gaps on quick and non-invasive ways of calculating and regulating the reaction chemistry of non-thermal plasmas, a quantitative study of the discharge dynamics and the thorough interface of non-thermal plasma with food and microorganisms is required, together with knowledge of new reactive species that are

generated. To maintain uniformity in the plasma process, process monitoring tools need to be available. For monitoring continuous plasma processes, specific diagnostic tools should be applied. For example, spectroscopy – the study of the interaction between matter and electromagnetic radiation emitting from the plasma – is used to monitor the different RONS originating from the plasma processes, and can be used to control and maintain concentration at the beginning and in the entire process. Electrical limitations, such as the temperature of electrons and/or power input to discharge, may be estimated directly using electrical tool diagnostics such as voltage and current probes. The gas temperature of the plasma, which determines its application, can be estimated using a thermocouple. Various systematic tools exist for both implicit and explicit diagnostics and monitoring of plasma chemistry. However, issues such as interference from several gaseous species, humidity, and noisy factory environments all contraindicate the live use of plasma diagnostics while treating food. There remain some unanswered questions before this non-thermal technology can be applied to commercial food processing.

10.6.2 Design of a Plasma Source

The major objective of atmospheric non-thermal plasma processing of foods is to get a result with maximum microbiological decontamination without affecting the nutritional and sensory quality of foods. Issues for consideration in the design of plasma sources for industrial use include the following.

Knowledge of the plasma chemistry of atmospheric non-thermal plasma is essential. Air is the combination of oxygen, nitrogen, water vapor, and a few other components. Due to various components present in air plasma, it can produce numerous unique chemical species that absorb several hundred instantaneous chemical reactions at various times. Second, as it is a new technology, standard operating procedures needs to be set for food, including processing power dissipation, breakdown voltage, applied frequency, gas flow rate, types of gas, and plasma diagnostics, to ensure a common platform for research and study conducted globally and to accelerate developments in plasma application. Setting standards will reduce unevenness of outcomes by different research groups.

The rare gases used in plasma are costly, so machinery required for commercial development of these plasma sources should be inexpensive. Plasma technology will come to dominate the food industry if sufficient volumes of plasma can be sourced. It will become acceptable to consumers and the associated costs will be reduced if it is used on a commercial scale.

To reduce the cost, air can be used instead of rare gases in APNTP. But air plasma generates high temperatures that damage heat-sensitive surfaces or food products. The main challenge for researchers is to reduce plasma temperature (around room temperature), which would make APNTP a very cost-effective technology.

Low running costs is the major market advantage of a commercial plasma system; costs can be reduced by, for example, developing plasma sources operating at low frequency instead of using high-frequency RF power. New plasma sources are required if plasma science is to advance and innovative applications are to be developed.

Attention should be paid to other essential components rather than simply to the primary functional components of the machine (i.e., robust and food-compatible plasma device design). A safe working environment and the health and safety of the operators should be prioritized over the design of scaled-up plasma methods and systems.

10.7 FUTURE SCOPE AND CONCLUSION

In order to meet growing global food demands, commercial technologies, environmental feasibility, and food safety practices must be introduced. The future of the non-thermal plasma technology for the food-oriented business looks promising. APNTP processes can be used for surface decontamination of food and water, and to transform packaging materials and improve the functionality of food ingredients, among other things. They will also contribute to conservation of the environment as cold plasma treatment can be introduced to the market as a 'dry process', as opposed to water or moist environments. Atmospheric pressure cold plasma uses only natural air and electricity to produce reactive gas, making it a cost-efficient technology.

REFERENCES

Ahmed, S.A., O. Rouaud, and M. Havet. 2007. *Electrohydrodynamic enhancement of heat and mass transfer in food processes. 3rd International Symposium on Food and Agricultural Products*, Naples, Italy.

Albertos, I., et al. 2019. Shelf-life extension of herring (Clupeaharengus) using in-package atmospheric plasma technology. *Innovative Food Science & Emerging Technologies*, 53: 85–91.

Audic, J.L., F. Poncin, and J.C. Brosse. 2001. Cold plasma surface modification of conventionally and nonconventionally plasticized poly (vinyl chloride)-based flexible films: Global and specific migration of additives into isooctane. *Journal of Applied Polymer Science*, 79: 1384–1393.

Bauer, A., et al. 2017. The effects of atmospheric pressure cold plasma treatment on microbiological, physical-chemical and sensory characteristics of vacuum packaged beef loin. *Meat Science*, 128: 77–87.

Calvo, T., et al. 2017. Stress adaptation has a minor impact on the effectivity of Non-Thermal Atmospheric Plasma (NTAP) against Salmonella spp. *Food Research International*, 102: 519–525.

Chen, G., et al. 2019a. Improving functional properties of zein film via compositing with chitosan and cold plasma treatment. *Industrial Crops and Products*, 129: 318–326.

Chen, J., et al. 2019b. Effect of cold plasma on maintaining the quality of chub mackerel (Scomberjaponicus): Biochemical and sensory attributes. *Journal of the Science of Food and Agriculture*, 99.1: 39–46.

Choi, S., et al. 2017. Structural and functional analysis of lysozyme after treatment with dielectric barrier discharge plasma and atmospheric pressure plasma jet. *Scientific Reports*, 7.1: 1–10.

Conrads, H., and M. Schmidt. 2000. Plasma generation and plasma sources. *Plasma Sources Science and Technology*, 9.4: 441.

Dasan, B.G., and I.H. Boyaci. 2018. Effect of cold atmospheric plasma on inactivation of Escherichia coli and physicochemical properties of apple, orange, tomato juices, and sour cherry nectar. *Food and Bioprocess Technology*, 11.2: 334–343.

de Souza Silva, D.A., et al. 2019. Use of cold atmospheric plasma to preserve the quality of white shrimp (Litopenaeusvannamei). *Journal of Food Protection*, 82.7: 1217–1223.

Dobrynin, D., et al. 2009. Physical and biological mechanisms of direct plasma interaction with living tissue. *New Journal of Physics*, 11.11: 115020.

Ercan, U.K., et al. 2016. Chemical changes in nonthermal plasma-treated N-acetylcysteine (NAC) solution and their contribution to bacterial inactivation. *Scientific Reports*, 6: 20365.

Farhoodi, M., and S. Beikzadeh. 2020. Effect of using cold plasma treatment on the surface and physicochemical properties of starch-chitosan composite film. *Iranian Journal of Nutrition Sciences & Food Technology*, 15.1: 103–111.

Fernandez, A., E. Noriega, and A. Thompson. 2013. Inactivation of Salmonella entericaserovar Typhimurium on fresh produce by cold atmospheric gas plasma technology. *Food Microbiology*, 33.1: 24–29.

Fridman, A., A. Chirokov, and A. Gutsol. 2005. Non-thermal atmospheric pressure discharges. *Journal of Physics D: Applied Physics*, 38.2: R1.

Gök, V., et al. 2019. The effects of atmospheric cold plasma on inactivation of Listeria monocytogenes and Staphylococcus aureus and some quality characteristics of pastırma: A dry-cured beef product. *Innovative Food Science & Emerging Technologies*, 56: 102188.

Han, L., et al. 2016. Mechanisms of inactivation by high-voltage atmospheric cold plasma differ for Escherichia coli and Staphylococcus aureus. *Applied and Environmental Microbiology*, 82.2: 450–458.

Han, Y., J.H. Cheng, and D.W. Sun. 2019. Activities and conformation changes of food enzymes induced by cold plasma: A review. *Critical Reviews in Food Science and Nutrition*, 59.5: 794–811.

Held, S., C.E. Tyl, and G.A. Annor. 2019. Effect of radio frequency cold plasma treatment on intermediate wheatgrass (Thinopyrumintermedium) flour and dough properties in comparison to hard and soft wheat (Triticumaestivum L.). *Journal of Food Quality*.

Hofmann, S. 2013. Atmospheric pressure plasma jets: Characterisation and interaction with human cells and bacteria. https://research.tue.nl/nl/publications/atmospheric-pressure-plasma-jets-characterisation-and-interaction.

Honarvar, Z., et al. 2017. Application of cold plasma to develop carboxymethyl cellulose-coated polypropylene films containing essential oil. *Carbohydrate Polymers*, 176: 1–10.

Jaffee, S., et al. 2018. *The safe food imperative: Accelerating progress in low-and middle-income countries*. The World Bank. http://openknowledge.worldbank.org/handle/10986/30568.

Jahid, I.K., N. Han, and S.D. Ha. 2014. Inactivation kinetics of cold oxygen plasma depend on incubation conditions of Aeromonashydrophila biofilm on lettuce. *Food Research International*, 55: 181–189.

Jayasena, D.D., et al. 2015. Flexible thin-layer dielectric barrier discharge plasma treatment of pork butt and beef loin: Effects on pathogen inactivation and meat-quality attributes. *Food Microbiology*, 46: 51–57.

Kang, J.H., S.H. Roh, and S.C. Min. 2019. Inactivation of potato polyphenol oxidase using microwave cold plasma treatment. *Journal of Food Science*, 84.5: 1122–1128.

Karam, L., et al. 2016. Optimization of cold nitrogen plasma surface modification process for setting up antimicrobial low density polyethylene films. *Journal of the Taiwan Institute of Chemical Engineers*, 64: 299–305.

Khamsen, N., et al. 2016. Rice (Oryza sativa L.) seed sterilization and germination enhancement via atmospheric hybrid nonthermal discharge plasma. *ACS Applied Materials & Interfaces*, 8.30: 19268–19275.

Kim, H.J., et al. 2015. Microbial safety and quality attributes of milk following treatment with atmospheric pressure encapsulated dielectric barrier discharge plasma. *Food Control*, 47: 451–456.

Kim, J.E., D.U. Lee, and S.C. Min. 2014. Microbial decontamination of red pepper powder by cold plasma. *Food Microbiology*, 38: 128–136.

Klämpfl, T.G., et al. 2012. Cold atmospheric air plasma sterilization against spores and other microorganisms of clinical interest. *Applied and Environmental Microbiology*, 78.15: 5077–5082.

Kogelschatz, U. 2002. Filamentary, patterned, and diffuse barrier discharges. *IEEE Transactions on plasma science*, 30.4: 1400–1408.

Lacombe, A., et al. 2015. Atmospheric cold plasma inactivation of aerobic microorganisms on blueberries and effects on quality attributes. *Food Microbiology*, 46: 479–484.

Lai, A.C.K., et al. 2016. Evaluation of cold plasma inactivation efficacy against different airborne bacteria in ventilation duct flow. *Building and Environment*, 98: 39–46.

Lee, H.J., et al. 2011. Inactivation of Listeria monocytogenes on agar and processed meat surfaces by atmospheric pressure plasma jets. *Food Microbiology*, 28.8: 1468–1471.

Lee, H.J., et al. 2012. Evaluation of a dielectric barrier discharge plasma system for inactivating pathogens on cheese slices. *Journal of Animal Science and Technology*, 54: 191–198.

Lin, L., et al. 2020. Inhibitory effect of cold nitrogen plasma on Salmonella Typhimurium biofilm and its application on poultry egg preservation. *LWT - Food Science and Technology*: 109340.

Los, A., et al. 2018. Improving microbiological safety and quality characteristics of wheat and barley by high voltage atmospheric cold plasma closed processing. *Food Research International*, 106: 509–521.

Lunov, O., et al. 2016. The interplay between biological and physical scenarios of bacterial death induced by non-thermal plasma. *Biomaterials*, 82: 71–83.

Ma, R., et al. 2015. Non-thermal plasma-activated water inactivation of food-borne pathogen on fresh produce. *Journal of Hazardous Materials*, 300: 643–651.

Mahnot, N.K., et al. 2020. In-package cold plasma decontamination of fresh-cut carrots: Microbial and quality aspects. *Journal of Physics D: Applied Physics*, 53.15: 154002.

Miao, H., and C. Jierong. 2009. Inactivation of Escherichia coli and properties of medical poly (vinyl chloride) in remote-oxygen plasma. *Applied Surface Science*, 255.11: 5690–5697.

Miao, H., and G. Yun. 2011. The sterilization of Escherichia coli by dielectric-barrier discharge plasma at atmospheric pressure. *Applied Surface Science*, 257.16: 7065–7070.

Misra, N.N., et al. 2014. In-package atmospheric pressure cold plasma treatment of strawberries. *Journal of Food Engineering*, 125: 131–138.

Moutiq, R., et al. 2020. In-package decontamination of chicken breast using cold plasma technology: Microbial, *quality and storage studies. Meat Science*, 159: 107942.

Muhammad, A.I., et al. 2018. Effects of nonthermal plasma technology on functional food components. *Comprehensive Reviews in Food Science and Food Safety*, 17.5: 1379–1394.

Oh, Y.A., S.H. Roh, and S.C. Min. 2016. Cold plasma treatments for improvement of the applicability of defatted soybean meal-based edible film in food packaging. *Food Hydrocolloids*, 58: 150–159.

Olatunde, O.O., S. Benjakul, and K. Vongkamjan. 2019. Dielectric barrier discharge high voltage cold atmospheric plasma: An innovative nonthermal technology for extending the shelflife of asian sea bass slices. *Journal of Food Science*, 84.7: 1871–1880.

Pankaj, S.K., N.N. Misra, and P.J. Cullen. 2013. Kinetics of tomato peroxidase inactivation by atmospheric pressure cold plasma based on dielectric barrier discharge. *Innovative Food Science & Emerging Technologies*, 19: 153–157.

Patil, S., T. Moiseev, N.N. Misra, P.J. Cullen, J.P. Mosnier, K.M. Keener, and P. Bourke, 2014. Influence of high voltage atmospheric cold plasma process parameters and role of relative humidity on inactivation of Bacillus atrophaeus spores inside a sealed package. *Journal of Hospital Infection*, 88(3): 162–169.

Perinban, S., O. Valerie, and V. Raghavan. 2019. Nonthermal plasma–liquid interactions in food processing: A review. *Comprehensive Reviews in Food Science and Food Safety*, 18, 1985–2008.

Pina-Perez, M.C., et al. 2020. Low-energy short-term cold atmospheric plasma: Controlling the inactivation efficacy of bacterial spores in powders. *Food Research International*, 130: 108921.

Prasad, P., et al. 2017. Effect of atmospheric cold plasma (ACP) with its extended storage on the inactivation of Escherichia coli inoculated on tomato. *Food Research International*, 102: 402–408.

Ragni, L., et al. 2010. Non-thermal atmospheric gas plasma device for surface decontamination of shell eggs. *Journal of Food Engineering*, 100.1: 125–132.

Raizer, Y.P., and J.E. Allen. 2017. *Gas discharge physics*, vol. 2. Berlin: Springer.

Ramazzina, I., et al. 2015. Effect of cold plasma treatment on physico-chemical parameters and antioxidant activity of minimally processed kiwifruit. *Postharvest Biology and Technology*, 107: 55–65.

Rodríguez, Ó., et al. 2017. Effect of indirect cold plasma treatment on cashew apple juice (Anacardiumoccidentale L.). *LWT - Food Science and Technology*, 84: 457–463.

Romani, V.P., et al. 2019. Improvement of fish protein films properties for food packaging through glow discharge plasma application. *Food Hydrocolloids*, 87: 970–976.

Sarangapani, C., et al. 2015. Effect of low-pressure plasma on physico-chemical properties of parboiled rice. *LWT - Food Science and Technology*, 63.1: 452–460.

Sarangapani, C., et al. 2017. Atmospheric cold plasma dissipation efficiency of agrochemicals on blueberries. *Innovative Food Science & Emerging Technologies*, 44: 235–241.

Schutze, A., et al. 1998. The atmospheric-pressure plasma jet: A review and comparison to other plasma sources. *IEEE Transactions on Plasma Science*, 26.6: 1685–1694.

Shah, U., et al. 2019. Effects of cold plasma treatments on spot-inoculated Escherichia coli O157: H7 and quality of baby kale (Brassica oleracea) leaves. *Innovative Food Science & Emerging Technologies*, 57: 102104.

Shen, J., et al. 2016. Bactericidal effects against *s. aureus* and physicochemical properties of plasma activated water stored at different temperatures. *Scientific Reports*, 6, 28505. doi:10.1038/srep28505

Song, A.Y., et al. 2016. Cold oxygen plasma treatments for the improvement of the physicochemical and biodegradable properties of polylactic acid films for food packaging. *Journal of Food Science*, 81.1: E86–E96.

Surowsky, B., et al. 2013. Cold plasma effects on enzyme activity in a model food system. *Innovative Food Science & Emerging Technologies*, 19: 146–152.

Surowsky, B., et al. 2014. Impact of cold plasma on Citrobacterfreundii in apple juice: Inactivation kinetics and mechanisms. *International Journal of Food Microbiology*, 174: 63–71.

Takai, E., et al. 2012. Protein inactivation by low temperature atmospheric pressure plasma in aqueous solution. *Plasma Processes and Polymers*, 9.1: 77–82.

Takamatsu, T. et al. 2015. Microbial inactivation in the liquid phase induced by multigas plasma jet. *PLoS One*, 10.7: E0132381.

Thirumdas, R., Saragapani, C., Ajinkya, M.T., Deshmukh, R.R., and Annapure, U.S. 2016. Influence of low pressure cold plasma on cooking and textural properties of brown rice. *Innovative Food Science & Emerging Technologies*, 37, Part A: 53–60.

Tolouie, H., et al. 2018. The impact of atmospheric cold plasma treatment on inactivation of lipase and lipoxygenase of wheat germs. *Innovative Food Science & Emerging Technologies*, 47: 346–352.

Tseng, S., et al. 2012. Gas discharge plasmas are effective in inactivating Bacillus and Clostridium spores. *Applied Microbiology and Biotechnology*, 93.6: 2563–2570.

Ulbin-Figlewicz, N., E. Brychcy, and A. Jarmoluk. 2015. Effect of low-pressure cold plasma on surface microflora of meat and quality attributes. *Journal of Food Science and Technology*, 52.2: 1228–1232.

Wan, Z., et al. 2017. High voltage atmospheric cold plasma treatment of refrigerated chicken eggs for control of Salmonella Enteritidis contamination on eggshell. *LWT - Food Science and Technology*, 76: 124–130.

Winter, J., R. Brandenburg, and K.D. Weltmann. 2015. Atmospheric pressure plasma jets: An overview of devices and new directions. *Plasma Sources Science and Technology*, 24.6: 064001.

Wiseman, H., and B. Halliwell. 1996. Damage to DNA by reactive oxygen and nitrogen species: Role in inflammatory disease and progression to cancer. *Biochemical Journal*, 313.1: 17–29.

Xiaohu, L., et al. 2013. Sterilization of Staphylococcus Aureus by an atmospheric non-thermal plasma jet. *Plasma Science and Technology*, 15.5: 439.

Xu, L., et al. 2017. Microbial inactivation and quality changes in orange juice treated by high voltage atmospheric cold plasma. *Food and Bioprocess Technology*, 10.10: 1778–1791.

Yong, H.I., et al. 2019. Color development, physiochemical properties, and microbiological safety of pork jerky processed with atmospheric pressure plasma. *Innovative Food Science & Emerging Technologies*, 53: 78–84.

Yost, A.D., and S.G. Joshi. 2015. Atmospheric nonthermal plasma-treated PBS inactivates Escherichia coli by oxidative DNA damage. *PLoS One*, 10.10: E0139903.

Zahoranová, A., et al. 2016. Effect of cold atmospheric pressure plasma on the wheat seedlings vigor and on the inactivation of microorganisms on the seeds surface. *Plasma Chemistry and Plasma Processing*, 36.2: 397–414.

Zhai, Y., et al. 2019. Effect of plasma-activated water on the microbial decontamination and food quality of thin sheets of bean curd. *Applied Sciences*, 9: 4223–4233.

Zhang, Y., et al. 2019. Bactericidal effect of cold plasma on microbiota of commercial fish balls. *Innovative Food Science & Emerging Technologies*, 52: 394–405.

Zhou, R. et al. 2020. Plasma-activated water: Generation, originof reactive species and biological applications. *Journal of Physics D: Applied Physics*, 53.30. doi:10.1088/1361-6463/ab81cf

Zhuang, H., et al. 2019. In-package air cold plasma treatment of chicken breast meat: Treatment time effect. *Journal of Food Quality*.

Ziuzina, D., et al. 2015. Cold plasma inactivation of bacterial biofilms and reduction of quorum sensing regulated virulence factors. *PLoS One*, 10.9: E0138209.

Ziuzina, D., et al. 2020. Investigation of a large gap cold plasma reactor for continuous in-package decontamination of fresh strawberries and spinach. *Innovative Food Science & Emerging Technologies*, 59: 102229.

Zou, J.J., C.J. Liu, and B. Eliasson. 2004. Modification of starch by glow discharge plasma. *Carbohydrate Polymers*, 55.1: 23–26.

Zouelm, F., et al. 2019. The effects of cold plasma application on quality and chemical spoilage of pacific white shrimp (litopenaeusvannamei) during refrigerated storage. *Journal of Aquatic Food Product Technology*, 28.6: 624–636.

Index

A

acoustics, 40
agitation, 73
agricultural commodities, 166
antinutritional factors, 50
antioxidant capacity, 108
APNTP, 224
APNTP generation, 228
ascorbic acid, 108
atmospeheric pressure cold plasma, 224
atmospeheric pressure plasma jets (APPJ), 228

B

bactericidal inactivation, 231
beverage industries, 137
bioaccesibility, 13
bioactive compounds, 53
biological factors of PEF, 32
breakdown of cell membrane, 33

C

clean label, 3
carotenoids, 13
cavitation, 42, 79
cheese processing, 134
chromosomes, 172
Clostridium botulinum bacteria, 172
cold pateurization, 133
colour, 108
combined processing, 109
concentation polarization, 147
concentration, 199
concentration of milk, 134
contamination, 217
corona discharge, 228
cryo freezing, 184
cryogenic grinding, 191
cryogens, 182
cryo-mechanical freezers, 189
crystalization of lipids, 10

D

dehydration and shrinkage, 190
dehydro-freezing, 81
demethylation, 56
dielectric barrier, 229
diffusion model, 210
diffusion rate, 72
drying kinetics, 71

E

electric field srength, 32
electro-permeabilization, 232
electrodialysis, 127
electron beams, 176
engineering properties of food, 6
enzyme inactivation, 107

F

flow cytometry, 43
fluence, 96
flux, 209
food freezing, 183
forward osmosis, 127
fouling mechanism, 148
fractionation of milk, 134
free flowing freezers, 189
freeze-cracking, 190
freezing rate, 186
freezing time, 186
fruit juice, 94
functionality of cold plasma, 231

G

gamma irradiation, 80
gamma rays, 169
grain hydration, 15

H

halomethanes, 59
heat transfer, 187
high-energy electrons, 169
high-pressure extraction, 12
high-pressure pasteurization, 5
high-voltage capacitor, 29
high-voltage pulse generator, 30
HIPEF, 28
homogenization, 53
hypertonic solution, 71

I

immersion freezers, 188
immersion freezing, 49
impingement freezers, 188
inactivation of microorganism, 33
inactivation rate, 33
individual quick freezers (IQF), 189
infusion, 14, 81

inpackage plasma technology, 245
International Atomic Energy Agency (IAEA), 173
ions, 198
irradiation, 165
isostatic principle, 3

L

Le Chatelier-Braun principle, 3
legislation, 172
liquid food, 102

M

macromolecules, 13
macronutrients, 167
mass spectrometric analysis, 178
mass transfer, 14, 70
media factors of PEF, 32
membrane distillation, 127
membrane filtration, 196
membrane fouling, 148
membrane processing, 124
microbial inactivation, 102
microbiological activity, 190
microfiltration, 125
micronutrients, 167
microscopic arrangement, 4
microstreaming, 56
microwave radiation, 77
microwave-assisted extraction, 46
module, 204
moisture diffusivity, 80
monosonication, 42
muscle tissues, 14

N

nano-emulsions, 54
nanofiltration, 1, 26, 217
non thermal, 79, 92
nucleation, 48

O

ohmic heating, 78
osmotic dehydration, 14, 68
osmotic distillation, 127

P

Paschen curve, 227
pathogenic safety, 179
PEF processing unit, 29
permeability, 203
peroxidase catalized oxidation, 56
pervaporation, 127
phase transition, 69
photochemical effect, 100
photo-reactivation effect, 101
photo-sanitization, 171
photothermal effect, 100
plasma, 225
plasma chemistry, 246
plasma parameters, 234
plasma source design, 246
plate freezers, 189
pores, 209
power supply, 29
pressure, 216
pressure-assisted freezing, 6
pressure-assisted thawing, 7
pressure-assisted thermal processing, 5
pressure-shifted freezing, 6
process freezers, 189
process parameters, 94
processing of eggs, 35
processing of juice and, 35
processing of milk, 34
product conveyors and control mechanisms, 174
pulse electric field, 27, 79
pulse wave shape, 30
pulsed light, 92
pulsed UV, 92

R

radiation, 169
radio-pasteurization, 171
raw milk concentratiom, 130
reactive oxygen and nitrogen species (RONS), 224
rectrystallization, 190
reduction of fouling, 149
reverse osmosis, 126

S

salt, 213
semi-permeable membrane, 69
sensory, 108
solute, 199
solution, 206
sono-emulsification, 54
sorption, 217
sorption isotherms, 74
spectroscopy, 246
spectrum, 166
statutory bodies, 166
sterilization, 166
superconductivity, 183
supercritical fluid extraction, 48
superfluidity, 183
survival RATE, 33
switches of PEF system, 30

T

technological factors of PEF, 32
thermal treatment, 76
thermosonication, 44
total phenol, 108
toxicological risk, 166
transition state theory, 4
transport, 212
treatment, 205
treatment chamber, 31
treatment of waste effluents, 144
tunnel and spiral freezers, 188

U

ultrafiltration, 126
ultrasound, 40, 79
ultrasound assisted drying, 52

V

vaccum impregnation, 78

W

whey processing, 135
white light, 92
World Health Organization, 169

X

X-rays, 169
xenon flash lamp, 92